Convection
Heat Transfer

Convection Heat Transfer

Adrian Bejan

*Department of Mechanical Engineering
and Materials Science
Duke University
Durham, North Carolina*

A WILEY-INTERSCIENCE PUBLICATION
JOHN WILEY & SONS
New York · Chichester · Brisbane · Toronto · Singapore

Readers should note that masculine pronouns are used throughout the book for succinctness and that they are intended to refer to both males and females.

Library of Congress Cataloging in Publication Data:
Bejan, Adrian, 1948–
Convection heat transfer.

"A Wiley-Interscience publication."
Includes bibliographical references and index.
1. Heat–Convection. I. Title.
QC327.B48   1984   536'.25   84-3583
ISBN 0-471-89612-8

Printed in the United States of America

10 9 8 7 6 5 4 3

To
Mary Bejan,
to whom I owe
my style

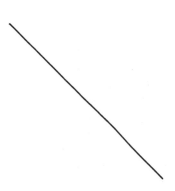

Preface

My main reason for writing a convection textbook is to place the field's past 100 years of exponential growth in perspective. This book is intended for the educator who wants to present his students with more than a review of the generally accepted "classical" methods and conclusions. Through this book I hope to encourage the convection student to question what is known and to think freely and creatively about what is unknown.

There is no such thing as "unanimous agreement" on any topic. The history of scientific progress shows clearly that our present knowledge and understanding—contents of today's textbooks—are the direct result of conflict and controversy. By encouraging our students to question authority, we encourage them to make discoveries on their own. We can all only benefit from the scientific progress that results.

In writing this book, I sought to make available a textbook alternative that offers something new on two other fronts: (1) content, or the selection of topics, and (2) method, or the approach to solving problems in convection heat transfer.

Regarding content, this textbook reflects the relative change in the priorities set by our technological society over the past two decades. Historically, the field of convective heat transfer grew out of great engineering pursuits such as energy conversion (power plant technology), the aircraft, and the exploration of extraterrestrial space. Today, we are forced to face additional challenges, primarily in the areas of "energy" and "ecology." Briefly stated, engineering education today places a strong emphasis on man's need to coexist with the environment. This new emphasis is reflected in the topics assembled in this book. Important areas covered for the first time in a convection textbook are: (1) natural convection on *an equal footing* with forced convection, with application to energy conservation in buildings and to geophysical dynamics, (2) convection through porous media saturated with fluid, with application to geothermal and thermal insulation engineering, and (3) turbulent mixing in free-stream flow, with application to the dispersion of pollutants in the atmosphere and the hydrosphere.

Regarding method, in this book I made a consistent effort to teach problem solving (a *Solutions Manual* is available from the publisher or from me). This book is a textbook to be used for teaching a course, not a handbook. Of course, important engineering results are listed; however, the emphasis is placed on the thinking that leads to these results. A unique feature of this book is that it stresses the importance of correct scale analysis as an eligible and cost-effective method of solution, and as a precondition for more refined methods of solution. It also stresses the need for correct scaling in the graphic reporting of more refined analytical results and of experimental and numerical data. The cost and the "return on investment" associated with a possible method of solution are issues that each student-researcher should examine critically: these issues are stressed throughout the text. The place of computer-aided solutions in convection is the object of an entire chapter contributed jointly by Dimos Poulikakos and Shigeo Kimura. I am very grateful to them for this contribution.

I wrote this book during the academic year 1982–1983, in our mountain-side house on the greenbelt of North Boulder. This project turned out to be a highly rewarding intellectual experience for me, because it forced upon me the rare opportunity to think about an entire field, while continuing my own research on special topics in convection and other areas (specialization usually inhibits the ability to enjoy a bird's-eye-view of anything). It is a cliché in engineering education and research for the author of a new book to end the preface by thanking his family for the "sacrifice" that allowed completion of the work. My experience with writing *Convection Heat Transfer* has been totally different (i.e., much more enjoyable!), to the point that I must thank this book for making me work at home and for triggering so many inspiring conversations with Mary. Convection can be entertaining .

ADRIAN BEJAN

Boulder, Colorado
July 1984

Acknowledgments

This is my opportunity to thank those individuals who, through their actions, have supported my activity and my morale during the past 2 years. By identifying these individuals, I continue the list started at the end of the Preface to *Entropy Generation through Heat and Fluid Flow*:

Prof. Andreas Acrivos, Stanford University
Prof. Vedat S. Arpaci, University of Michigan
Prof. Win Aung, National Science Foundation and Howard University
Prof. Arthur E. Bergles, Iowa State University
Mr. Frank J. Cerra, John Wiley and Sons
Prof. Warren M. Rohsenow, Massachusetts Institute of Technology
Prof. Clifford Truesdell, The Johns Hopkins University
Prof. T. Nejat Veziroglu, University of Miami

A. B.

Contents

APPENDIX   460

INDEX   471

1

Fundamental Principles

Convective heat transfer or, simply, convection is the study of heat transport processes effected by the flow of fluids. The very word *convection* has its roots in the Latin verb *convehere* [1], which means *to bring together* or *to carry into one place* [2]. Convective heat transfer has grown to the status of a contemporary science because of man's desire to understand and predict the extent to which a fluid flow will act as "carrier" or "conveyor belt" for energy and matter. Convective heat transfer, clearly, is a field at the interface between two older fields—heat transfer and fluid mechanics. For this reason the study of any convective heat transfer problem must rest on a solid understanding of basic heat transfer and fluid mechanics principles. The objective of this first chapter is to review these principles in order to establish a common language to debate the more specific issues addressed in later chapters.

Before reviewing the foundations of convective heat transfer methodology, it is worth reexamining the historic relationship between fluid mechanics and heat transfer at the interface we call *convection*. Especially during the past 100 years, heat transfer and fluid mechanics have enjoyed a symbiotic relationship in their parallel development, a relationship where one field was stimulated by the curiosity in the other field. Examples of this symbiosis abound in the history of boundary layer theory and natural convection (see Chapters 2 and 4). The field of convection heat transfer grew out of this symbiosis and, if we are to learn anything from history, important advances in convection will continue to result from this symbiosis. Thus, the student and the future researcher would be well advised to devote equal attention to fluid mechanics and heat transfer literature.

MASS CONSERVATION

The first principle to review is undoubtedly the oldest: it has to do with the conservation of mass in a closed system or the "continuity" of mass through a flow (open) system. From engineering thermodynamics, we recall the mass

1

conservation statement for a control volume [3]

$$\frac{\partial M_{cv}}{\partial t} = \sum_{\substack{\text{inlet} \\ \text{ports}}} \dot{m} - \sum_{\substack{\text{outlet} \\ \text{ports}}} \dot{m} \tag{1}$$

where M_{cv} is the mass instantaneously trapped inside the control volume (cv), while the $\dot{m}$'s are the mass flowrates associated with the flow into and out of the control volume. In convective heat transfer we are usually interested in the velocity and temperature distributions in some flow field near a solid wall; hence, the control volume to consider is the infinitesimally small $\Delta x \Delta y$ box drawn around a fixed location (x, y) in a flow field. In Fig. 1.1, as in most of the problems analyzed in this textbook, the flow field is two-dimensional (i.e., the same in any plane parallel to the plane of Fig. 1.1); in a three-dimensional flow field, the control volume of interest would be the parallelepiped $\Delta x \Delta y \Delta z$. Taking u and v as the local velocity components at point (x, y), the mass conservation equation (1) requires

$$\frac{\partial}{\partial t}(\rho \Delta x \Delta y) = \rho u \Delta y + \rho v \Delta x - \left[\rho u + \frac{\partial(\rho u)}{\partial x} \Delta x\right] \Delta y$$

$$- \left[\rho v + \frac{\partial(\rho v)}{\partial y} \Delta y\right] \Delta x \tag{2}$$

or, dividing through the constant size of the control volume $(\Delta x \Delta y)$,

$$\frac{\partial \rho}{\partial t} + \frac{\partial(\rho u)}{\partial x} + \frac{\partial(\rho v)}{\partial y} = 0 \tag{3}$$

In a three-dimensional flow, an analogous argument yields

$$\frac{\partial \rho}{\partial t} + \frac{\partial(\rho u)}{\partial x} + \frac{\partial(\rho v)}{\partial y} + \frac{\partial(\rho w)}{\partial z} = 0 \tag{4}$$

where w is the velocity component in the z direction.

The local mass conservation statement (4) can also be written as

$$\frac{\partial \rho}{\partial t} + u\frac{\partial \rho}{\partial x} + v\frac{\partial \rho}{\partial y} + w\frac{\partial \rho}{\partial z} + \rho\left(\frac{\partial u}{\partial x} + \frac{\partial v}{\partial y} + \frac{\partial w}{\partial z}\right) = 0 \tag{5}$$

or

$$\frac{D\rho}{Dt} + \rho \nabla \cdot \mathbf{v} = 0 \tag{6}$$

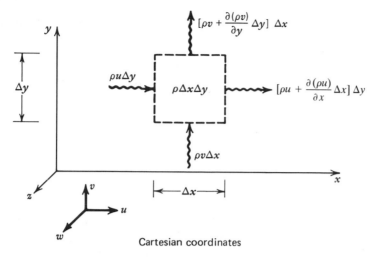

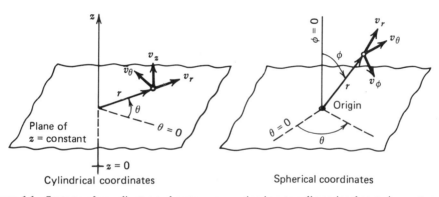

Figure 1.1 Systems of coordinates and mass conservation in a two-dimensional cartesian system.

In this last expression **v** is the velocity vector (u, v, w) while D/Dt represents the "material derivative" operator encountered frequently in convective heat and mass transfer,

$$\frac{D}{Dt} = \frac{\partial}{\partial t} + u\frac{\partial}{\partial x} + v\frac{\partial}{\partial y} + w\frac{\partial}{\partial z} \tag{7}$$

Of particular interest to the *classroom* treatment of the convection problem is the wide class of flows in which the temporal and spatial variations in density are negligible relative to the local variations in velocity. For this class, the mass conservation statement reads

$$\frac{\partial u}{\partial x} + \frac{\partial v}{\partial y} + \frac{\partial w}{\partial z} = 0 \tag{8}$$

The equivalent forms of eq. (8) in cylindrical and spherical coordinates are (Fig. 1.1)

$$\frac{\partial v_r}{\partial r} + \frac{v_r}{r} + \frac{1}{r}\frac{\partial v_\theta}{\partial \theta} + \frac{\partial v_z}{\partial z} = 0 \tag{9}$$

and

$$\frac{1}{r}\frac{\partial}{\partial r}\left(r^2 v_r\right) + \frac{1}{\sin\phi}\frac{\partial}{\partial\phi}\left(v_\phi \sin\phi\right) + \frac{1}{\sin\phi}\frac{\partial v_\theta}{\partial\theta} = 0 \tag{10}$$

It is tempting to regard eqs. (8)–(10) as valid only for incompressible fluids; in fact, their derivation shows that they apply to flows (not fluids) where the density and velocity gradients are such that the $D\rho/Dt$ terms are negligible relative to the $\rho\nabla \cdot \mathbf{v}$ terms in eq. (6). Most of the gas flows encountered in heat exchangers, heated enclosures, and porous media obey the simplified version of the mass conservation principle [eqs. (8)–(10)].

FORCE BALANCES (MOMENTUM EQUATIONS)

From the dynamics of thrust or propulsion systems, we recall that the instantaneous force balance on a control volume requires [4]

$$\frac{\partial}{\partial t}\left(Mv_n\right)_{cv} = \sum F_n + \sum_{\substack{\text{inlet}\\\text{ports}}} \left(\dot{m}v_n\right) - \sum_{\substack{\text{outlet}\\\text{ports}}} \dot{m}v_n \tag{11}$$

where n is the direction chosen for analysis and (v_n, F_n) are the projections of fluid velocity and forces on the n direction. Equation (11) is recognized in the literature as the *momentum principle* or the *momentum theorem*: in essence, eq. (11) is the control volume formulation of Newton's Second Law of Motion where, in addition to terms accounting for *forces* and *mass* × *acceleration*, we now have the *impact* due to the flow of momentum into the control volume, plus the *reaction* associated with the flow of momentum out of the control volume. In the two-dimensional flow situation of Fig. 1.2, we can write two force balances of type (11), one for the x direction and the other for the y direction.

Consider now the special form taken by eq. (11) when applied to the finite-size control volume $\Delta x \Delta y$ drawn around point (x, y) in Fig. 1.2. Consider first the balance of forces in the x direction. In the top drawing of the $\Delta x \Delta y$ control volume, we see the sense of the impact and reaction forces associated with the flow of momentum through the control volume. In the bottom drawing, we see the more classical forces represented by the normal stress (σ_x), tangential stress (τ_{xy}), and the x body force per unit volume (X).

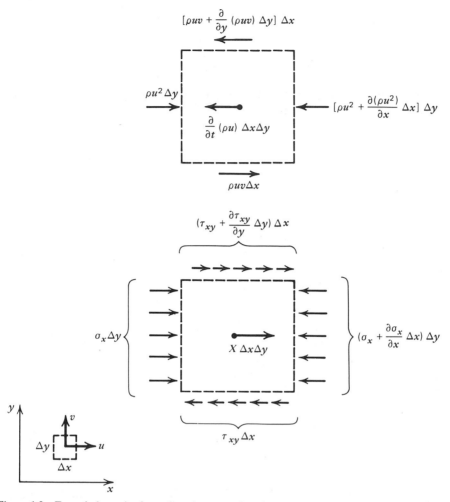

Figure 1.2 Force balance in the x direction on an imaginary control volume in two-dimensional flow.

Projecting all these forces on the x axis, we obtain

$$-\frac{\partial}{\partial t}(\rho u \, \Delta x \Delta y) + \rho u^2 \, \Delta y - \left[\rho u^2 + \frac{\partial}{\partial x}(\rho u^2) \, \Delta x\right] \Delta y$$

$$+ \rho u v \, \Delta x - \left[\rho u v + \frac{\partial}{\partial y}(\rho u v) \, \Delta y\right] \Delta x$$

$$+ \sigma_x \Delta y - \left(\sigma_x + \frac{\partial \sigma_x}{\partial x} \, \Delta x\right) \Delta y - \tau_{xy} \Delta x$$

$$+ \left(\tau_{xy} + \frac{\partial \tau_{xy}}{\partial y} \, \Delta y\right) \Delta x + X \Delta x \Delta y = 0 \qquad (12)$$

or, dividing by $\Delta x \Delta y$ in the limit $\Delta x, \Delta y \to 0$,

$$\rho \frac{Du}{Dt} + u\left[\frac{D\rho}{Dt} + \rho\left(\frac{\partial u}{\partial x} + \frac{\partial v}{\partial y}\right)\right] = -\frac{\partial \sigma_x}{\partial x} + \frac{\partial \tau_{xy}}{\partial y} + X \tag{13}$$

According to the mass conservation equation (6), the quantity in the square brackets is equal to zero; hence

$$\rho \frac{Du}{Dt} = -\frac{\partial \sigma_x}{\partial x} + \frac{\partial \tau_{xy}}{\partial y} + X \tag{14}$$

Next, we relate the stresses σ_x and τ_{xy} to the local flow field by recalling the constitutive relations [5]

$$\sigma_x = P - 2\mu\frac{\partial u}{\partial x} + \frac{2}{3}\mu\left(\frac{\partial u}{\partial x} + \frac{\partial v}{\partial y}\right) \tag{15}$$

$$\tau_{xy} = \mu\left(\frac{\partial u}{\partial y} + \frac{\partial v}{\partial x}\right) \tag{16}$$

These relations are of empirical origin: they summarize the experimental observation that a fluid packet offers no resistance to a change of shape, but resists the time rate of a change of shape. Equations (15) and (16) are the definition of the measurable coefficient of *viscosity* μ. Combining eqs. (14)–(16) yields the *Navier–Stokes* equation

$$\rho \frac{Du}{Dt} = -\frac{\partial P}{\partial x} + \frac{\partial}{\partial x}\left[2\mu\frac{\partial u}{\partial x} - \frac{2\mu}{3}\left(\frac{\partial u}{\partial x} + \frac{\partial v}{\partial y}\right)\right] + \frac{\partial}{\partial y}\left[\mu\left(\frac{\partial u}{\partial y} + \frac{\partial v}{\partial x}\right)\right] + X \tag{17}$$

Of particular interest is the case when the flow may be treated as incompressible (see Problem 1) and the viscosity μ may be regarded as constant. Then the *x momentum equation* reduces to

$$\rho\left(\frac{\partial u}{\partial t} + u\frac{\partial u}{\partial x} + v\frac{\partial u}{\partial y}\right) = -\frac{\partial P}{\partial x} + \mu\left(\frac{\partial^2 u}{\partial x^2} + \frac{\partial^2 u}{\partial y^2}\right) + X \tag{18}$$

A similar equation can be derived from the force balance in the y direction. For a three-dimensional flow in the (x, y, z), (u, v, w) Cartesian system, the three momentum equations for $\rho, \mu \cong$ constant flows are

$$\rho\left(\frac{\partial u}{\partial t} + u\frac{\partial u}{\partial x} + v\frac{\partial u}{\partial y} + w\frac{\partial u}{\partial z}\right) = -\frac{\partial P}{\partial x} + \mu\left(\frac{\partial^2 u}{\partial x^2} + \frac{\partial^2 u}{\partial y^2} + \frac{\partial^2 u}{\partial z^2}\right) + X \tag{19a}$$

$$\rho\left(\frac{\partial v}{\partial t} + u\frac{\partial v}{\partial x} + v\frac{\partial v}{\partial y} + w\frac{\partial v}{\partial z}\right) = -\frac{\partial P}{\partial y} + \mu\left(\frac{\partial^2 v}{\partial x^2} + \frac{\partial^2 v}{\partial y^2} + \frac{\partial^2 v}{\partial z^2}\right) + Y$$

(19b)

$$\rho\left(\frac{\partial w}{\partial t} + u\frac{\partial w}{\partial x} + v\frac{\partial w}{\partial y} + w\frac{\partial w}{\partial z}\right) = -\frac{\partial P}{\partial z} + \mu\left(\frac{\partial^2 w}{\partial x^2} + \frac{\partial^2 w}{\partial y^2} + \frac{\partial^2 w}{\partial z^2}\right) + Z$$

(19c)

Alternative forms of eqs. (19) are [6]:

Vectorial notation

$$\rho\frac{D\mathbf{v}}{Dt} = -\nabla P + \mu\nabla^2\mathbf{v} + \mathbf{F}$$ 　　(20)

where $\mathbf{F}$ is the body force per unit volume vector (X, Y, Z),

Cylindrical coordinates (Fig. 1.1)

$$\rho\left(\frac{\partial v_r}{\partial t} + v_r\frac{\partial v_r}{\partial r} + \frac{v_\theta}{r}\frac{\partial v_r}{\partial \theta} - \frac{v_\theta^2}{r} + v_z\frac{\partial v_r}{\partial z}\right)$$

$$= -\frac{\partial P}{\partial r} + \mu\left(\frac{\partial^2 v_r}{\partial r^2} + \frac{1}{r}\frac{\partial v_r}{\partial r} - \frac{v_r}{r^2} + \frac{1}{r^2}\frac{\partial^2 v_r}{\partial \theta^2} - \frac{2}{r^2}\frac{\partial v_\theta}{\partial \theta} + \frac{\partial^2 v_r}{\partial z^2}\right) + F_r$$

(21a)

$$\rho\left(\frac{\partial v_\theta}{\partial t} + v_r\frac{\partial v_\theta}{\partial r} + \frac{v_\theta}{r}\frac{\partial v_\theta}{\partial \theta} + \frac{v_r v_\theta}{r} + v_z\frac{\partial v_\theta}{\partial z}\right)$$

$$= -\frac{1}{r}\frac{\partial P}{\partial \theta} + \mu\left(\frac{\partial^2 v_\theta}{\partial r^2} + \frac{1}{r}\frac{\partial v_\theta}{\partial r} - \frac{v_\theta}{r^2} + \frac{1}{r^2}\frac{\partial^2 v_\theta}{\partial \theta^2} + \frac{2}{r^2}\frac{\partial v_r}{\partial \theta} + \frac{\partial^2 v_\theta}{\partial z^2}\right) + F_\theta$$

(21b)

$$\rho\left(\frac{\partial v_z}{\partial t} + v_r\frac{\partial v_z}{\partial r} + \frac{v_\theta}{r}\frac{\partial v_z}{\partial \theta} + v_z\frac{\partial v_z}{\partial z}\right)$$

$$= -\frac{\partial P}{\partial z} + \mu\left(\frac{\partial^2 v_z}{\partial r^2} + \frac{1}{r}\frac{\partial v_z}{\partial r} + \frac{1}{r^2}\frac{\partial^2 v_z}{\partial \theta^2} + \frac{\partial^2 v_z}{\partial z^2}\right) + F_z$$ 　(21c)

where (v_r, v_θ, v_z) and (F_r, F_θ, F_z) are the velocity and body force vectors.

Spherical coordinates (Fig. 1.1)

$$\rho\left(\frac{Dv_r}{Dt} - \frac{v_\phi^2 + v_\theta^2}{r}\right)$$

$$= -\frac{\partial P}{\partial r} + \mu\left(\nabla^2 v_r - \frac{2v_r}{r^2} - \frac{2}{r^2}\frac{\partial v_\phi}{\partial \phi} - \frac{2v_\phi\cot\phi}{r^2} - \frac{2}{r^2\sin\phi}\frac{\partial v_\theta}{\partial\theta}\right) + F_r$$

(22a)

$$\rho\left(\frac{Dv_\phi}{Dt} + \frac{v_r v_\phi}{r} - \frac{v_\theta^2\cot\phi}{r}\right)$$

$$= -\frac{1}{r}\frac{\partial P}{\partial \phi} + \mu\left(\nabla^2 v_\phi + \frac{2}{r^2}\frac{\partial v_r}{\partial \phi} - \frac{v_\phi}{r^2\sin^2\phi} - \frac{2\cos\phi}{r^2\sin^2\phi}\frac{\partial v_\theta}{\partial\theta}\right) + F_\phi \quad (22b)$$

$$\rho\left(\frac{Dv_\theta}{Dt} + \frac{v_\theta v_r}{r} + \frac{v_\phi v_\theta\cot\phi}{r}\right)$$

$$= -\frac{1}{r\sin\phi}\frac{\partial P}{\partial \theta} + \mu\left(\nabla^2 v_\theta - \frac{v_\theta}{r^2\sin^2\phi} + \frac{2}{r^2\sin\phi}\frac{\partial v_r}{\partial\theta} + \frac{2\cos\phi}{r^2\sin^2\phi}\frac{\partial v_\phi}{\partial\theta}\right) + F_\theta$$

(22c)

where (v_r, v_ϕ, v_θ) and (F_r, F_ϕ, F_θ) are the velocity and body force vectors, and

$$\frac{D}{Dt} = \frac{\partial}{\partial t} + v_r\frac{\partial}{\partial r} + \frac{v_\phi}{r}\frac{\partial}{\partial \phi} + \frac{v_\theta}{r\sin\phi}\frac{\partial}{\partial \theta}$$

(23)

$$\nabla^2 = \frac{1}{r^2}\frac{\partial}{\partial r}\left(r^2\frac{\partial}{\partial r}\right) + \frac{1}{r^2\sin\phi}\frac{\partial}{\partial \phi}\left(\sin\phi\frac{\partial}{\partial \phi}\right) + \frac{1}{r^2\sin^2\phi}\frac{\partial^2}{\partial\theta^2}$$

(24)

are the material derivative and Laplacian operators in spherical coordinates.

THE FIRST LAW OF THERMODYNAMICS

The preceding two principles—mass conservation and force balance—are in many cases sufficient for solving the flow part of the convective heat transfer problem: note at this juncture the availability of four equations (mass convection plus three force balances) for determining four unknowns (three velocity components plus pressure). The exception to this statement is the subject of Chapter 4, where the natural flow is driven by the heat administered to the flowing fluid. In all cases, however, the heat transfer part of the convection problem requires a solution for the temperature distribution through the flow, especially in the close vicinity of the solid walls bathed by the heat-carrying fluid stream (Chapter 2). The additional equation for accomplishing this ultimate objective is the First Law of Thermodynamics or the *energy equation*.

For the control volume of finite size $\Delta x \Delta y$ in Fig. 1.3, the First Law of Thermodynamics requires that

$$\left\{ \begin{array}{l} \text{The rate of energy} \\ \text{accumulation in the} \\ \text{control volume} \end{array} \right\}_1$$

$$= \left\{ \begin{array}{l} \text{The net transfer of} \\ \text{energy by fluid flow} \end{array} \right\}_2 + \left\{ \begin{array}{l} \text{The net heat transfer} \\ \text{by conduction} \end{array} \right\}_3$$

$$+ \left\{ \begin{array}{l} \text{The rate of internal} \\ \text{heat generation (e.g.,} \\ \text{electrical power} \\ \text{dissipation)} \end{array} \right\}_4 - \left\{ \begin{array}{l} \text{The net work transfer} \\ \text{from the control} \\ \text{volume to its} \\ \text{environment} \end{array} \right\}_5$$

According to the energy flow diagrams sketched in Fig. 1.3 the above groups of terms are

$$\{ \ \}_1 = \Delta x \Delta y \frac{\partial}{\partial t}(\rho e)$$

$$\{ \ \}_2 = -(\Delta x \Delta y)\left[\frac{\partial}{\partial x}(\rho u e) + \frac{\partial}{\partial y}(\rho v e)\right]$$

$$\{ \ \}_3 = -(\Delta x \Delta y)\left(\frac{\partial q_x''}{\partial x} + \frac{\partial q_y''}{\partial y}\right)$$

$$\{ \ \}_4 = (\Delta x \Delta y)q'''$$

$$\{ \ \}_5 = (\Delta x \Delta y)\left(\sigma_x \frac{\partial u}{\partial x} - \tau_{xy}\frac{\partial u}{\partial y} + \sigma_y \frac{\partial v}{\partial y} - \tau_{yx}\frac{\partial v}{\partial x}\right)$$

$$+ (\Delta x \Delta y)\left(u\frac{\partial \sigma_x}{\partial x} - u\frac{\partial \tau_{xy}}{\partial y} + v\frac{\partial \sigma_y}{\partial y} - v\frac{\partial \tau_{yx}}{\partial x}\right)_* \qquad (25)$$

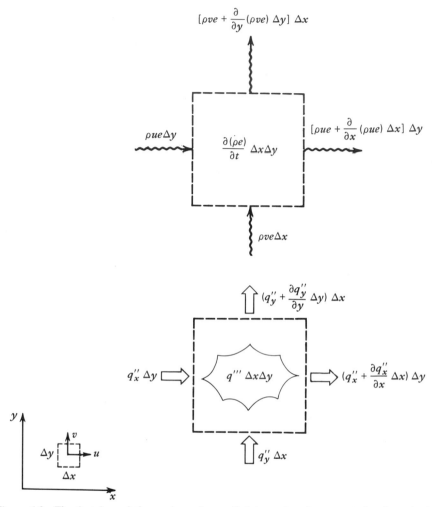

Figure 1.3 The first law of thermodynamics applied to an imaginary control volume in two-dimensional flow (for work transfer interactions see Fig. 1.2).

where e, q_x'', q_y'', and q''' are the specific internal energy, the heat flux in the x direction, the heat flux in the y direction, and the dissipation rate or rate of internal heat generation. The origin of the dissipation rate term $\{\;\}_5$ lies in the work transfer effected by the normal and tangential stresses sketched in the lower half of Fig. 1.2. For example, the work done per unit time by the normal stresses σ_x on the left side of the $\Delta x\,\Delta y$ element is negative and equal to the force acting on the boundary $(\sigma_x\Delta y)$ times the boundary displacement per unit time (u), which yields $-u\sigma_x\Delta y$. Likewise, the work transfer rate associated with normal stresses acting on the right side of the element is positive and equal to $[\sigma_x + (\partial\sigma_x/\partial x)\Delta x][u + (\partial u/\partial x)\Delta x]\Delta y$. The net work transfer rate due to these two contributions is $[\sigma_x(\partial u/\partial x) + u(\partial\sigma_x/\partial x)](\Delta x \Delta y)$, as shown in the { }$_5$ term of eq. (25). Three more work transfer rates can be calculated in the same manner by examining the effect of the remaining three stresses, τ_{xy} in the x direction, and σ_y and τ_{yx} in the y direction. In the { }$_5$ expression above, the eight terms have been separated into two groups. It can be shown that the group denoted as ( )$_*$ reduces to $-\rho(D/Dt)(u^2 + v^2)/2$, which represents the change in the kinetic energy of the fluid packet: in the present treatment, this change is considered negligible relative to the internal energy change $\partial(\rho e)/\partial t$ appearing in { }$_1$ (see Problem 7).

Assembling expressions (25) into the energy conservation statement that preceded them, and using the constitutive relations (15) and (16) we obtain

$$\rho\frac{De}{Dt} + e\left(\frac{D\rho}{Dt} + \rho\nabla\cdot\mathbf{v}\right) = -\nabla\cdot\mathbf{q}'' + q''' - P\nabla\cdot\mathbf{v} + \mu\Phi \qquad (26)$$

where $\mathbf{q}''$ is the heat flux vector (q_x'', q_y'') and Φ is the viscous dissipation function for incompressible two-dimensional flow,

$$\Phi = 2\left[\left(\frac{\partial u}{\partial x}\right)^2 + \left(\frac{\partial v}{\partial y}\right)^2\right] + \left(\frac{\partial u}{\partial y} + \frac{\partial v}{\partial x}\right)^2 \qquad (27)$$

Note that the quantity between parentheses on the left-hand side of eq. (26) is equal to zero [see eq. (6)].

In order to express eq. (26) in terms of enthalpy, we use the thermodynamics definition $h = e + (1/\rho)P$; hence

$$\frac{Dh}{Dt} = \frac{De}{Dt} + \frac{1}{\rho}\frac{DP}{Dt} - \frac{P}{\rho^2}\frac{D\rho}{Dt} \qquad (28)$$

In addition, we can express the directional heat fluxes (q_x'', q_y'') in terms of the local temperature gradients, that is, we invoke the *Fourier law* of heat conduction

$$\mathbf{q}'' = -k\nabla T \qquad (29)$$

Then, combining eqs. (26), (28), and (29) in the desired manner we obtain

$$\rho\frac{Dh}{Dt} = \nabla\cdot(k\nabla T) + q''' + \frac{DP}{Dt} + \mu\Phi - \frac{P}{\rho}\left(\frac{D\rho}{Dt} + \rho\nabla\cdot\mathbf{v}\right) \qquad (30)$$

Finally, we learn again from the mass conservation equation (6) that the last terms in the parentheses in eq. (30) add up to zero; in conclusion, the First Law of Thermodynamics reduces to

$$\rho\frac{Dh}{Dt} = \nabla\cdot(k\nabla T) + q''' + \frac{DP}{Dt} + \mu\Phi \qquad (31)$$

In order to express the energy equation (31) in terms of temperature, it is tempting to replace the specific enthalpy on the left-hand side by the product of specific heat × temperature. This move is correct only in cases where the fluid behaves like an ideal gas (see the ideal gas model, Table 1.1). In general, the change in specific enthalpy is expressed by the canonical relation for enthalpy [7],

$$dh = T \, ds + \frac{1}{\rho} \, dP \tag{32}$$

where T is the absolute temperature and ds the specific entropy change

$$ds = \left(\frac{\partial s}{\partial T} \right)_P dT + \left(\frac{\partial s}{\partial P} \right)_T dP \tag{33}$$

From the last of Maxwell's relations [8], we have

$$\left(\frac{\partial s}{\partial P} \right)_T = -\left[\frac{\partial (1/\rho)}{\partial T} \right]_P = \frac{1}{\rho^2} \left(\frac{\partial \rho}{\partial T} \right)_P = -\frac{\beta}{\rho} \tag{34}$$

where β is the coefficient of thermal expansion

$$\beta = -\frac{1}{\rho} \left(\frac{\partial \rho}{\partial T} \right)_P \tag{35}$$

Table 1.1 shows also that

$$\left(\frac{\partial s}{\partial T} \right)_P = \frac{c_P}{T} \tag{36}$$

Together, eqs. (32)–(36) state

$$dh = c_P dT + \frac{1}{\rho} (1 - \beta T) \, dP \tag{37}$$

in other words, the left-hand side of the energy equation (31) is

$$\rho \frac{Dh}{Dt} = \rho c_P \frac{DT}{Dt} + (1 - \beta T) \frac{DP}{Dt} \tag{38}$$

The temperature-formulation of the First Law of Thermodynamics is therefore

$$\rho c_P \frac{DT}{Dt} = \nabla \cdot (k \nabla T) + q''' + \beta T \frac{DP}{Dt} + \mu \Phi \tag{39}$$

Table 1.1. Summary of Thermodynamic Relations[a] and Models [9]

	Internal Energy $du = T\,ds - P\,dv$	Enthalpy $dh = T\,ds + v\,dP$	Entropy $ds = \dfrac{1}{T}du + \dfrac{P}{T}dv$
Pure substance	$du = c_v\,dT + \left[T\left(\dfrac{\partial P}{\partial T}\right)_v - P\right]dv$	$dh = c_P\,dT + \left[-T\left(\dfrac{\partial v}{\partial T}\right)_P + v\right]dP$	$ds = \dfrac{c_P}{T}dT - \left(\dfrac{\partial v}{\partial T}\right)_P dP$ $= \dfrac{c_v}{T}dT + \left(\dfrac{\partial P}{\partial T}\right)_v dv$
Ideal gas	$du = c_v\,dT$	$dh = c_P\,dT$	$ds = c_P\dfrac{dT}{T} - R\dfrac{dP}{P}$ $= c_v\dfrac{dT}{T} + R\dfrac{dv}{v}$ $= c_v\dfrac{dP}{P} + c_P\dfrac{dv}{v}$
Incompressible liquid	$du = c\,dT$	$dh = c\,dT + v\,dP$	$ds = c\dfrac{dT}{T}$

[a] According to the classical thermodynamics notation, v is the specific volume, $v = 1/\rho$, and u is the internal energy (e in the text).

13

with the two special forms

Ideal gas ($\beta T = 1$)

$$\rho c_P \frac{DT}{Dt} = \nabla \cdot (k \nabla T) + q''' + \frac{DP}{Dt} + \mu \Phi \tag{40}$$

Incompressible liquid ($\beta = 0$)

$$\rho c \frac{DT}{Dt} = \nabla \cdot (k \nabla T) + q''' + \mu \Phi \tag{41}$$

Most of the convection problems addressed in this textbook obey an even simpler model, namely, constant fluid conductivity k, zero internal heat generation q''', negligible viscous dissipation $\mu \Phi$, and negligible compressibility effect $\beta T \, DP/Dt$. The energy equation for this model is

$$\rho c_P \frac{DT}{Dt} = k \nabla^2 T \tag{42}$$

or, in terms of specific coordinate systems (Fig. 1.1),

Cartesian (x, y, z)

$$\rho c_P \left(\frac{\partial T}{\partial t} + u \frac{\partial T}{\partial x} + v \frac{\partial T}{\partial y} + w \frac{\partial T}{\partial z} \right) = k \left(\frac{\partial^2 T}{\partial x^2} + \frac{\partial^2 T}{\partial y^2} + \frac{\partial^2 T}{\partial z^2} \right) \tag{43a}$$

Cylindrical (r, θ, z)

$$\rho c_P \left(\frac{\partial T}{\partial t} + v_r \frac{\partial T}{\partial r} + \frac{v_\theta}{r} \frac{\partial T}{\partial \theta} + v_z \frac{\partial T}{\partial z} \right)$$

$$= k \left[\frac{1}{r} \frac{\partial}{\partial r} \left(r \frac{\partial T}{\partial r} \right) + \frac{1}{r^2} \frac{\partial^2 T}{\partial \theta^2} + \frac{\partial^2 T}{\partial z^2} \right] \tag{43b}$$

Spherical (r, ϕ, θ)

$$\rho c_P \left(\frac{\partial T}{\partial t} + v_r \frac{\partial T}{\partial r} + \frac{v_\phi}{r} \frac{\partial T}{\partial \phi} + \frac{v_\theta}{r \sin \phi} \frac{\partial T}{\partial \theta} \right)$$

$$= k \left[\frac{1}{r^2} \frac{\partial}{\partial r} \left(r^2 \frac{\partial T}{\partial r} \right) + \frac{1}{r^2 \sin \phi} \frac{\partial}{\partial \phi} \left(\sin \phi \frac{\partial T}{\partial \phi} \right) + \frac{1}{r^2 \sin^2 \phi} \frac{\partial^2 T}{\partial \theta^2} \right] \tag{43c}$$

If the fluid can be modeled thermodynamically as an incompressible liquid then, as in eq. (41), the specific heat at constant pressure c_P is replaced by the lone specific heat of the incompressible liquid, c (Table 1.1).

When dealing with extremely viscous flows of the type encountered in lubrication problems or the piping of crude oil, the above model is improved by taking into account the internal heating due to viscous dissipation,

$$\rho c_P \frac{DT}{Dt} = k \nabla^2 T + \mu \Phi \tag{44}$$

In three dimensions, the viscous dissipation function can be expressed as

Cartesian (x, y, z)

$$\Phi = 2 \left[\left(\frac{\partial u}{\partial x} \right)^2 + \left(\frac{\partial v}{\partial y} \right)^2 + \left(\frac{\partial w}{\partial z} \right)^2 \right]$$

$$+ \left[\left(\frac{\partial u}{\partial y} + \frac{\partial v}{\partial x} \right)^2 + \left(\frac{\partial v}{\partial z} + \frac{\partial w}{\partial y} \right)^2 + \left(\frac{\partial w}{\partial x} + \frac{\partial u}{\partial z} \right)^2 \right]$$

$$- \frac{2}{3} \left(\frac{\partial u}{\partial x} + \frac{\partial v}{\partial y} + \frac{\partial w}{\partial z} \right)^2 \tag{45a}$$

Cylindrical (r, θ, z)

$$\Phi = 2 \left[\left(\frac{\partial v_r}{\partial r} \right)^2 + \left(\frac{1}{r} \frac{\partial v_\theta}{\partial \theta} + \frac{v_r}{r} \right)^2 + \left(\frac{\partial v_z}{\partial z} \right)^2 \right.$$

$$+ \frac{1}{2} \left(\frac{\partial v_\theta}{\partial r} - \frac{v_\theta}{r} + \frac{1}{r} \frac{\partial v_r}{\partial \theta} \right)^2 + \frac{1}{2} \left(\frac{1}{r} \frac{\partial v_z}{\partial \theta} + \frac{\partial v_\theta}{\partial z} \right)^2$$

$$\left. + \frac{1}{2} \left(\frac{\partial v_r}{\partial z} + \frac{\partial v_z}{\partial r} \right)^2 - \frac{1}{3} (\nabla \cdot \mathbf{v})^2 \right] \tag{45b}$$

Spherical (r, ϕ, θ)

$$\Phi = 2 \left\{ \left[\left(\frac{\partial v_r}{\partial r} \right)^2 + \left(\frac{1}{r} \frac{\partial v_\phi}{\partial \phi} + \frac{v_r}{r} \right)^2 + \left(\frac{1}{r \sin\phi} \frac{\partial v_\theta}{\partial \theta} + \frac{v_r}{r} + \frac{v_\phi \cot \phi}{r} \right)^2 \right] \right.$$

$$+ \frac{1}{2} \left[r \frac{\partial}{\partial r} \left(\frac{v_\phi}{r} \right) + \frac{1}{r} \frac{\partial v_r}{\partial \phi} \right]^2 + \frac{1}{2} \left[\frac{\sin\phi}{r} \frac{\partial}{\partial \phi} \left(\frac{v_\theta}{r \sin\phi} \right) + \frac{1}{r \sin\phi} \frac{\partial v_\phi}{\partial \theta} \right]^2$$

$$\left. + \frac{1}{2} \left[\frac{1}{r \sin\phi} \frac{\partial v_r}{\partial \theta} + r \frac{\partial}{\partial r} \left(\frac{v_\theta}{r} \right) \right]^2 \right\} - \frac{2}{3} (\nabla \cdot \mathbf{v})^2 \tag{45c}$$

Note that in deriving the general energy equation (39) as well as the simplified models (42) and (44), the flow density has not been regarded as uniform. If the density is regarded as practically constant, then $\nabla \cdot \mathbf{v} = 0$ [eq. (6)] and the last term in each of expressions (45) vanishes.

THE SECOND LAW OF THERMODYNAMICS

Any discussion of the basic principles of convective heat transfer must include the Second Law of Thermodynamics, not because the second law is necessary for determining the flow and temperature field (it is not, because it is not an equation), but because the second law is the basis for much of the engineering *motive* for formulating and solving convection problems. For example, in the development of know-how for the heat exchanger industry, we strive for improved thermal contact (enhanced heat transfer) *and* reduced pump power loss in order to improve the *thermodynamic* efficiency of the heat exchanger. Good heat exchanger design means, ultimately, efficient thermodynamic performance, that is, the least generation of entropy or least destruction of available work (exergy) in the power/refrigeration system incorporating the heat exchanger [10]. For this reason, it is pedagogically necessary to review the second law and, in this way, to explain the common-sense origin of the engineering questions that led to today's field of convective heat transfer.

The Second Law of Thermodynamics states that all real-life processes are irreversible: in the case of a control volume, as in Fig. 1.1, this statement is [11]

$$\frac{\partial S_{cv}}{\partial t} \geq \sum \frac{q_i}{T_i} + \underbrace{\sum \dot{m}s}_{\substack{\text{inlet} \\ \text{ports}}} - \underbrace{\sum \dot{m}s}_{\substack{\text{outlet} \\ \text{ports}}} \tag{46}$$

where S_{cv} is the instantaneous entropy inventory of the control volume, $\dot{m}s$ is the entropy flows (streams) into and out of the control volume, and T_i is the absolute temperature of the boundary crossed by the heat transfer interaction q_i.[†] The irreversibility of the process is measured by the strength of the inequality sign in eq. (46) or by the entropy generation rate S_{gen} defined as

$$S_{gen} = \frac{\partial S_{cv}}{\partial t} - \sum \frac{q_i}{T_i} - \underbrace{\sum \dot{m}s}_{\substack{\text{inlet} \\ \text{ports}}} + \underbrace{\sum \dot{m}s}_{\substack{\text{outlet} \\ \text{ports}}} \geq 0 \tag{47}$$

It is easy to show that the rate of one-way destruction of useful work in an engineering system, W_{lost}, is directly proportional to the rate of entropy generation

$$W_{lost} = T_0 S_{gen} \tag{48}$$

[†] Defined as positive *into* the control volume.

where T_0 is the absolute temperature of the ambient temperature reservoir (T_0 = constant) [10]. Equation (48) stresses the engineering importance of estimating the irreversibility or entropy generation rate of convective heat transfer processes: if not used wisely, these processes effect the deplorable waste of precious fuel resources.

Based on an analysis similar to the analyses presented for mass conservation, force balances, and the First Law of Thermodynamics, the second law (47) may be applied to a finite-size control volume $\Delta x \Delta y \Delta z$ at an arbitrary point (x, y, z) in a flow field. Thus, the rate of entropy generation per unit time and per unit volume S_{gen}''' is [12]

$$S_{gen}''' = \underbrace{\frac{k}{T^2}(\nabla T)^2}_{\geq 0} + \underbrace{\frac{\mu}{T}\Phi}_{\geq 0} \geq 0 \tag{49}$$

where k and μ are assumed constant. In a two-dimensional convection situation such as in Figs. 1.1–1.3, the local entropy generation rate (69) yields

$$S_{gen}''' = \frac{k}{T^2}\left[\left(\frac{\partial T}{\partial x}\right)^2 + \left(\frac{\partial T}{\partial y}\right)^2\right]$$

$$+ \frac{\mu}{T}\left\{2\left[\left(\frac{\partial u}{\partial x}\right)^2 + \left(\frac{\partial v}{\partial y}\right)^2\right] + \left(\frac{\partial u}{\partial y} + \frac{\partial v}{\partial x}\right)^2\right\} \geq 0 \tag{50}$$

In the last two equations, T represents the *absolute* temperature of the point where S_{gen}''' is being evaluated. The two-dimensional expression (50) illustrates the cooperation between viscous dissipation and imperfect thermal contact (finite-temperature gradients) in the generation of entropy via convective heat transfer.

Equations (48) and (50) constitute the bridge between two research activities, namely, fundamental convection heat transfer and applied heat transfer (thermodynamic design). Beginning with the next chapter we focus on the fundamental problems of determining the flow and temperature fields in a given convection heat transfer configuration. However, through eq. (50), we are invited to keep in mind that these fields contribute hand-in-hand to downgrading the thermodynamic merit of the engineering device that ultimately employs the convection process under consideration. The art of adjusting the convection process so that it destroys the least available work (subject to various system constraints) is the focus of the applied field of entropy generation minimization: this activity has been outlined recently in a textbook [10].

SCALE ANALYSIS

This section is designed to familiarize the student with the common-sense intellectual exercise that in the fields of heat transfer and fluid mechanics goes by the name of *scale analysis* or *scaling*. This section is necessary because scale analysis is used extensively throughout this textbook, in fact, scale analysis is recommended as the premier method for obtaining the most information per unit of intellectual effort. Furthermore, this section is necessary because scale analysis is not discussed in the heat transfer and fluid mechanics textbooks of our time, despite the fact that it is a precondition for good analysis in dimensionless form. Scale analysis is often confused with dimensional analysis or the often arbitrary nondimensionalization of the governing equations before performing a perturbation analysis or a numerical simulation on the computer.

The object of scale analysis is to use the basic principles of convective heat transfer in order to produce order-of-magnitude estimates for the quantities of interest. This means that if one of the quantities of interest is the thickness of the boundary layer in forced convection, the object of scale analysis is to determine whether the boundary layer thickness is measured in millimeters or meters. Note that scale analysis goes beyond dimensional analysis (whose objective is to determine the dimension of boundary layer thickness, namely, length): when done properly, scale analysis anticipates within a factor of order one (or within percentage points) the expensive results produced by "exact" analyses. The value of scale analysis is remarkable, particularly when we realize that the notion of "exact analysis" is as false and ephemeral as the notion of "experimental fact."

As the first example of scale analysis, consider a problem from the field of conduction heat transfer. In Fig. 1.4, we see a metal plate plunged at $t = 0$ into a highly conducting fluid, such that the surfaces of the plate instantaneously assume the fluid temperature $T_\infty = T_0 + \Delta T$. Suppose that we are interested in estimating the time needed by the thermal front to penetrate through the plate, that is, the time until the center plane of the plate "feels" the heating imposed on the outer surfaces.

To answer the above question, we focus on the half-plate of thickness $D/2$ and the energy equation for pure conduction in one direction.

$$\rho c_P \frac{\partial T}{\partial t} = k \frac{\partial^2 T}{\partial x^2} \tag{51}$$

Next, we estimate the order of magnitude of each of the terms appearing in eq. (51). On the left-hand side, we have

$$\rho c_P \frac{\partial T}{\partial t} \sim \rho c_P \frac{\Delta T}{t} \tag{52}$$

in other words, the scale of temperature changes in the chosen space and in a time of order t is ΔT. On the right-hand side, we obtain

$$k \frac{\partial^2 T}{\partial x^2} = k \frac{\partial}{\partial x} \left(\frac{\partial T}{\partial x} \right) \sim \frac{k}{D/2} \frac{\Delta T}{D/2} = \frac{k \Delta T}{(D/2)^2} \tag{53}$$

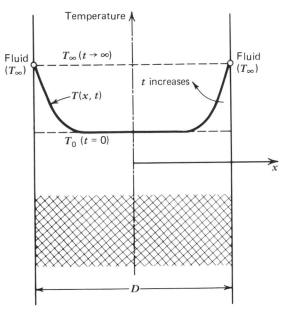

Figure 1.4 Transient heat conduction in a one-dimensional conducting slab with a sudden temperature change at the boundary.

Equating the two orders of magnitude (52) and (53), as required by the energy equation (51), we find the answer to the problem

$$t \sim \frac{(D/2)^2}{\alpha} \tag{54}$$

where α is the thermal diffusivity of the medium, $k/(\rho c_p)$. The penetration time (54) compares well with any interpretation of the exact solution to this classical problem [13]. However, the time and effort associated with deriving eq. (54) do not compare with the labor required by Fourier analysis and the graphical presentation of Fourier series.

Based on the above example,[†] the following rules of scale analysis are worth stressing.

Rule 1. Always define the spatial extent of the region in which you perform the scale analysis. In the example of Fig. 1.4, the size of the region of interest is $D/2$. In other problems such as boundary layer flow, the size of the region of interest is unknown: as shown in Chapter 2, the scale analysis begins by selecting the region and by labeling the unknown thickness of this region δ. Any scale analysis of a flow or a flow region that is not uniquely defined is pure nonsense.

[†]More examples of scale analysis are presented in Problems 11 and 12 at the end of this chapter.

Rule 2. Any equation constitutes an equivalence between the scales of two dominant terms appearing in the equation. In the transient conduction example of Fig. 1.4, the left-hand side of eq. (51) could only be the same order of magnitude as the right-hand side. The two terms appearing in eq. (51) are the dominant terms (considering that the discussion referred to pure conduction): in general, the energy equation can contain many more terms [eq. (39)], not all of them important. The art of selecting the dominant scales from many scales is condensed in rules 3–5.

Rule 3. If in the sum of two terms

$$c = a + b \tag{55}$$

the order of magnitude of one term is greater than the order of magnitude of the other term

$$O(a) > O(b) \tag{56}$$

then the order of magnitude of the sum is dictated by the dominant term

$$O(c) \sim O(a) \tag{57}$$

The same conclusion holds if, instead of eq. (55), we have the difference $c = a - b$ or $c = -a + b$.

Rule 4. If in the sum of two terms

$$c = a + b \tag{55}$$

the two terms are of the same order of magnitude

$$O(a) \sim O(b) \tag{58}$$

then the sum is also of the same order of magnitude

$$O(c) \sim O(a) \sim O(b) \tag{59}$$

Rule 5. In any product

$$p = ab \tag{60}$$

the order of magnitude of the product is equal to the product of the orders of magnitude of the two factors

$$O(p) \sim O(a)O(b) \tag{61}$$

If, instead of eq. (60), we have the ratio

$$r = \frac{a}{b} \tag{62}$$

then

$$O(r) \sim \frac{O(a)}{O(b)} \tag{63}$$

Notation

In addition to having its own set of rules, scale analysis requires special care with regard to notation. In rules 1–5 we used the following symbols:

$\sim$ is of the same order of magnitude as

$O(a)$ the order of magnitude of a

$>$ greater than, in an order-of-magnitude sense

For brevity, the scale analyses included in this textbook employ the language of expressions (56), (57), (61) and (63) without the repetitive notation $O(\ )$ for order of magnitude.

STREAMLINES AND HEATLINES

The opportunity to actually "see" the solution to a problem is essential to a problem-solver's ability to learn from his experience and, in this way, to improve his technique. In convection problems, it is important to visualize the flow of fluid and, riding on this, the flow of energy. For example, in the two-dimensional Cartesian configuration of Fig. 1.1 it has become common practice to define a streamfunction $\psi(x, y)$ as

$$u = \frac{\partial \psi}{\partial y}, \qquad v = -\frac{\partial \psi}{\partial x} \tag{64}$$

such that the mass continuity equation for incompressible flow

$$\frac{\partial u}{\partial x} + \frac{\partial v}{\partial y} = 0 \tag{65}$$

is satisfied identically. It is easy to verify that the actual flow is locally parallel to the $\psi = $ constant line passing through the point of interest. Therefore, although there are no substitutes for (u, v) as bearers of precise information regarding the local flow, the family of $\psi = $ constant streamlines provides a much needed bird's-eye view of the entire flow field and its main characteristics.

In convection, the transport of energy through the flow field is a combination of both thermal diffusion and enthalpy flow [eq. (42)]. For any such field we can define a new function $H(x, y)$ such that the net flow of energy (thermal diffusion and enthalpy flow) is zero across each $H = \text{constant}$ line. The mathematical definition of the *heatfunction H* follows in the steps of eqs. (64) if, this time, the aim is to satisfy the energy equation. For steady state two-dimensional convection through a constant-property homogeneous fluid, eq. (42) becomes

$$u\frac{\partial T}{\partial x} + v\frac{\partial T}{\partial y} = \alpha\left(\frac{\partial^2 T}{\partial x^2} + \frac{\partial^2 T}{\partial y^2}\right) \tag{66}$$

or

$$\frac{\partial}{\partial x}\left(\rho c_p u T - k\frac{\partial T}{\partial x}\right) + \frac{\partial}{\partial y}\left(\rho c_p v T - k\frac{\partial T}{\partial y}\right) = 0 \tag{67}$$

The heatfunction is defined as [14]

$$\frac{\partial H}{\partial y} = \rho c_p u T - k\frac{\partial T}{\partial x}, \quad \text{net energy flow in the } x \text{ direction}$$

$$-\frac{\partial H}{\partial x} = \rho c_p v T - k\frac{\partial T}{\partial y}, \quad \text{net energy flow in the } y \text{ direction} \tag{68}$$

so that $H(x, y)$ satisfies eq. (66) identically. Note that the above definition also applies to convection through a fluid-saturated porous medium, where eqs. (66) and (67) account for energy conservation (see Chapter 10).

Figure 12.7, which appears at the end of Chapter 12, shows the merits of heatline visualization relative to the classical method of plotting isotherms. The comparison between heatlines and isotherms is done by presenting the solution to the classical problem of natural convection in a square cavity heated from the side (Chapter 5), where the Rayleigh number is high enough so that the side-to-side heat transfer rate is dominated by convection. The heatlines shown vividly that "heat rises" and that the true energy corridor consists of two vertical boundary layers connected through an energy tube positioned along the upper wall. The heatlines are parallel to the top and bottom walls which are adiabatic. Along the two isothermal vertical walls, the heatlines are normal to the wall because the near-wall regions are dominated by conduction (both u and v vanish at the wall). One interesting contribution of the heatline pattern is that it shows graphically the magnitude of the Nusselt number: note that the conduction-referenced Nusselt number appears on Fig. 12.7c as the value of maximum H on the top heatline of the heatfunction plot. Note further that the heatline pattern shows graphically the flow of energy downward through the core.

It is worth noting that if the fluid flow subsides ($u = v = 0$), the heatlines become identical to the *heat-flux lines* employed frequently in the study of conduction phenomena. Therefore, as a heat transfer visualization technique, the use of heatlines is the convection counterpart or the generalization of a standard technique (heat-flux lines) used in conduction. It is interesting to also point out that the contemporary use of $T =$ constant lines is not a proper way to visualize heat transfer in the field of convection: isotherms are a proper heat transfer visualization tool only in the field of conduction (where, in fact, they have been invented), because only there they are locally orthogonal to the true direction of energy flow. The use of $T =$ constant lines to visualize convection heat transfer makes as little sense as using $P =$ constant lines to visualize fluid flow.

SYMBOLS

c	specific heat of incompressible substance
c_P	specific heat at constant pressure
c_v	specific heat at constant volume
e	specific energy (labeled u in Table 1.1)
F	force
h	specific enthalpy
H	heatfunction, defined via eqs. (68)
k	thermal conductivity
$\dot{m}$	mass flowrate
M	mass
P	pressure
q	heat transfer rate [W/m]
q''	heat flux [W/m^2]
q'''	rate of internal heat generation [W/m^3]
R	ideal gas constant
r, θ, z	cylindrical coordinates (Fig. 1.1)
r, ϕ, θ	spherical coordinates (Fig. 1.1)
s	specific entropy
S	entropy
S_{gen}	entropy generation
t	time
T	absolute temperature
T_0	the absolute temperature of the ambient
u, v, w	velocity components in the x, y, z system of coordinates (Fig. 1.1)
x, y, z	cartesian coordinates (Fig. 1.1)
X, Y, Z	body force terms [eqs. (19)]
α	thermal diffusivity
β	coefficient of thermal expansion [eq. (35)]
μ	viscosity

ρ density (labeled $1/v$ in Table 1.1, where v is the specific volume)

σ normal stress

τ shear stress

Φ viscous dissipation function [eqs. (45)]

ψ streamfunction [eqs. (64)]

$(\ )_{cv}$ property of the control volume

$(\ )_{r,\theta,z}$ components of a vector in cylindrical coordinates

$(\ )_{r,\phi,\theta}$ components of a vector in spherical coordinates

REFERENCES

1. D. B. Guralnik, ed., *Webster's New World Dictionary*, Second College Edition, World Publishing Company, New York, 1970, p. 310.
2. D. P. Simpson, *Cassell's Latin Dictionary*, Macmillan, New York, 1978, p. 150.
3. E. G. Cravalho and J. L. Smith, Jr., *Engineering Thermodynamics*, Pitman, Boston, MA, 1981, pp. 347–350.
4. W. C. Reynolds and H. C. Perkins, *Engineering Thermodynamics*, 2nd ed., McGraw-Hill, New York, 1977, p. 348.
5. W. M. Rohsenow and H. Y. Choi, *Heat, Mass and Momentum Transfer*, Prentice-Hall, Englewood Cliffs, NJ, 1961, p. 48.
6. S. W. Yuan, *Foundations of Fluid Mechanics*, Prentice-Hall, Englewood Cliffs, NJ, 1967.
7. W. M. Rohsenow and H. Y. Choi, *op. cit.*, p. 170.
8. E. G. Cravalho and J. L. Smith, Jr., *op. cit.*, p. 282.
9. A. Bejan and H. M. Paynter, *Solved Problems in Thermodynamics*, p. 10-5. Issued since 1976 in the Mechanical Engineering Department, Massachusetts Institute of Technology, Cambridge, MA.
10. A. Bejan, *Entropy Generation through Heat and Fluid Flow*, Wiley, New York, 1982, Chapter 2.
11. E. G. Cravalho and J. L. Smith, Jr., *op. cit.*, p. 359.
12. A. Bejan, *op. cit.*, Chapter 5.
13. V. S. Arpaci, *Conduction Heat Transfer*, Addison-Wesley, Reading, MA, 1966, Chapter 5.
14. S. Kimura and A. Bejan, The "heatline" visualization of convective heat transfer, *J. Heat Transfer*, Vol. 105, 1983, pp. 916–919.

PROBLEMS

1. Consider the unsteady mass conservation equation (5) as it might describe the flow accelerating through a duct with a variable cross-section. If the largest velocity gradient measured locally is du/dx, and if the largest density gradient is $d\rho/dx$, then what relationship must exist between du/dx and $d\rho/dx$ for the simplified eq. (8) to be applicable?

2. Derive the mass conservation equation in cylindrical coordinates [eq. (9)] by applying the general principle (1) to an elementary control volume of size $(\Delta r)(r\Delta\theta)(\Delta z)$ in the bottom-left frame of Fig. 1.1 (assume $\rho = $ constant).

3. Derive the mass conservation statement for spherical coordinates [eq. (10)] by writing eq. (1) for the elementary control volume $(\Delta r)(r \sin\phi \Delta\theta)(r\Delta\phi)$ around point (r, θ, ϕ) in the bottom-right drawing of Fig. 1.1 (assume $\rho =$ constant).

4. Consider the flow where ρ and μ may be regarded as constant. Show that the x momentum equation (18) follows from eq. (17) through the proper use of the mass conservation principle.

5. Imagine a certain flow described by eqs. (9) and (21) in cylindrical coordinates. If the flow itself is situated on one side of and infinitely far from the $r = 0$ origin of the coordinate system, then the local three-directional increments $\Delta r, r\Delta\theta, \Delta z$ become analogous to three Cartesian increments $\Delta x, \Delta y, \Delta z$ measured away from the local point (r, θ, z) in the flow field. Show that in the limit $r \to \infty$, the transformation $\Delta r \to \Delta x, r\Delta\theta \to \Delta y, \Delta z \to \Delta z$ leads to the collapse of eqs. (9) and (21) into their (x, y, z) Cartesian equivalents [eqs. (8) and (19)].

6. Consider the conservation of mass and the three force balances in spherical coordinates [eqs. (10) and (22)]. If the flow described by these equations is situated infinitely far from the $r = 0$ origin of the spherical system, then the following transformation is applicable (Fig. 1.1): $\Delta r \to \Delta x, r \sin\phi \Delta\theta \to \Delta y, r\Delta\phi \to \Delta z$. Show that in the limit $r \to \infty$ through this transformation eqs. (10) and (22) become the same as eqs. (8) and (19).

7. Implicit in the derivation of the energy equation (39) is the assumption that changes in kinetic energy $V^2/2$ are negligible relative to changes in internal energy e [see expressions (25), where e should, in general, be replaced by $e + V^2/2$]. Retrace the path leading to eq. (39) by taking into account changes in kinetic energy; show that the result of this more rigorous analysis is identical to eq. (39).

8. Demonstrate that lost work is always proportional to entropy generation [eq. (48)], where $W_{lost} = W_{maximum} - W_{actual}$, and where $W_{maximum}$ corresponds to the fictitious reversible limit ($S_{gen} = 0$). Write the First Law of Thermodynamics for a control volume, first for the actual (real) process and, next, for the reversible process. Then use the definition of W_{lost} and S_{gen} to prove eq. (48). (Details of this derivation are presented in Ref. 10.)

9. Derive the formula for the local rate of entropy generation [eq. (49)]. Begin with translating the general statement (47) into the language of the two-dimensional control volume $\Delta x \Delta y$. Combine the resulting expression with the First Law of Thermodynamics as given by eq. (26), plus the canonical relation for internal energy (Table 1.1). (For details, consult Ref. 10.)

10. Consider the Couette flow between two parallel plates separated by a gap of width D and moving relative to one another with a speed U. The temperature difference ΔT is imposed between the two plates. Estimate the rate of entropy generation per unit volume in this flow. What relationship must exist between $D, U, \Delta T$ and fluid properties (μ, k) in order for S_{gen}''' to be dominated by the irreversibility due to fluid friction?

11. According to the one-dimensional (longitudinal) conduction model of a fin, the temperature distribution along the fin, $T(x)$, obeys the energy equation (see Ref. 13, Sec. 3-8)

$$\underbrace{kA\frac{d^2T}{dx^2}}_{\substack{\text{Longitudinal} \\ \text{conduction}}} - \underbrace{hP(T - T_0)}_{\substack{\text{Lateral} \\ \text{convection}}} + \underbrace{q'''A}_{\substack{\text{Internal} \\ \text{heat} \\ \text{generation}}} = 0$$

where A, h, P, and q''' are the fin cross-sectional area, the fin-fluid heat transfer coefficient, the perimeter of the fin cross-section (the so-called *wetted* perimeter), and the volumetric rate of heat generation. Consider the semiinfinite fin that, as shown in the sketch, is bathed by a fluid of temperature T_0 and is attached to a solid wall of temperature T_0. The heat generated by the fin is absorbed either by the fluid or the solid wall.

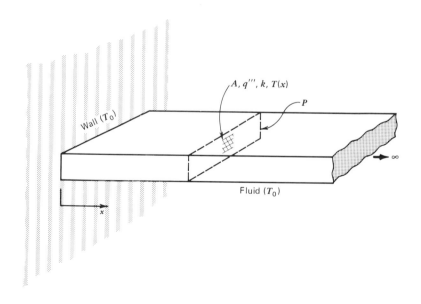

(a) As a system for scale analysis, select the fin section of length x, where x is measured away from the wall. Let T_∞ be the fin temperature sufficiently far from the wall. Show that if x is large enough, the longitudinal conduction term becomes negligible in the energy equation.

(b) Invoking the balance between lateral convection and internal heat generation, determine the fin temperature sufficiently far from the wall, T_∞.

(c) Determine the fin section of length δ near the wall where the heat transfer is ruled by the balance between longitudinal conduction and internal heat generation.

(d) Determine the heat transfer rate into the wall, through the base of the fin.

12. Consider the laminar flow near a flat, solid wall, as illustrated in Fig. 2.1. The momentum equation for this flow involves the competition between three effects: inertia, pressure gradient, and friction [see eq. (26) of Chapter 2]. For the purpose of scale analysis, consider the flow region of length L and thickness L. Show that in this region the ratio (inertia)/(friction) is of order Re_L, where Re_L is the Reynolds number based on wall length. (Note that the region selected for analysis is not the boundary layer region discussed in Chapter 2.) In a certain flow the value of Re_L is 10^3. What force balance rules the $L \times L$ region: inertia ~ pressure, inertia ~ friction, or pressure–friction?

Laminar Boundary Layer Flow

2

To begin a course in convective heat transfer with a chapter on "boundary layers" is to recognize the origins of the field. In this chapter, we will take a close look at the meaning of the boundary layer theory and at the revolution this theory triggered not only in convective heat transfer but in fluid mechanics as well. What today is a universally accepted viewpoint and language was, at the turn of the century, only one man's revolutionary idea.

Boundary layer theory was proposed by Prandtl shortly after the completion of his doctoral dissertation in 1904 [1]. To say that it was not immediately accepted by its creator's contemporaries[†] is an understatement. To appreciate Prandtl's enormous accomplishment in converting the establishment, the reader has only to examine the 1932 edition of Lamb's *Hydrodynamics* [3]. This treatise of 385 articles devotes only one article to boundary layer theory and its pre-1932 results. It took three or four decades of persistent exposition by Prandtl for his theory to become the common language we speak today.

However, it is unfortunate and pedagagically inaccurate for today's teachers of convective heat transfer to present the boundary layer theory as dogma to be unreflectively applied to solve a very long list of engineering problems. By definition, a theory is never perfect [4], and, there is little exact about the similarity solutions to Prandtl's approximate boundary layer equations. As students and researchers, we can learn important lessons from the history of boundary layer theory. For example:

1. No theory is perfect and forever, not even the boundary layer theory.

2. It is legal and, indeed, desirable to question any accepted theory.

3. Any theory is better than no theory at all.

[†]As in Dryden's preface to Schlichting's *Boundary Layer Theory* [2].

4. It is legal to propose a new theory or a new idea in place of any accepted theory.

5. Lack of immediate acceptance of a new theory does not mean that the new theory is not better.

6. It is crucial to persevere to prove the worth of a new theory.

The history of scientific progress is full of episodes of the type exemplified by Prandtl. Since knowledge of this history and its lessons is a precondition for good research, the reader is urged to read Feyerabend's brilliant book on the origins of scientific progress [4].

THE FUNDAMENTAL PROBLEM IN CONVECTIVE HEAT TRANSFER

Consider the basic questions that engineers ask in connection with heat transfer from a solid object to a fluid stream in external flow. Think, for example, of a flat plate of temperature T_0 suspended in a uniform stream of velocity U_∞ and temperature T_∞, as is shown in Fig. 2.1. If this flat plate is the plate-fin protruding from a heat-exchanger surface into the stream that bathes it, we want to know:

1. The net force exerted by the stream on the plate.

2. The resistance to the transfer of heat from the plate to the stream.

We must answer question no. 1 in order to predict the total drag force exerted

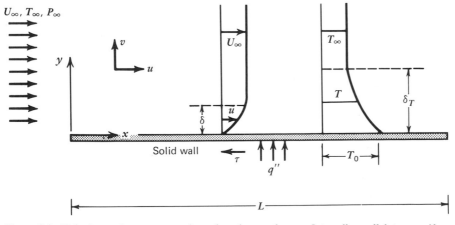

Figure 2.1 Velocity and temperature boundary layers along a flat wall parallel to a uniform stream.

by the stream on the heat-exchanger surface: from a simple force balance around the duct through which the stream flows (Chapter 3), we learn that the drag force felt by the solid surface translates into the pressure drop; hence, the pumping power or exergy payment [5] required to keep the stream flowing. We must also answer question no. 2 in order to predict the heat transfer rate between solid and fluid. In fact, question no. 2 is the fundamental question in the field of heat transfer, while question no. 1 is the fundamental question in fluid mechanics as it applies to heat transfer engineering.

Referring to Fig. 2.1, we are interested in calculating the total force

$$F = \int_0^L \tau W \, dx \tag{1}$$

and the total heat transfer rate

$$q = \int_0^L q'' W \, dx \tag{2}$$

Symbols τ, q'', and W stand for skin friction (shear stress experienced by the wall)

$$\tau = \mu \left(\frac{\partial u}{\partial y} \right)_{y=0} \tag{3}$$

wall heat-flux

$$q'' = h(T_0 - T_\infty) \tag{4}$$

and the width of the flat plate in the direction perpendicular to the plane of Fig. 2.1, respectively. In eqs. (3) and (4) we recognize two ideas inherited from Newton—definitions for the concepts of viscosity μ and heat-transfer coefficient h.

In this treatment we accept empirically, that is, as a matter of repeated physical observation, that the fluid layer situated at $y = 0^+$ is, in fact, stuck to the solid wall. This is the so-called *no-slip* hypothesis on which the bulk of modern convective heat transfer research is based: it acknowledges the observation that from heat exchangers to the honey in a jar, a fluid *wets* the solid surface with which it makes contact. The no-slip condition implies that since the $0 < y < 0^+$ fluid layer is motionless, the transfer of heat from the wall to the fluid is first by pure conduction. Therefore, in place of eq. (4), we can write *pure conduction* through the fluid layer immediately adjacent to the wall,

$$q'' = -k \left(\frac{\partial T}{\partial y} \right)_{y=0} \tag{5}$$

Note at this point the sign convention defined in Fig. 2.1: the heat flux q'' is defined as positive when the wall releases energy into the stream. Combining eqs. (4) and (5) we find a way to express and, eventually, calculate the heat-transfer coefficient

$$h = \frac{-k(\partial T/\partial y)_{y=0}}{T_0 - T_\infty} \tag{6}$$

To summarize, the two key questions in the field of convective heat transfer, the questions of friction and thermal resistance, boil down to carrying out the calculations dictated by eqs. (1) and (2). However, eqs. (3) and (5) demonstrate that in order to be able to calculate F and q, we must first determine the flow and temperature fields in the vicinity of the solid wall. Thus, it is the engineering demand for F and q that leads to the mathematical problem of solving for the flow (u, v) and temperature (T) in the fluid space outlined in Fig. 2.1. Modeling the flow as incompressible and constant-property (Chapter 1), the complete mathematical statement of this problem is—Solve four equations:

$$\frac{\partial u}{\partial x} + \frac{\partial v}{\partial y} = 0 \tag{7}$$

$$u\frac{\partial u}{\partial x} + v\frac{\partial u}{\partial y} = -\frac{1}{\rho}\frac{\partial P}{\partial x} + \nu\left(\frac{\partial^2 u}{\partial x^2} + \frac{\partial^2 u}{\partial y^2}\right) \tag{8}$$

$$u\frac{\partial v}{\partial x} + v\frac{\partial v}{\partial y} = -\frac{1}{\rho}\frac{\partial P}{\partial y} + \nu\left(\frac{\partial^2 v}{\partial x^2} + \frac{\partial^2 v}{\partial y^2}\right) \tag{9}$$

$$u\frac{\partial T}{\partial x} + v\frac{\partial T}{\partial y} = \alpha\left(\frac{\partial^2 T}{\partial x^2} + \frac{\partial^2 T}{\partial y^2}\right) \tag{10}$$

for four unknowns (u, v, P, T), subject to the following boundary conditions:

(i)	No slip	$u = 0$	
(ii)	Impermeability	$v = 0$	At the solid wall
(iii)	Wall temperature	$T = T_0$	

$$\tag{11}$$

(iv)	Uniform flow	$u = U_\infty$	Infinitely far from
(v)	Uniform flow	$v = 0$	the solid, in both
(vi)	Uniform temperature	$T = T_\infty$	directions (y and x)

In eqs. (7)–(10) we recognize, in order, statements accounting for the steady state conservation of mass, momentum, and energy at every point in the two-dimensional flow field. Conditions (i) and (ii) apply to the horizontal surfaces in Fig. 2.1; along the short leading and trailing surfaces ($x = 0, L$), the no-slip condition reads $v = 0$ and the impermeable wall condition reads $u = 0$.

THE CONCEPT OF BOUNDARY LAYER

The nonlinear partial differential problem stated as eqs. (7)–(11) has served as one of the central stimuli in the development of the field of applied mathematics during the past 200 years. The most remarkable feature of this problem is that, despite all this time and effort, it has not been solved. It is this feature that makes the boundary layer idea so special: it is a clever way to think and, a way to solve many historically unsolvable engineering problems. As with any great theory, it is a way to see simplicity in the complexity of unsolved problems.

Referring once again to Fig. 2.1 and the complete problem statement (7)–(11), we have the freedom to think that the velocity change from $u = 0$ to $u = U_\infty$ and the temperature change from $T = T_0$ to $T = T_\infty$ occurs in a *space* situated relatively close to the solid wall. How close is close is the object of the scale analysis presented later in this section. The important thing to understand at this stage is the revolutionary step taken by Prandtl in thinking of the region close to the wall (the boundary layer) as a region *distinct* from the immense domain in which the difficult mathematical problem (7)–(11) was formulated by his contemporaries. Prandtl's decision is equivalent to mentally carving out of the entire flow field only that region that is truly relevant to answering the engineering questions formulated in the preceding section. Outside the boundary layer, he imagines a *free-stream*, that is, a flow region not affected by the obstruction and heating effect introduced by the solid object. The free stream is characterized by

$$u = U_\infty, \qquad v = 0, \qquad P = P_\infty, \qquad T = T_\infty \tag{12}$$

Let δ be the order of magnitude of the distance in which u changes from 0 to roughly U_∞. Thus, in a space of height δ and length L on Fig. 2.1, we identify the following scales for changes in x, y and u:

$$x \sim L, \qquad y \sim \delta, \qquad u \sim U_\infty \tag{13}$$

In the $\delta \times L$ region, the longitudinal momentum equation (8) accounts for the competition between three types of forces,

$$
\begin{array}{ccc}
\text{Inertia} & \text{Pressure} & \text{Friction} \\[4pt]
U_\infty \dfrac{U_\infty}{L}, \; v\dfrac{U_\infty}{\delta} & \dfrac{P}{\rho L} & \nu\dfrac{U_\infty}{L^2}, \; \nu\dfrac{U_\infty}{\delta^2}
\end{array} \tag{14}
$$

In the above expression, each term represents the scale of each of the five terms appearing in eq. (8). Since the mass continuity equation (7) requires

$$\frac{U_\infty}{L} \sim \frac{v}{\delta} \tag{15}$$

we learn that the inertia terms in eq. (14) are *both* of order U_∞^2/L; hence, neither can be neglected at the expense of the other. However, if the boundary layer region $\delta \times L$ is *slender* such that

$$\delta \ll L \tag{16}$$

then the last scale in eq. (14) is the scale most representative of the friction force in that region. Thus, neglecting the $\partial^2 u/\partial x^2$ term at the expense of the $\partial^2 u/\partial y^2$ term in the x momentum equation (8) yields

$$u\frac{\partial u}{\partial x} + v\frac{\partial u}{\partial y} = -\frac{1}{\rho}\frac{\partial P}{\partial x} + v\frac{\partial^2 u}{\partial y^2} \tag{17}$$

Invoking the same scaling argument—the *slenderness* of the boundary layer region—the y momentum equation reduces to

$$u\frac{\partial v}{\partial x} + v\frac{\partial v}{\partial y} = -\frac{1}{\rho}\frac{\partial P}{\partial y} + v\frac{\partial^2 v}{\partial y^2} \tag{18}$$

Equation (18) is not usually discussed in connection with the boundary layer analysis of specific laminar flow problems. However, it is the basis for another important result, namely, the replacement of $\partial P/\partial x$ by a known quantity (dP_∞/dx) in eq. (17). To show how this is done, consider answering the following question: In a slender region $\delta \times L$, is the pressure variation in the y direction negligible when compared with the pressure variation in the x direction? Intuitively, we suspect that the answer must be "yes", because the region of interest $(\delta \times L)$ is by definition slender.

In general, the pressure at any point in the fluid of Fig. 2.1 is a function of both x and y; hence, the total derivative

$$dP = \frac{\partial P}{\partial x}\,dx + \frac{\partial P}{\partial y}\,dy \tag{19}$$

Dividing by dx, the question formulated in the preceding paragraph amounts to whether or not the last term is negligible in the expression

$$\frac{\partial P}{dx} = \frac{\partial P}{\partial x} + \frac{\partial P}{\partial y}\frac{dy}{dx} \tag{20}$$

The orders of magnitude of the two pressure gradients can be deduced from eqs. (17) and (18) by recognizing a balance between pressure forces and *either* friction or inertia [eq. (14)]. For the present argument it is not crucial which balance we invoke, as long as the same balance is invoked in both eqs. (17) and (18). For instance, a pressure ~ friction balance in eq. (17) suggests

$$\frac{\partial P}{\partial x} \sim \frac{\mu U_\infty}{\delta^2} \tag{21}$$

whereas the same balance in eq. (18) yields

$$\frac{\partial P}{\partial y} \sim \frac{\mu v}{\delta^2} \tag{22}$$

Now, turning our attention to the right-hand side of eq. (20), the ratio of the second term divided by the first term scales as

$$\frac{(\partial P/\partial y)(dy/dx)}{\partial P/\partial x} \sim \frac{v\delta}{U_\infty L} \sim \left(\frac{\delta}{L}\right)^2 \ll 1 \tag{23}$$

Note that in order to complete this last statement, we had to use the mass continuity scaling [eq. (15)] and the slenderness postulate [eq. (16)]. In conclusion, the last term in eq. (20) is less significant as the $\delta \times L$ region becomes more slender,

$$\frac{dP}{dx} = \frac{\partial P}{\partial x} \tag{24}$$

This means that in the boundary layer the pressure varies chiefly in the longitudinal direction, in other words, at any x the pressure inside the boundary layer region is practically the same as the pressure immediately outside it,

$$\frac{\partial P}{\partial x} = \frac{dP_\infty}{dx} \tag{25}$$

Making this last substitution in x momentum equation (17), we obtain finally

$$u\frac{\partial u}{\partial x} + v\frac{\partial u}{\partial y} = -\frac{1}{\rho}\frac{dP_\infty}{dx} + \nu\frac{\partial^2 u}{\partial y^2} \tag{26}$$

This is the *boundary layer equation* for momentum and, keeping in mind how it was derived, it is a statement of momentum conservation in *both* the x and y directions.

The boundary layer equation for energy follows from eq. (10), where we neglect the term accounting for thermal diffusion in the x direction,

$$u\frac{\partial T}{\partial x} + v\frac{\partial T}{\partial y} = \alpha\frac{\partial^2 T}{\partial y^2} \tag{27}$$

With this statement, we finish rewriting the original flow and heat transfer problem [eqs. (7)–(11)] in the language of boundary layer theory. We now have only three equations to solve [eqs. (7), (26), and (27)] for three unknowns (u, v, T). Compare this with the four equations—the four unknowns problem contemplated originally. In addition, the disappearance of the $\partial^2/\partial x^2$ diffusion terms from the momentum and energy equations makes this new problem solvable in a variety of ways. In the next section, we begin with the most cost-effective method of solution: scale analysis.

VELOCITY AND THERMAL BOUNDARY LAYER THICKNESSES

The boundary layer equations (26) and (27) are based on the thought that the significant variations in velocity and temperature occur in a slender region near the solid wall. This thought does not in any way imply that u and T reach their free-stream values within the same distance δ. Indeed, we have the freedom to think not of one but of an infinity of slender flow regions adjacent to the wall. Let δ be the thickness of the region in which u varies from 0 at the wall to U_∞ in the free stream. Let δ_T be the thickness of another slender region in which T varies from T_0 at the wall to T_∞ in the free stream. Keeping up with tradition, in the present treatment we refer to δ and δ_T as the velocity boundary layer thickness and the thermal boundary layer thickness, respectively. These scales are shown schematically in Fig. 2.1; in general, $\delta \neq \delta_T$.

In scaling terms, the flow friction question (3) is answered by writing

$$\tau \sim \mu\frac{U_\infty}{\delta} \tag{28}$$

Thus, in order to estimate the wall frictional-shear stress we must evaluate the extent δ of this imaginary slender wall region. Consider the simplest free stream possible, namely, a free stream with uniform pressure P_∞. (This is a very good approximation for the flow around a plate fin in a heat-exchanger passage, because the pressure drop in the direction of flow is not significant over the longitudinal length L dictated by the plate fin.) With $dP_\infty/dx = 0$ in eq. (26), the boundary layer momentum equation implies

$$\text{Inertia} \quad \sim \quad \text{Friction}$$

$$\frac{U_\infty^2}{L}, \frac{vU_\infty}{\delta} \quad \sim \quad v\frac{U_\infty}{\delta^2} \tag{29}$$

Referring once again to the mass continuity scaling (15), we conclude that the two inertia terms are of the same order of magnitude. Therefore, eq. (29) requires

$$\delta \sim \left(\frac{\nu L}{U_\infty} \right)^{1/2} \tag{30}$$

in other words,

$$\frac{\delta}{L} \sim \mathrm{Re}_L^{-1/2} \tag{31}$$

where Re_L is the Reynolds number based on the longitudinal dimension of the boundary layer region.

Equation (31) is an important result: it states that the slenderness postulate on which the boundary layer theory is based ($\delta \ll L$) is valid provided $\mathrm{Re}_L^{1/2} \gg 1$. Thus, eq. (31) is a test of whether a given external flow situation lends itself to boundary layer analysis, as Re_L can easily be calculated beforehand. Furthermore, even when $\mathrm{Re}_L^{1/2} \gg 1$, eq. (31) can be used to assess the limitations of the boundary layer analysis: for example, the boundary layer solution will fail in the tip region of length l, short enough so that $\mathrm{Re}_l^{1/2}$ is not considerably greater than unity.

Returning to the engineering question at hand [eq. (28)], the wall shear stress scales as

$$\tau \sim \mu \frac{U_\infty}{L} \mathrm{Re}_L^{1/2} \sim \rho U_\infty^2 \mathrm{Re}_L^{-1/2} \tag{32}$$

Therefore, the dimensionless *skin-friction coefficient* $C_f = \tau/(\frac{1}{2}\rho U_\infty^2)$ depends on the Reynolds number.

$$C_f \sim \mathrm{Re}_L^{-1/2} \tag{33}$$

At this point, the question of wall friction has been answered in an order of magnitude sense. The scaling analysis on which eq. (32) is based assures us that the real (measured or calculated) value of τ will differ from $\rho U_\infty^2 \mathrm{Re}_L^{-1/2}$ by only a factor of order unity. This prediction is amply verified by more exact analyses, as is shown later in this chapter.

The heat transfer engineering question [eq. (6)] is answered by focusing on the thermal boundary layer of thickness δ_T,

$$h \sim \frac{k(\Delta T/\delta_T)}{\Delta T} \sim \frac{k}{\delta_T} \tag{34}$$

where $\Delta T = (T_0 - T_\infty)$ is the temperature variation scale in the region $\delta_T \times L$.

The boundary layer energy equation (27) states that there is always a balance between conduction from the wall into the stream and convection (enthalpy flow) parallel to the wall,

$$\text{Convection} \quad \sim \quad \text{Conduction}$$

$$u\frac{\Delta T}{L}, \; v\frac{\Delta T}{\delta_T} \quad \sim \quad \alpha\frac{\Delta T}{\delta_T^2} \tag{35}$$

Mass continuity in a layer of thickness δ_T dictates $u/L \sim v/\delta_T$; hence, the two convection scales in eq. (35) are of comparable magnitude ($\sim u\,\Delta T/L$). To proceed with the evaluation of δ_T, it is very tempting to take $u \sim U_\infty$, as is done by Schlichting [6]. However, to do this would be erroneous. As is shown in Fig. 2.2, the actual velocity scale in the δ_T layer depends on the *relative* size of δ_T and δ. Let us first assume that

$$\textbf{(i)} \;\; \delta \ll \delta_T, \quad \text{hence } u \sim U_\infty \tag{36}$$

With this assumption (whose range of applicability we do not know yet), eq. (35) yields

$$\frac{\delta_T}{L} \sim \text{Pe}_L^{-1/2} \sim \text{Pr}^{-1/2}\text{Re}_L^{-1/2} \tag{37}$$

where $\text{Pe}_L = U_\infty L/\alpha$ is the Peclet number. Comparing eq. (37) with eq. (31), we find the most interesting result that the relative size of δ_T and δ depends on the Prandtl number $\text{Pr} = \nu/\alpha$,

$$\frac{\delta_T}{\delta} \sim \text{Pr}^{-1/2} \gg 1 \tag{38}$$

The first assumption, (i) $\delta \ll \delta_T$, is therefore valid strictly in the limit $\text{Pr}^{1/2} \ll 1$, which represents the range occupied by liquid metals.

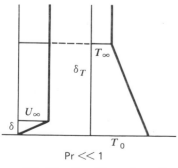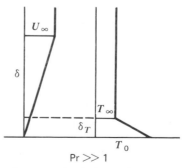

Figure 2.2 The Prandtl number effect on the joint development of velocity and temperature boundary layers.

The heat transfer coefficient corresponding to assumption (i) is

$$h \sim \frac{k}{L} Pr^{1/2} Re_L^{1/2}, \qquad Pr \ll 1 \tag{39}$$

or, expressed as a Nusselt number $Nu = hL/k$:

$$Nu \sim Pr^{1/2} Re_L^{1/2} \tag{40}$$

Of considerable practical interest is the case of fluids with a Prandtl number of order unity (e.g., air) or greater than unity (e.g., water, or oils). As is shown on the right side of Fig. 2.2, let us assume that

$$\textbf{(ii)} \ \delta \gg \delta_T \tag{41}$$

Geometrically, it is clear that the scale of u in the δ_T layer is not U_∞ but $(\delta_T/\delta)U_\infty$. Substituting this scale into the convection $\sim$ conduction balance (35) yields

$$\frac{\delta_T}{L} \sim Pr^{-1/3} Re_L^{-1/2} \tag{42}$$

which means

$$\frac{\delta_T}{\delta} \sim Pr^{-1/3} \ll 1 \tag{43}$$

Thus assumption (ii) is valid in the case of $Pr^{1/3} \gg 1$ fluids. The heat-transfer coefficient and Nusselt number vary as

$$h \sim \frac{k}{L} Pr^{1/3} Re_L^{1/2}, \qquad Pr \gg 1 \tag{44}$$

$$Nu \sim Pr^{1/3} Re_L^{1/2}, \qquad Pr \gg 1 \tag{45}$$

These scaling results agree within a factor of order unity with the classical results discussed next.

An important observation concerns eq. (31) which is the first place we encounter the Reynolds number in external flow, $Re_L = U_\infty L/\nu$. In most introductory treatments of fluid mechanics, the Reynolds number is interpreted as the order of magnitude of the ratio inertia/friction in a particular flow.[†] This interpretation is not always correct because, at least in the boundary layer region examined above, there is always a balance between inertia and friction, whereas Re_L can reach as high as 10^4 before the transition

---

[†]See Problem 12 at the end of Chapter 1.

to turbulent flow (Table 6.1). The only physical interpretation of the Reynolds number in boundary layer flow appears to be

$$\mathrm{Re}_L^{1/2} = \frac{\text{the wall length}}{\text{the boundary layer thickness}}$$

In other words, it is not Re_L but the square root of Re_L that means something: $\mathrm{Re}_L^{1/2}$ is *a geometric parameter of the flow region—the slenderness ratio*.

It is also worth noting that according to eq. (30), δ must be proportional to $L^{1/2}$. More refined analyses described later in this chapter confirm that along the wall $(0 < x < L)$ the boundary layer thickness increases as $x^{1/2}$. Now, one particular property of the $x^{1/2}$ function is that its slope is infinite at $x = 0$, as is shown schematically in Figs. 2.8 and 2.9. This geometric feature of the boundary layer is unexplicably absent from the graphics employed by most texts that teach boundary layer theory.

INTEGRAL SOLUTIONS

The next step in the act of refining our answers to the friction and heat transfer questions (3) and (6) amounts to determining the numerical coefficients (factors) missing from the scaling laws (32), (39), and (44). So far, the above scaling laws tell us the *manner* in which various flow and geometric parameters affect τ and h. For example, we know now that both τ and h are proportional to $L^{-1/2}$, meaning that the skin friction and heat flux are more intense near the leading edge of the flat plate. In the realm of scaling analysis, we made no distinction between the *local* values of τ an h (the values right at $x = L$) and the *average* values τ_{0-L} and h_{0-L} defined as

$$\tau_{0-L} = \frac{1}{L} \int_0^L \tau\, dx, \qquad h_{0-L} = \frac{1}{L} \int_0^L h\, dx \qquad (46)$$

The reason for such treatment is that the average quantities $(\tau, h)_{0-L}$ have exactly the same scale as the τ and h evaluated at $x = L$; this scaling conclusion is easily drawn from eqs. (46) and the specific results developed in the remainder of this chapter.

The *integral* approach to solving the boundary layer equations is an important piece of analysis developed by Pohlhausen and von Karman in the first decades of this century in Germany. The engineering philosophy on which this approach is based is the same philosophy that allowed Prandtl to separate from an immense and complicated flow field only the region *most relevant* to answering the practical question at hand. In the integral method, we look at the definitions of τ and h [eqs. (3) and (6)] and recognize that what we need is not a complete solution for the velocity $u(x, y)$ and temperature $T(x, y)$ near the wall, but only the gradients $\partial(u, T)/\partial y$ evaluated at $y = 0$. Since the

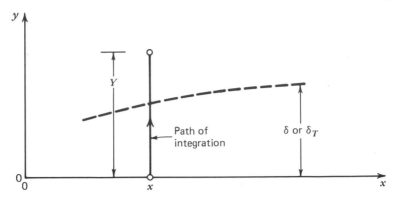

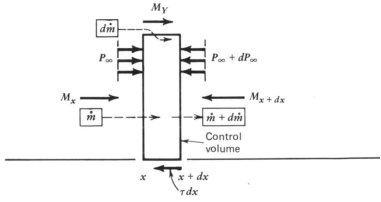

Figure 2.3 Derivation of the integral boundary layer equations, or force balance on a control volume of thickness dx.

$y > 0$ variation of u and T is not the most relevant to evaluating τ and h, we have the opportunity to simplify the boundary layer equations (26) and (27) by eliminating y as a variable. As is shown in Fig. 2.3, this is accomplished by integrating each equation term-by-term from $y = 0$ to $y = Y$, where $Y > \max(\delta, \delta_T)$ is situated in the free stream. Before performing the integrals, it is useful to rewrite eqs. (26) and (27) as

$$\frac{\partial}{\partial x}(u^2) + \frac{\partial}{\partial y}(uv) = -\frac{1}{\rho}\frac{dP_\infty}{dx} + \nu\frac{\partial^2 u}{\partial y^2} \tag{47}$$

$$\frac{\partial}{\partial x}(uT) + \frac{\partial}{\partial y}(vT) = \alpha\frac{\partial^2 T}{\partial y^2} \tag{48}$$

Form (47) is obtained by multiplying the left-hand side of the mass conservation equation (7) by u and adding it to the left-hand side of eq. (26); form (48) is obtained in similar fashion, that is, by multiplying eq. (7) by T.

Integrating eqs. (47) and (48) from $y = 0$ to $y = Y$ yields

$$\frac{d}{dx} \int_0^Y u^2 \, dy + u_Y v_Y - u_0 v_0 = -\frac{1}{\rho} Y \frac{dP_\infty}{dx} + \nu \left(\frac{\partial u}{\partial y} \right)_Y - \nu \left(\frac{\partial u}{\partial y} \right)_0 \quad (49)$$

$$\frac{d}{dx} \int_0^Y uT \, dy + v_Y T_Y - v_0 T_0 = \alpha \left(\frac{\partial T}{\partial y} \right)_Y - \alpha \left(\frac{\partial T}{\partial y} \right)_0 \quad (50)$$

in which indices Y and 0 indicate the level y where the respective quantities are to be evaluated. Since the free stream is uniform, $(\partial/\partial y)_Y = 0$ and $u_Y = U_\infty$, $T_Y = T_\infty$. Also, since the wall is impermeable, $v_0 = 0$; we evaluate v_Y by performing the same integral on the continuity equation (7),

$$\frac{d}{dx} \int_0^Y u \, dy + v_Y - v_0 = 0 \quad (51)$$

Substituting v_Y into eqs. (49) and (50) and rearranging the resulting expressions we obtain finally

$$\frac{d}{dx} \int_0^Y u(U_\infty - u) \, dy = \frac{1}{\rho} Y \frac{dP_\infty}{dx} + \frac{dU_\infty}{dx} \int_0^Y u \, dy + \nu \left(\frac{\partial u}{\partial y} \right)_0 \quad (52)$$

$$\frac{d}{dx} \int_0^Y u(T_\infty - T) \, dy = \alpha \left(\frac{\partial T}{\partial y} \right)_0 \quad (53)$$

These are the integral boundary layer equations for momentum and energy: they account for the conservation of momentum and energy not at every point (x, y) as eqs. (26) and (27), but in every *slice* of thickness dx and height Y (see bottom of Fig. 2.3). It is worth noting that eqs. (52) and (53) can also be derived by invoking the x momentum theorem and the First Law of Thermodynamics (Chapter 1) for the control volume of size $(Y) \times (dx)$ shown in the bottom-half of Fig. 2.3. For example, the momentum equation (52) represents the following force balance:

1. Forces acting from left to right on the control volume (Fig. 2.3, bottom):

$$M_x = \int_0^Y \rho u^2 \, dy \qquad \text{Impulse due to the flow of a stream } into \text{ the control volume.}$$

$M_Y = U_\infty \, d\dot{m}$

Impulse due to the flow of fast fluid (U_∞) into the control volume, at a rate $d\dot{m}$, where $\dot{m} = \int_0^Y \rho u \, dy$ is the mass flowrate through the slice of height Y.

$P_\infty Y$

Pressure force.

2. Forces acting from right to left on the control volume (Fig. 2.3, bottom):

$M_{x+dx} = M_x + (dM_x/dx) \, dx$

Reaction force due to flow of a stream *out* of the control volume.

$\tau \, dx$

Tangential force due to friction.

$Y[P_\infty + (dP_\infty/dx) \, dx]$

Pressure force.

Setting the resultant of all these forces equal to zero, we derive eq. (52). The integral energy equation (53) can be obtained similarly by summing up all the heat transfer and enthalpy flowrates around the control surface. The details of this analytical work are presented in most textbooks on heat transfer (e.g., in Refs. 7–10).

Consider now the simplest laminar boundary layer problem—the uniform flow (U_∞, P_∞ = constants) analyzed in the preceding section. In order to solve for the wall shear stress appearing now explicitly in eq. (52), we must make an assumption as a substitute for the information we gave up when we integrated the original boundary layer equation (26): the y variation of the flow. Let us assume that the *shape* of the longitudinal velocity profile is described by

$$u = U_\infty m(n), \qquad 0 \le n \le 1$$

$$u = U_\infty, \qquad 1 \le n \tag{54}$$

where m is an unspecified profile shape function that varies from 0 to 1 and $n = y/\delta$ (see Fig. 2.4). Substituting this assumption into eq. (52) with $dP_\infty/dx = 0$ yields a first-order ordinary differential equation for the velocity boundary layer thickness $\delta(x)$,

$$\delta \frac{d\delta}{dx} \left[\int_0^1 m(1 - m) \, dn \right] = \frac{\nu}{U_\infty} \left(\frac{dm}{dn} \right)_{n=0} \tag{55}$$

The resulting expressions for *local* boundary layer thickness and skin-friction coefficient are

$$\frac{\delta}{x} = a_1 \mathrm{Re}_x^{-1/2} \tag{56}$$

$$C_{f,x} = \frac{\tau}{\frac{1}{2}\rho U_\infty^2} = a_2 \mathrm{Re}_x^{-1/2} \tag{57}$$

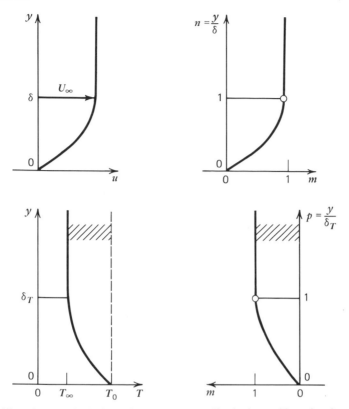

Figure 2.4 The selection of velocity and temperature profiles for integral boundary layer analysis.

with the following notation

$$a_1 = \left[\frac{2(dm/dn)_{n=0}}{\int_0^1 m(1-m)\,dn} \right]^{1/2}$$

$$a_2 = \left[2\left(\frac{dm}{dn} \right)_{n=0} \int_0^1 m(1-m)\,dn \right]^{1/2}$$

Results (56) and (57) agree in order of magnitude sense with the earlier conclusions [eqs. (31) and (33)]. The numerical coefficients a_1 and a_2 depend on the particular guess made for the profile shape function m: Table 2.1 shows that as long as this guess is reasonable,[†] the choice of $m(n)$ does not influence the skin friction result appreciably.

[†]A function that increases from $n = 0$ to $n = 1$ monotonically and smoothly, and has a finite slope at $n = 0$.

Table 2.1   The Impact of Profile Shape on the Integral Solution to the Laminar Boundary Layer Friction and Heat Transfer Problem

Profile Shapes $m(n)$ or $m(p)$ (Fig. 2.4)	$\frac{\delta}{x}\mathrm{Re}_x^{1/2}$ (Ref. 7)	$C_{f,x}\mathrm{Re}_x^{1/2}$ (Ref. 7)	$\mathrm{Nu}\,\mathrm{Re}_x^{-1/2}\mathrm{Pr}^{-1/3}$ Uniform Temperature [7] (Pr > 1)	Uniform Heat Flux[17] (Pr > 1)
$m = n$	3.46	0.577	0.289	0.364
$m = \frac{n}{2}(3 - n^2)$	4.64	0.646	0.331	0.417
$m = \sin(\pi n/2)$	4.8	0.654	0.337	0.424
Similarity solution	4.92^a	0.664	0.332	0.458^b

[a] Thickness defined as the y value corresponding to $u/U_\infty = 0.99$.
[b] From Ref. 23.

Heat-transfer coefficient information is extracted in a similar fashion from eq. (53). Thus we assume the temperature profile shapes

$$T_0 - T = (T_0 - T_\infty)m(p), \qquad 0 \le p \le 1$$

$$T = T_\infty, \qquad 1 < p \tag{58}$$

with $p = y/\delta_T$; educated by the scale analysis discussed earlier, we assume that

$$\frac{\delta_T}{\delta} = \Delta \tag{59}$$

where Δ is a function of Prandtl number only and δ is given by eq. (56). Based on these assumptions and $\delta_T < \delta$ (high Pr fluids), the integral energy equation (53) reduces to

$$\mathrm{Pr} = \frac{2(dm/dp)_{p=0}}{(a_1\Delta)^2}\left[\int_0^1 m(p\Delta)[1 - m(p)]\,dp\right]^{-1} \tag{60}$$

This result is an implicit expression for a thickness ratio function $\Delta(\mathrm{Pr})$, thus confirming the validity of the scaling arguments on which eq. (59) was based. Assuming the simplest temperature profile, $m = p$, expression (60) becomes

$$\Delta = \mathrm{Pr}^{-1/3} \tag{61}$$

which is numerically identical to the scaling law for $\mathrm{Pr} \gg 1$ fluids [eq. (43)]. As hinted by Table 2.1, other choices of profile shape $m(p)$ will change the proportionality factor in eq. (61) by only percentage points. The results usually listed in the literature correspond to the cubic profile $m = (p/2)(3 - p^2)$:

$$\Delta = \frac{\delta_T}{\delta} = 0.977\,\mathrm{Pr}^{-1/3} \tag{62}$$

$$h = 0.323\frac{k}{x}\mathrm{Pr}^{1/3}\mathrm{Re}_x^{1/2} \tag{63}$$

$$\mathrm{Nu} = \frac{hx}{k} = 0.323\,\mathrm{Pr}^{1/3}\mathrm{Re}_x^{1/2} \tag{64}$$

The local heat transfer results listed above are anticipated correctly by the scale analysis [eq. (44) and (45)].

In the case of liquid metals ($\Delta \gg 1$), instead of eq. (60) we obtain

$$\mathrm{Pr} = \frac{2(dm/dp)_{p=0}}{(a_1\Delta)^2}\left[\int_0^{1/\Delta} m(p\Delta)[1 - m(p)]\,dp + \int_{1/\Delta}^1 [1 - m(p)]\,dp\right]^{-1} \tag{65}$$

The sum of two integrals stems from the fact that when $\delta_T \gg \delta$, immediately next to the wall ($0 < y < \delta$) the velocity is described by the assumed shape $U_\infty m$, whereas for $\delta < y < \delta_T$ the velocity is uniform, $u = U_\infty$ [eq. (54)]. Since Δ is much greater than unity, the second integral dominates in eq. (65). Taking again the simplest profile $m = p$, we obtain

$$\Delta = \frac{\delta_T}{\delta} = (3\,\mathrm{Pr})^{-1/2}, \qquad \mathrm{Pr} \ll 1 \tag{66}$$

in other words,

$$\frac{\delta_T}{x} = 2\,\mathrm{Pr}^{-1/2}\mathrm{Re}_x^{-1/2}, \qquad \mathrm{Pr} \ll 1 \tag{67}$$

From eq. (6), the heat-transfer coefficient is

$$h = \frac{k}{\delta_T} = \frac{1}{2}\frac{k}{x}\mathrm{Pr}^{1/2}\mathrm{Re}_x^{1/2}, \qquad \mathrm{Pr} \ll 1 \tag{68}$$

or

$$\mathrm{Nu} = \frac{hx}{k} = \frac{1}{2}\mathrm{Pr}^{1/2}\mathrm{Re}_x^{1/2}, \qquad \mathrm{Pr} \ll 1 \tag{69}$$

These results compare favorably with the scaling laws [eqs. (37)–(40)]. As shown in the next section, they also compare favorably with more exact (and expensive) solutions.

SIMILARITY SOLUTIONS

In this section we review the exact solutions to the boundary layer problem of Fig. 2.1, solutions due to Blasius [11] for the flow problem and Pohlhausen [12] for the heat transfer problem. Relative to the integral solutions presented in the preceding section, the Blasius–Pohlhausen solutions have the added benefit that they describe the y variation of the flow and temperature fields in the boundary layer regions. The basic idea in the construction of these solutions is the feeling that from one location x to another, then u and T profiles look *similar* (hence, the name of *similarity solutions*). Figure 2.5 shows that although more and more fluid slows down near the wall as x increases, the longitudinal velocity is always $u = 0$ at the wall and $u = U_\infty$ sufficiently far from the wall. It is possible to think (to suppose) that the two profiles $u_1(y)$ and $u_2(y)$ were drawn by an artist who used the master profile shown in Fig. 2.5: like the elastic metal band of a wristwatch, this master profile can be stretched appropriately at x_1 and x_2, so as to fit the actual velocity profiles.

Mathematically, the appropriate stretching of a master profile amounts to writing

$$\frac{u}{U_\infty} = \text{function}(\eta)$$

where the *similarity variable* η is proportional to y and the proportionality

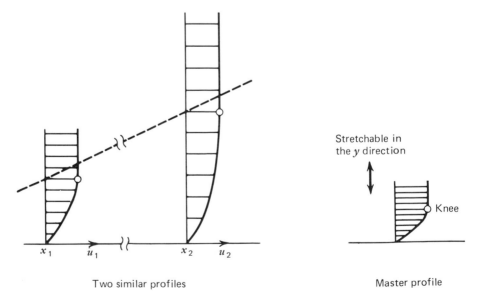

Two similar profiles   Master profile

Figure 2.5 The construction of similar profiles and similarity solutions in the analysis of boundary layers.

factor depends on x. Based on the scaling laws we already know, it is fairly obvious that η must be proportional to $y/\delta(x)$, with $\delta \sim x\,\mathrm{Re}_x^{-1/2}$. We assume therefore

$$\frac{u}{U_\infty} = f'(\eta) \tag{70}$$

and

$$\eta = \frac{y}{x}\,\mathrm{Re}_x^{1/2} \tag{71}$$

Function $f' = df/d\eta$ is presently unknown and accounts for the shape of the master profile: this function is the object of the analysis constructed by Blasius.

The flow problem can be restated here as the conservation of mass and momentum at every point in a $P_\infty =$ constant boundary layer

$$\frac{\partial u}{\partial x} + \frac{\partial v}{\partial y} = 0 \tag{72}$$

$$u\frac{\partial u}{\partial x} + v\frac{\partial u}{\partial y} = \nu\frac{\partial^2 u}{\partial y^2} \tag{73}$$

subject to only three boundary conditions

$$u = v = 0 \quad \text{at } y = 0 \tag{74}$$

$$u \to U_\infty \quad \text{as } y \to \infty \tag{75}$$

A useful bit of shorthand is the introduction of a streamfunction $\psi(x, y)$ defined as

$$u = \frac{\partial \psi}{\partial y}, \qquad v = -\frac{\partial \psi}{\partial x} \tag{76}$$

so that the continuity equation (72) is satisfied identically (see the last section of Chapter 1). In terms of the streamfunction, the problem consists of solving

$$\frac{\partial \psi}{\partial y}\frac{\partial^2 \psi}{\partial x\partial y} - \frac{\partial \psi}{\partial x}\frac{\partial^2 \psi}{\partial y^2} = \nu\frac{\partial^3 \psi}{\partial y^3} \tag{77}$$

subject to

$$\frac{\partial \psi}{\partial y} = 0, \qquad \psi = 0 \quad \text{at } y = 0 \tag{78}$$

$$\frac{\partial \psi}{\partial y} \to U_\infty \quad \text{as } y \to \infty \tag{79}$$

This problem is finally placed in the language of the similarity transformation (70) and (71) by evaluating ψ and its derivatives. For example, from the first of eqs. (76) we obtain

$$\psi = (U_\infty \nu x)^{1/2} f(\eta) \tag{80}$$

and from the second of eqs. (76)

$$v = \frac{1}{2}\left(\frac{\nu U_\infty}{x}\right)^{1/2}(\eta f' - f) \tag{81}$$

Expressions for the partial derivatives of ψ appearing in eq. (77) are obtained by keeping in mind that according to eq. (80), ψ depends on x directly and via $\eta(x, y)$. The similarity statement of the problem is

$$2f''' + ff'' = 0 \tag{82}$$

with the following boundary conditions

$$f' = f = 0 \quad \text{at } \eta = 0 \tag{83}$$

$$f' \to 1 \quad \text{as } \eta \to \infty \tag{84}$$

Equation (82) is nonlinear; Blasius solved it approximately by the method of matched asymptotic expansions (see Problem 2 at the end of this chapter). Blasius' method as well as a number of more recent solutions are discussed in Schlichting [13].

The resulting velocity profile f' is shown in Fig. 2.6: u reaches U_∞ asymptotically as η tends to infinity. Unlike in the integral solutions developed earlier, there is no clear *knee* in the u/U_∞ curve to mark the boundary layer thickness δ. For this reason, in the Blasius profile, δ is defined based on *convention*. Numerically, it is found that $u = 0.99U_\infty$ at $\eta = 4.92$; the boundary layer thickness is taken as equal to the value of y corresponding to this condition

$$\frac{\delta}{x} = 4.92 \, \text{Re}_x^{-1/2} \tag{85}$$

To get around the need for convention in defining δ, two other thicknesses have been in use in the field of boundary layer theory,

$$\delta^* = \int_0^\infty \left(1 - \frac{u}{U_\infty}\right) dy, \quad \text{displacement thickness} \tag{86}$$

$$\theta = \int_0^\infty \frac{u}{U_\infty}\left(1 - \frac{u}{U_\infty}\right) dy, \quad \text{momentum thickness} \tag{87}$$

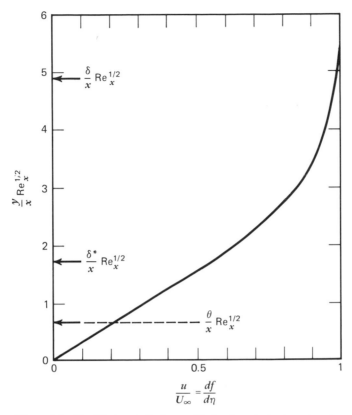

Figure 2.6 The Blasius similarity profile for the velocity boundary layer.

As is shown in Fig. 2.7, the displacement thickness is a measure of the fraction of the original free stream slowed down viscously by the wall

$$\delta^* U_\infty = \int_0^\infty U_\infty \, dy - \int_0^\infty u \, dy \tag{88}$$

The dotted line on Fig. 2.7 shows that at any x the free stream appears to be displaced away from the wall so that it can avoid and flow past the fluid viscously stuck to the wall. The momentum thickness θ is based on a similar argument: it is a measure of the longitudinal momentum missing at any x relative to the original ($x = 0$) amount

$$\theta U_\infty^2 = \underbrace{\int_0^\infty U_\infty^2 \, dy}_{\substack{x \text{ momentum} \\ \text{at } x = 0}} - \underbrace{\int_0^\infty u^2 \, dy}_{\substack{x \text{ momentum} \\ \text{at any } x}} - \underbrace{U_\infty \int_0^\infty (U_\infty - u) \, dy}_{\substack{x \text{ momentum of the} \\ \text{fluid displaced out of the} \\ \text{boundary layer region}}} \tag{89}$$

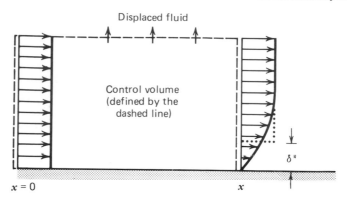

Figure 2.7   The displacement thickness δ^* and its physical interpretation.

In this regard, note that expression (87) is a restatement of the left-hand side of the integral momentum equation (52). The displacement and momentum thicknesses for the Blasius similarity solution (Fig. 2.6) are

$$\frac{\delta^*}{x} = 1.73 \, \mathrm{Re}_x^{-1/2}, \qquad \frac{\theta}{x} = 0.664 \, \mathrm{Re}_x^{-1/2} \tag{90}$$

Finally, the skin-friction coefficient predicted by the similarity solution is

$$C_{f,x} = \frac{\mu(\partial u/\partial y)_0}{\frac{1}{2}\rho U_\infty^2} = 2(f'')_{\eta=0}\mathrm{Re}_x^{-1/2} \tag{91}$$

Numerically it is found that $(f'')_{y=0} = 0.332$ [14]; hence

$$C_{f,x} = 0.664 \, \mathrm{Re}_x^{-1/2} \tag{92}$$

This result is not far off from any of the considerably less laborious predictions based on the integral method (Table 2.1).

The heat transfer part of the problem was solved along similar lines [12]: introducing the dimensionless similarity temperature profile

$$\theta(\eta) = \frac{T - T_0}{T_\infty - T_0} \tag{93}$$

the boundary layer energy equation (27) assumes the form

$$\theta'' + \frac{\mathrm{Pr}}{2} f\theta' = 0 \tag{94}$$

This equation must be solved subject to the known wall and free-stream temperature conditions

$$\theta = 0 \quad \text{at } \eta = 0 \tag{95}$$

$$\theta \to 1 \quad \text{as } \eta \to \infty \tag{96}$$

An interesting observation is the fact that if $\Pr = 1$ and $\theta = f'$, then the heat transfer problem (94)–(96) becomes identical to the Blasius problem (82)–(84). This means that for $\Pr = 1$ fluids, the similarity temperature profile is already known and plotted in Fig. 2.6.

In general (for any $\Pr$), eq. (94) can be integrated, keeping in mind that $f(\eta)$ is a known function available in tabular form [14]. Via separation of variables, we integrate eq. (94) and obtain

$$\theta'(\eta) = \theta'(0)\exp\left[-\frac{\Pr}{2}\int_0^\eta f(\beta)\,d\beta\right] \tag{97}$$

Integrating again from 0 to η and using the wall condition (95) yields

$$\theta(\eta) = \theta'(0)\int_0^\eta \exp\left[-\frac{\Pr}{2}\int_0^\gamma f(\beta)\,d\beta\right]d\gamma \tag{98}$$

where β and γ are two dummy variables. The above solution for $\theta(\eta)$ depends on an unknown constant of integration, $\theta'(0)$, because the free-stream condition (96) has not been used yet; using it, we find

$$\theta'(0) = \left\{\int_0^\infty \exp\left[-\frac{\Pr}{2}\int_0^\gamma f(\beta)\,d\beta\right]d\gamma\right\}^{-1} \tag{99}$$

The solution is now complete and, as one may have expected, the value of $\theta'(0)$ is all important in calculating the heat-transfer coefficient: from eq. (6) we learn that

$$h = \frac{k}{x}\,\mathrm{Re}_x^{1/2}\theta'(0) \tag{100}$$

hence,

$$\mathrm{Nu} = \frac{hx}{k} = \theta'(0)\mathrm{Re}_x^{1/2} \tag{101}$$

As is shown by eq. (99), $\theta'(0)$ is a function of the Prandtl number which accounts for the relationship between Nu and Pr predicted on scaling grounds early in this chapter. Pohlhausen [12] calculated a number of $\theta'(0)$ values which, if $\Pr > 0.5$, are correlated accurately by

$$\theta'(0) = 0.332\,\Pr^{1/3} \tag{102}$$

The theoretical basis for this so far empirical correlation is contained in the discussion leading to eq. (45). The similarity solution for the heat-transfer coefficient (Nusselt number) is therefore

$$\mathrm{Nu} = 0.332\,\Pr^{1/3}\mathrm{Re}_x^{1/2}, \qquad \Pr > 0.5 \tag{103}$$

A different correlation must be used below Pr < 0.5 or, if the Prandtl number of a particular liquid metal is given, eq. (99) should be solved once for that particular Pr. It is better still to develop an analytical replacement for the Nu formula (103) in the limit Pr → 0. According to the left side of Fig. 2.2, in highly conductive fluids the velocity boundary layer is much thinner than the thermal layer. Therefore, in this limit it is permissible to set $f' = 1$ in the region occupied by the thermal boundary layer $\theta(\eta)$. Differentiating eq. (94) once

$$\frac{d}{d\eta}\left(\frac{\theta''}{\theta'}\right) = -\frac{\text{Pr}}{2}f' \tag{104}$$

This equation paves the way to an explicit solution for $\theta(\eta)$ in the limit Pr → 0. The ensuing analysis is proposed as an exercise to the student; its chief results are

$$\theta(\eta) = \text{erf}\left(\frac{\eta}{2}\sqrt{\text{Pr}}\right)$$

$$\theta'(0) = \left(\frac{\text{Pr}}{\pi}\right)^{1/2} \tag{105}$$

$$\text{Nu} = \frac{hx}{k} = 0.564\,\text{Pr}^{1/2}\,\text{Re}_x^{1/2}, \qquad \text{Pr} \to 0$$

This limiting heat transfer result compares favorably with the scaling law (40).

In concluding this section, it is worth commenting further on the imperfect character of the boundary layer theory and the approximation built into the so-called *exact similarity solution*. Examination of the Blasius solution for the velocity normal to the wall shows that v tends to a finite value, $0.86U_\infty\,\text{Re}_x^{-1/2}$, as η tends to infinity [14]. This feature distinguishes the boundary layer problem from the complete problem stated in eqs. (7)–(11) where v must clearly vanish sufficiently far from the wall [condition (v) eq. (11)]. Since in the boundary layer theory $v/U_\infty \sim \text{Re}_x^{-1/2}$ as $\eta \to \infty$, this theory becomes "better" as $\text{Re}_x^{1/2}$ increases, that is, as the boundary layer region becomes more slender. Other limitations of the theory have been discussed earlier in connection with the breakdown of the *slenderness* feature in the region near the tip [see the discussion following eq. (31)].

OTHER WALL HEATING CONDITIONS

What we have seen so far is open competition between three methodologies (scaling, integral, similarity) in the search for engineering answers to the basic questions of convective heat transfer. The laminar boundary layer near an isothermal flat plate was the simplest and, historically, oldest setting in which to witness this competition. Of course, despite what the pure scientists and pure engineers among us may want us to believe, there can be no official winner in such a competition. However, an individual researcher with a personal mathematics background and, most important, with a personal supply of curiosity and time can and should judiciously evaluate the worthiness of any of these methodologies relative to his ability and taste. And he is free to choose.

The engineering problems we encounter in the field are diverse and, quite often, demand models that differ from the isothermal flat plate problem of Fig. 2.1. Although in each case the model and engineering answers (C_f, Nu) are different, the conceptual basis is the same, as defined by Prandtl's boundary layer theory. There have been very many advances made along the lines of this theory and the most important of these are reviewed in the handbooks [6, 9]. In this section, we mention only a few examples.

Unheated Starting Length

In cases such as the forced convection cooling of an electric circuit board, the heating effect is distributed discretely along the flat plate. The simplest question to formulate in relation to this practical situation is sketched in Fig. 2.8: what is the heat transfer rate from the wall to the fluid stream if the leading segment $0 < x < x_0$ is unheated ($T = T_\infty$). An answer is possible based on the integral method [7]. Assuming the temperature profile shape $m = (p/2)(3 - p^2)$ and the velocity profile shape $m = (n/2)(3 - n^2)$, the integral energy equation (53) yields

$$\Delta^3 + 4\Delta^2 x \frac{d\Delta}{dx} = \frac{0.932}{\Pr} \tag{106}$$

with the general solution

$$\Delta^3 = \frac{0.932}{\Pr} + Cx^{-3/4} \tag{107}$$

Constant C follows from the condition that heating, that is, a thermal boundary layer starts at $x = x_0$; hence

$$\Delta = 0.977 \,\Pr^{-1/3}\left[1 - \left(\frac{x_0}{x}\right)^{3/4}\right]^{1/3} \tag{108}$$

which is the same as eq. (62) if $x_0 = 0$. The local Nusselt number is

$$\mathrm{Nu} = \frac{hx}{k} = 0.323 \,\Pr^{1/3}\mathrm{Re}_x^{1/2}\left[1 - \left(\frac{x_0}{x}\right)^{3/4}\right]^{-1/3} \tag{109}$$

Laminar Boundary Layer Flow

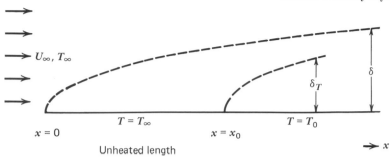

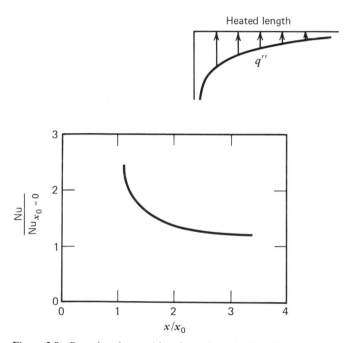

Figure 2.8 Boundary layer with unheated starting length.

As is shown in the bottom of Fig. 2.8, the effect of the unheated length x_0 on the local Nusselt number drops below 20 percent if x is beyond $3x_0$ from the leading edge of the flat plate.

Arbitrary Wall Temperature

The integral solution for heat transfer with an unheated starting length presented above is the central ingredient in the construction of heat transfer results for more complicated situations. Consider, for example, the heat transfer from the heated spot $x_1 < x < x_2$ shown in Fig. 2.9c: the wall

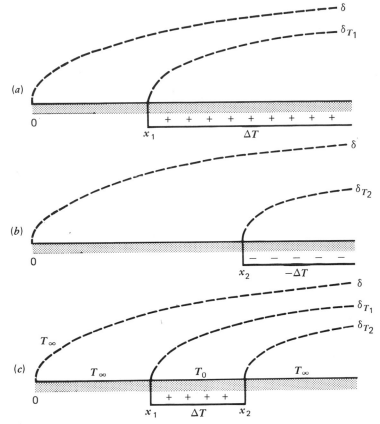

Figure 2.9 The principle of superposition in the construction of integral solutions for boundary layers with finite heated length.

temperature upstream and downstream from the heated spot is T_∞, while the spot temperature is T_0. Since the integral energy equation (53) is linear in temperature, the thermal boundary layer generated by the T_0 spot can be reconstructed as the superposition of two thermal boundary layers of type (108) and (109). The first thermal boundary layer $\delta_{T,1}$(Fig. 2.9a) is the fingerprint of wall heating $(T_\infty + \Delta T)$ downstream from $x = x_1$. The second thermal boundary layer (Fig. 2.9b) is the result of wall cooling $(T_\infty - \Delta T)$ downstream from $x = x_2$. The superposition of the two thermal layers (Fig. 2.9c) constitutes the thermal boundary layer due to spot heating. Of interest is the heat flux q'' from the wall to the fluid: to calculate q'' we identify three distinct wall regions:

(i) $0 < x < x_1$, the unheated started length, where $q'' = 0$ because the wall is in thermal equilibrium with the free stream.

(ii) $x_1 < x < x_2$, the heated spot, where eq. (109) applies unchanged

$$q'' = 0.323 \frac{k}{x} \Pr^{1/3} \mathrm{Re}_x^{1/2} \left\{ \frac{\Delta T}{\left[1 - (x_1/x)^{3/4} \right]^{1/3}} \right\} . \qquad (110)$$

(iii) $x > x_2$, the unheated trailing section, where q'' is the superposition of two effects of type (110)

$$q'' = 0.323 \frac{k}{x} \Pr^{1/3} \mathrm{Re}_x^{1/2} \left\{ \frac{\Delta T}{\left[1 - (x_1/x)^{3/4} \right]^{1/3}} + \frac{-\Delta T}{\left[1 - (x_2/x)^{3/4} \right]^{1/3}} \right\}$$

$$(111)$$

Note that since $x_2 > x_1$, the heat flux q'' in region (iii) is negative. This means that in the trailing section the wall reabsorbs part of the heat released earlier in region (ii).

Result (111) can be generalized [15, 16]. The heat flux from the wall to the fluid, downstream from N step changes ΔT_i in wall temperature, is given by

$$q'' = 0.323 \frac{k}{x} \Pr^{1/3} \mathrm{Re}_x^{1/2} \sum_{i=1}^{N} \frac{\Delta T_i}{\left[1 - (x_i/x)^{3/4} \right]^{1/3}} \qquad (112)$$

where x_i is the longitudinal position of each temperature step change ΔT_i. If the wall temperature varies smoothly, $T_0(x)$, then formula (112) is replaced by its integral limit (the limit of infinitesimally small steps)

$$q'' = 0.323 \frac{k}{x} \Pr^{1/3} \mathrm{Re}_x^{1/2} \int_0^x \frac{(dT_0/d\xi)\, d\xi}{\left[1 - (\xi/x)^{3/4} \right]^{1/3}} \qquad (113)$$

Uniform Heat Flux

In many problems, particularly those involving the cooliing of electrical and nuclear components, the wall heat-flux q'' is known. In such problems, overheating, burnout, and meltdown are very important issues; therefore, the object of heat transfer analysis is the prediction of the wall temperature variation $T_0(x)$. Technically, the heat transfer problem is still the calculation of heat-transfer coefficient $h = q''/[T_0(x) - T_\infty]$ [eq. (6)].

The integral method and profile shapes used to generate eqs. (62)–(64) can be applied to the calculation of $T_0(x) - T_\infty$ when $q'' = $ constant is specified. This analysis is recommended as an exercise to the student (see Table 2.1). One such result is [17]

$$\mathrm{Nu} = \frac{q''}{T_0(x) - T_\infty} \frac{x}{k} = 0.458 \Pr^{1/3} \mathrm{Re}_x^{1/2}, \qquad (\Pr > 1) \qquad (114)$$

The more general result corresponding to the case of nonuniform wall heat-flux $q''(x)$ is [18]

$$T_0(x) - T_\infty = \frac{0.623}{k} \mathrm{Pr}^{-1/3} \mathrm{Re}_x^{-1/2} \int_{\xi=0}^{x} \left[1 - \left(\frac{\xi}{x} \right)^{3/4} \right]^{-2/3} q''(\xi) \, d\xi,$$

$$(\mathrm{Pr} > 1) \qquad (115)$$

THE EFFECT OF LONGITUDINAL PRESSURE GRADIENT

The preceding analytical results are all based on the assumption that the pressure gradient term is negligible relative to inertia and friction in the boundary layer momentum equation (26). This assumption applies to the case of a flat wall parallel to a uniform stream. If, as is shown in the sketch at the bottom of Fig. 2.10, the wall makes a positive angle $\beta/2$ with the free stream, then the free stream is accelerated in the x direction along the wall (x is measured away from the tip of the wedge). Graphically, the acceleration of the flow is indicated by the gradual increase in the density of streamlines. Neglect-

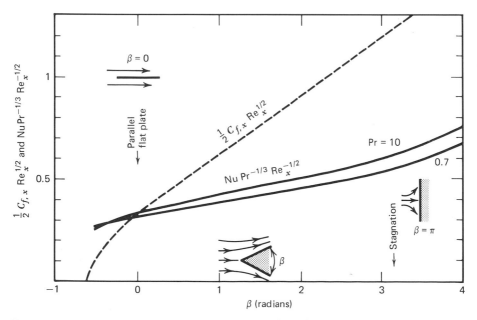

Figure 2.10 Local friction and heat transfer in laminar boundary layer flow along an isothermal wedge-shaped body (drawn based on data collected from Refs. 24 and 25).

ing the laminar boundary layer in which viscosity balances inertia, the flow engulfing the wedge of angle β may be treated as inviscid and may be determined analytically based on potential flow theory. Note that the inviscid flow residing outside the laminar boundary layer is governed by the balance between inertia and pressure gradients. The potential flow solution for the velocity variation along the wedge-shaped wall (i.e., along the boundary layer that coats the wall) is

$$U_\infty(x) = Cx^m \tag{116}$$

where C is a constant and m is related to the β angle of Fig. 2.10,

$$m = \frac{\beta}{2\pi - \beta} \tag{117}$$

Invoking the Bernoulli equation along the wall streamline,

$$\frac{1}{\rho}\frac{dP_\infty}{dx} = -U_\infty\frac{dU_\infty}{dx} \tag{118}$$

and using formula (116), the boundary layer equation for momentum becomes

$$u\frac{\partial u}{\partial x} + v\frac{\partial u}{\partial y} = \frac{m}{x}U_\infty^2 + \nu\frac{\partial^2 u}{\partial y^2} \tag{119}$$

Falkner and Skan [24] showed that eq. (119) admits a similarity solution with m as an additional parameter (note that the Blasius solution is the special case $m = 0$). The development of the similarity equation is left as an exercise for the student (see Problem 8). The skin-friction coefficient produced by the Falkner–Skan solution is shown graphically in Fig. 2.10. We learn that relative to the flat plate case ($\beta = 0$), flow acceleration brings about a substantial increase in the numerical coefficient appearing in the $C_{f,x} \sim \mathrm{Re}_x^{-1/2}$ scaling law.

The heat transfer similarity solution can be developed in the same way as Pohlhausen's solution, this time by substituting the Falkner–Skan similarity flow into the boundary layer energy equation (27). Eckert [25] performed this calculation; a few of his results are illustrated in Fig. 2.10. Compared with the skin-friction coefficient, the Nusselt number is considerably less sensitive to changes in the longitudinal pressure gradient. Note also that the $\mathrm{Nu} \sim \mathrm{Pr}^{1/3}\mathrm{Re}_x^{1/2}$ scaling law prevails over a wide range of wall angles. In the heat transfer literature (e.g., Ref. 9) Eckert's calculations are reported in a two-dimensional table as $\mathrm{Nu}\,\mathrm{Re}_x^{-1/2}$ versus Pr and β; Fig. 2.10 shows that when the proper scaling law is used, the group $\mathrm{Nu}\,\mathrm{Pr}^{-1/3}\mathrm{Re}_x^{-1/2}$ depends mainly on β.

SYMBOLS

$C_{f,x}$	local skin-friction coefficient [eq. (57)]
f	streamfunction similarity profile [eq. (80)]
F	wall friction force
h	heat transfer coefficient [eq. (4)]
k	thermal conductivity
L	wall length
m	profile shape function for integral analysis [eq. (54)]
M	impulse or reaction force due to fluid flow into or out of a control volume
n	dimensionless coordinate across the velocity boundary layer (y/δ)
Nu	local Nusselt number
p	dimensionless coordinate across the thermal boundary layer (y/δ_T)
P	pressure
P_∞	pressure outside the boundary layer
Pe_L	Peclet number ($U_\infty L/\alpha$)
Pr	Prandtl number (ν/α)
Re_L	Reynolds number ($U_\infty L/\nu$)
q	heat transfer rate [W]
q''	heat flux [W/m^2]
T	temperature
T_0	wall temperature
T_∞	free-stream temperature
ΔT	temperature difference ($T_0 - T_\infty$)
u	longitudinal velocity
U_∞	free-stream velocity
v	velocity normal to the wall
W	width
x, y	cartesian coordinates (Fig. 2.1)
x_0	unheated starting length (Fig. 2.8)
Y	the y coordinate of a point situated in the free stream
α	thermal diffusivity
δ	velocity boundary layer thickness
δ_T	thermal boundary layer thickness
δ^*	displacement thickness [eq. (86)]
η	similarity variable [eq. (71)]
θ	dimensionless temperature function [eq. (93)]
θ	momentum thickness [eq. (87)]
μ	viscosity
ν	kinematic viscosity
ρ	density

τ wall shear stress

ψ streamfunction [eq. (76)]

$(\ )_{0-L}$ quantity averaged from $x = 0$ to $x = L$

REFERENCES

1. L. Prandtl, *Über Flussigkeitsbewegung bei sehr kleiner Reibung*, Proc. 3rd Int. Math. Congr., Heidelberg, 1904; also NACA TM 452, 1928.

2. H. L. Dryden, in H. Schlichting, *Boundary Layer Theory*, 4th ed., McGraw-Hill, New York, 1960, p. V.

3. H. Lamb, *Hydrodynamics*, Dover, New York, 1945.

4. P. Feyerabend, *Against Method*, Verso, London, England, 1978.

5. A. Bejan, *Entropy Generation through Heat and Fluid Flow*, John Wiley & Sons, New York, 1982.

6. H. Schlichting, *Boundary Layer Theory*, 4th ed., McGraw-Hill, New York (1960), p. 299–301.

7. W. M. Rohsenow and H. Y. Choi, *Heat, Mass and Momentum Transfer*, Prentice-Hall, Englewood Cliffs, NJ, (1961), p. 149.

8. F. Kreith and W. Z. Black, *Basic Heat Transfer*, Harper & Row, New York, 1980, pp. 207, 208.

9. W. M. Kays and M. E. Crawford, *Convective Heat and Mass Transfer*, McGraw-Hill, New York, 1980, pp. 51–54.

10. A. J. Chapman, *Heat Transfer*, 3rd ed., Macmillan, New York, 1974, pp. 257–259.

11. H. Blasius, *Grenzschichten in Flussigkeiten mit kleiner Reibung*, Z. Math. Phys., Vol. 56, 1908, p. 1; also NACA TM 1256.

12. E. Pohlhausen, *Der Warmeaustausch zwischen festen Korpern und Flussigkeiten mit kleiner Reibung und kleiner Warmeleitung*, Z. Angew. Math. Mech., Vol. 1, 1921, pp. 115–121.

13. H. Schlichting, *op. cit.*, pp. 116–122.

14. L. Howarth, On the solution to the laminar boundary layer equations, *Proc. R. Soc. London Ser. A*, Vol. 164, 1938, p. 547.

15. R. J. Nickerson, Heat transfer to incompressible boundary layers, Ch. 2, in *Developments in Heat Transfer*, W. M. Rohsenow, ed., MIT Press, Cambridge, MA, 1964, p. 40.

16. E. R. G. Eckert, *Introduction to the Transfer of Heat and Mass*, McGraw-Hill, New York, 1959.

17. C. Prolhac, Term Project, Course ME 564, University of Colorado, Boulder, 1982.

18. W. M. Kays and M. E. Crawford, *op. cit.* p. 151.

19. H. Weyl, On the differential equations of the simplest boundary layer problems, *Ann. Math.*, Vol. 43, 1942, pp. 381–407.

20. M. Van Dyke, *Perturbation Methods in Fluid Mechanics*, Parabolic Press, Stanford, CA, 1975, p. 131.

21. S. Goldstein, ed., *Modern Developments in Fluid Dynamics*, Oxford University Press, London, 1938, p. 135.

22. L. Rosenhead, *Laminar Boundary Layers*, Oxford University Press, London, 1963, p. 223.

23. R. J. Nickerson and H. P. Smith, Term Project, Course 2.521, MIT, Cambridge, MA, 1958.

24. V. M. Falkner and S. W. Skan, Some approximate solutions of the boundary layer equations, *Philos. Mag.*, Vol. 12, 1931, p. 865.

25. E. R. G. Eckert, *VDI-Forschungsh.*, Vol. 416, 1942, pp. 1–24.

PROBLEMS

1. Derive eq. (53) by invoking the First Law of Thermodynamics in the open system defined by the control volume in the bottom-half of Fig. 2.3.

2. Develop a power series expression for the Blasius profile (Fig. 2.6), as a solution to eq. (82) subject to conditions (83) and (84). Assume that $f = \sum_{i=0}^{\infty} a_i \eta^i$. Show that for small η the expression

$$f = \frac{\alpha \eta^2}{2!} - \frac{\alpha^2 \eta^5}{(2)(5!)} + \frac{11 \alpha^3 \eta^8}{(4)(8!)} - \frac{375 \alpha^4 \eta^{11}}{(8)(11!)} + \cdots$$

satisfies the Blasius equation and the boundary conditions at $\eta = 0$. The curvature at the wall, $\alpha = f''(0)$ is unknown and would have to be determined from the condition (84) at $\eta = \infty$, where, unfortunately, the above series expression does not hold. To complete the solution, it is necessary to develop an asymptotic expansion valid for large values of η,

$$f = f_1 + f_2 + \cdots$$

where the higher order approximations must be small compared with the lower order (e.g., $f_2 \ll f_1$). Using [13] as a guide, show that for large values of η

$$f = \eta - \beta + \gamma \int_{\infty}^{\eta} d\eta \int_{\infty}^{\eta} \exp\left[-\frac{1}{4}(\eta - \beta)^2\right] d\eta + \cdots$$

where β and γ are two additional unknown constants. The three unknowns (α, β, γ) can be determined *matching* the two expansions, that is, by making f, f', and f'' equal at some $\eta = \eta_1 = O(1)$. Although this is how Blasius found $\alpha = 0.332$ (which is very close to the correct numerical result [14]), it has been pointed out that in this problem the matching of the two solutions is inappropriate because the convergence radii of the two expressions are too small [19, 20].

3. Determine the Blasius profile (Fig. 2.6) by solving eq. (82) numerically using a shooting scheme. First, divide the η domain into small intervals $\Delta \eta$ of size 0.01 or smaller. Write finite-difference approximations for the derivatives f'' and f''' [e.g., $f_i'' = (f_{i+1} + f_{i-1} - 2f_i)/(\Delta \eta)^2$, where i indicates the position of the ith node defined as $\eta_i = i \Delta \eta$], and substitute these expressions into eq. (82). The result of this operation is a formula for calculating the value of f at any node, based on the f values at the *three* preceding notes. The numerical integration starts from the wall, by first calculating f_3 based on $f_0 = f_1 = 0$ (eq. 83) and a $O(1)$ guess for the value of f_2. The calculation is repeated for $f_4, f_5, \ldots$, etc., until η becomes large, $O(10)$: in this range, the third boundary condition [eq. (84)] must be used as the test for how good the initial f_2 guess was. If eq. (84) is not satisfied, the integration sequence is repeated using an updated guess for the value of f_2 (synonymous to guessing $f''(0)$).

The need for performing the integration more than once is eliminated based on the observation that the Blasius equation (82) and the initial conditions (83) are invariant under the transformation [20–22]

$$f \to bf, \qquad \eta \to \frac{\eta}{b}$$

If eq. (82) is integrated to $\eta \sim 10$ using a certain initial curvature, say, $f''(0) = 1$, then the numerically calculated outer slope $a = f'(10)$ and eq. (84) imply that the correct guess for initial curvature $f''(0)$ must be $a^{-3/2}$.

4. Derive the expression for a local Nusselt number along a flat wall with uniform heat flux using the integral method. Assume one of the temperature profiles listed in Table 2.1, and you will take advantage of the fact that the flow part of the problem has already been solved (i.e., $\delta(x)$ is known). Keep in mind that in this problem $T_0(x)$ is an additional unknown: the necessary additional equation is the definition of q'' (known) [eq. (5)].

5. Consider the laminar boundary layer flow of an isothermal fluid (U_∞, T_∞) over a flat isothermal wall (T_0). At a certain distance x from the leading edge, the local skin-friction coefficient is $C_{f,x} = 0.0066$. What is the value of the local Nusselt number at the same location, if the Prandtl number is Pr $= 7$?

6. Your job is to design a plate fin for maximum heat transfer rate q_B subject to fixed volume ($V = bLt =$ constant). If the heat transfer through the fin can be described as one-dimensional, then the total heat transfer rate pulled by the fin from the wall T_B is

$$q_B = (T_B - T_\infty)(hpkA)^{1/2}\tanh\left[L\left(\frac{hp}{kA}\right)^{1/2}\right]$$

where h, p, k, and A are the average heat-transfer coefficient at location x along the fin (away from the wall), the wetted perimeter at $x =$ constant, the thermal conductivity of the fin, and the fin cross-sectional area at $x =$ constant. It is assumed also that $L \gg b \gg t$, where L, b, t are the dimensions of the plate fin. The fin is in contact with a uniform stream (U_∞, T_∞) in laminar boundary layer flow parallel to the b dimension of the fin; hence, the heat-transfer coefficient at any location x along the fin is

$$\frac{hb}{k_f} = 0.664 \, \mathrm{Pr}^{1/3}\left(\frac{U_\infty b}{\nu}\right)^{1/2}$$

where k_f, Pr, and ν are the fluid conductivity, the Prandtl number, and the kinematic viscosity.

Assuming all other design variables are given (including the plate thickness t), determine the optimum dimension b for maximum q_B and fixed V. Express your result in dimensionless form as

$$\frac{b_{opt}}{t} = \text{function}\left(\frac{V}{t^3}, \frac{k_f}{k}, \text{Pr}, \frac{U_\infty t}{\nu}\right)$$

where k is the thermal conductivity of the fin material.

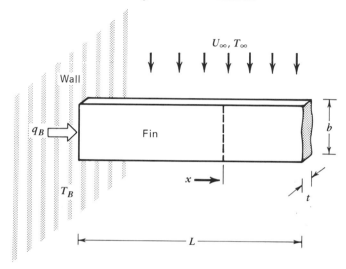

7. Consider the same plate fin shown in the above figure and discussed in the preceding problem statement, in the limit where the fin length L is large enough so that the base heat transfer rate q_B is no longer influenced by L.

(a) If all design variables except b are fixed, and if b increased by a factor of 2, then by what factor will the total base heat transfer rate q_B increase?

(b) When in contact with air, the fin shown in the figure experiences the heat transfer rate $q_{B,a}$. Immersed in a water stream with the same velocity and temperature as the original air stream (U_∞, T_∞), the same fin experiences a new heat transfer rate $q_{B,w}$. Calculate the ratio $q_{B,w}/q_{B,a}$, keeping in mind the following property ratios

$$\frac{k_w}{k_a} = 23, \qquad \frac{\nu_w}{\nu_a} = 0.07, \qquad \frac{\text{Pr}_w}{\text{Pr}_a} = \frac{7}{0.72}$$

where subscripts w and a indicate *water* and *air*, respectively.

8. Develop the similarity forms of the boundary layer momentum and energy equations for uniform flow past an inclined wall, as shown in the bottom sketch of Fig. 2.10. For the momentum equation, begin with eq. (119) and apply the similarity transformation contained in eqs. (71) and (80). Show that

Blasius' equation (82) is replaced now by [24]

$$2f''' + (m + 1)ff'' + 2m\left[1 - (f')^2\right] = 0$$

Apply the same transformation to the energy equation and show that Pohlhausen's equation (94) is replaced by the more general form [25]

$$2\theta'' + \Pr(m + 1)f\theta' = 0$$

Establish whether the angle of inclination has any effect on the boundary conditions to be used in conjunction with the above equations.

9. Consider the laminar boundary layer frictional heating [12] of an adiabatic wall parallel to a free stream (U_∞, T_∞; Fig. 2.1). Modeling the flow as one with temperature-independent properties and assuming that the Blasius velocity solution holds, use scaling arguments to show that the relevant boundary layer energy equation for this problem is

$$\rho c_P\left(u\frac{\partial T}{\partial x} + v\frac{\partial T}{\partial y}\right) = k\frac{\partial^2 T}{\partial y^2} + \mu\left(\frac{\partial u}{\partial y}\right)^2$$

and that the wall temperature rise scales as U_∞^2/c_P when $\Pr > 1$. Determine the wall temperature ($T_0 > T_\infty$), assuming that the wall is insulated ($\partial T/\partial y = 0$ at $y = 0$) and that $T \to T_\infty$ as $y \to \infty$. The suggested path is to develop the similarity solution for the dimensionless temperature profile

$$\theta_r(\eta) = \frac{T - T_\infty}{U_\infty^2/(2c_P)}$$

where the similarity transformation is the same as in eqs. (71) and (80) of Chapter 2. Show that the energy equation reduces to

$$\theta_r'' + \frac{\Pr}{2}f\theta_r' + 2\Pr(f'')^2 = 0$$

where $f(\eta)$ is the Blasius solution. Solving this equation subject to $\theta_r'(0) = 0$ and $\theta_r'(\infty) = 0$, prove that the temperature rise in the boundary layer is

$$\theta_r(\eta) = 2\Pr\int_\eta^\infty\left\{\int_0^p[f''(\beta)]^2\exp\left(\frac{\Pr}{2}\int_0^\beta f(\gamma)\,d\gamma\right)d\beta\right\}$$

$$\times\exp\left(-\frac{\Pr}{2}\int_0^p f(m)\,dm\right)dp$$

Calculate the wall temperature rise $\theta_r(0)$ as a function of Prandtl number using the above formula. (Evaluate the nested integrals numerically using an appropriate analytical approximation for the streamfunction profile $f(\eta)$.) Based on scale analysis show that $\theta_r(0)$ is of order $O(\text{Pr})$ and $O(1)$ in the two limits $\text{Pr} \to 0$ and $\text{Pr} \to \infty$, respectively.

10. Consider the development of a two-dimensional laminar jet discharging in the x-direction into a fluid reservoir that contains the same fluid as the jet. The reservoir pressure P_∞ is uniform. The jet is generated by a narrow slit of width D_0; the average fluid velocity through the slit is U_0.

Let $D(x)$ and $U(x)$ be the jet thickness scale and the centerline velocity scale at a sufficiently long distance x away from the nozzle (the slit). Relying on the mass and momentum conservation equations, on boundary layer theory $(D \ll x)$, and on scale analysis in a flow region of length x and thickness D, determine the order of magnitude of D and U in terms of (D_0, U_0, x, ν).

Hint: Integrate the momentum equation over an $x = $ constant plane (i.e., from $y = -\infty$ to $y = +\infty$) and show that the integral $\int_{-\infty}^{+\infty} u^2 dy$ is independent of x. This result is the basis for an additional scaling law necessary for determining the D and U scales uniquely.

11. Consider the development of a two-dimensional thermal jet if the velocity jet (D, U) determined in the preceding problem has an original temperature T_0 as it comes out through the slit. The reservoir temperature is uniform, $T_\infty = 0$, and buoyancy effects are negligible.

Let $D_T(x)$ and $T(x)$ be the thermal jet thickness scale and the centerline temperature scale at a sufficiently long distance x away from the slit. Again, based on boundary-layer scale analysis determine the order of magnitude of D_T and T in terms of $(D_0, U_0, x, \nu, T_0, \alpha)$. Find the relationship $D_T(x)/D(x)$ ~ function(Pr) and sketch the geometric implication of this law.

Hint: Integrate the energy equation over an $x = $ constant plane and show that the integral $\int_{-\infty}^{\infty} uT \, dy$ is independent of x. Consider the two possibilities $(D_T < D$ and $D_T > D)$ separately, as you interpret the scaling law implied by the x independent longitudinal enthalpy flow integral.

12. Consider the laminar flow of a two-dimensional liquid film on a flat wall inclined at an angle α relative to the horizontal direction. The film flow is driven by the gravitational acceleration component $(g \sin \alpha)$ acting parallel to the wall. Attach the Cartesian system of coordinates (x, y) and (u, v) to the wall, such that x and u point in the flow direction. In this notation, derive the terminal velocity distribution in the liquid film $u(y)$; in other words, determine the flow in the limit where the film inertia is negligible and the x momentum equation expresses a balance between film weight and wall friction. Let U be the undetermined free-surface velocity at $y = \delta$, where δ is the film thickness. (U is undetermined because the film flowrate can be varied at will by the individual who pours liquid on the incline.)

Consider next the heat transfer from the wall to the liquid film, in the case where the film and wall temperature is T_0 everywhere upstream of $x = 0$, and where the wall temperature alone is raised to $(T_0 + \Delta T)$ downstream of $x = 0$.

Let δ_T be the thermal boundary layer thickness of the thin liquid region in which the wall heating effect is felt. Based on scale analysis, demonstrate that immediately downstream from $x = 0$ (where δ_T is much smaller than δ), the thermal boundary layer thickness δ_T scales as $(\alpha \delta x / U)^{1/3}$.

Determine the temperature distribution in the film based on an *integral* analysis, assuming the following temperature profile:

$$\frac{T(x, y) - T_0}{\Delta T} = 1 - 2\frac{y}{\delta_T} + \left(\frac{y}{\delta_T}\right)^2; \qquad 0 \leq y \leq \delta_T$$

$$T(x, y) = T_0; \qquad \delta_T < y \leq \delta$$

Note that this integral analysis is valid as long as $\delta_T(x) \leq \delta$. At what distance $x = x_1$ will the free surface feel the heating effect of the wall (i.e., at what x will δ_T equal δ)? Devise an integral analysis to determine the film temperature field $T(x, y)$ downstream from the point $x = x_1$.

13. An infinitely long, flat plate is initially at rest immersed in a liquid pool with properties ν, α, T_∞. The plate is also in thermal equilibrium with the pool. At a certain instant, $t = 0$, the plate starts moving at constant velocity U along itself. Determine the time-dependent velocity distribution in the fluid, for times $t > 0$ in the immediate vicinity of the solid wall. At another point in time, $t = t_1$, the plate temperature is changed to a new temperature $T = T_0$. Determine the time-dependent temperature distribution in the fluid in the immediate vicinity of the wall and for times $t > t_1$. Based on the expressions obtained for the velocity and temperature fields, decide whether the solid plate is lined by boundary layer regions. Does or does not the temperature field depend on the velocity distribution (as in Pohlhausen's problem, eq. (98), Chapter 2)?

3

Laminar Duct Flow

In this chapter we consider the fluid friction and heat transfer between a stream and a solid object in internal flow, that is, when the solid surface is the duct that guides the stream. The fundamental questions are the same as in external flow (Chapter 2):

1. What is the friction force or the pressure drop in the longitudinal (flow) direction?

2. What is the heat-transfer coefficient or the thermal resistance to heat transfer in the direction normal to the flow?

We will see shortly that the theoretical view, that is, the idea that makes the answers to these questions most accessible to the analyst is the concept of *fully developed* flow and temperature fields. This is a powerful concept that, like Prandtl's boundary layer in external convective heat transfer is responsible for much of the language and results presently known in connection with laminar duct flow and heat transfer.

Traditionally, the concept of fully developed flow is taught as a self-standing topic, as a useful approximation one can make when confronted by the tough (complete) forms of the Navier–Stokes equations [1]. This traditional approach is not incorrect provided that the theoretical basis for the approximation is well understood. We find it more appropriate to introduce the concept of fully developed flow as a direct consequence of the concept of boundary layer encountered in external flow. That a close relationship between these two concepts must exist should not surprise anyone: both concepts are expressions of the same philosophical point of view, namely, the view that certain *finite-size* regions of a flow field possess special properties.[†] The concept of boundary

---

[†] The same point of view is exercised in Chapters 6 to 8, where the longitudinal buckling wavelength λ_B is a property of a flow region of *finite thickness.*

layer divides an external flow into a free stream and a boundary layer, just like the concept of *fully developed* divides a duct flow into a *developing* length succeeded by a fully developed length.

HYDRODYNAMIC ENTRANCE LENGTH

Consider the application of the integral method to the flow configuration sketched in Fig. 3.1: two parallel plates form a two-dimensional duct intaking the uniform stream U. We are interested in the friction force exerted by the flow on the two walls and the dependence of the velocity profile $u(x, y)$ on the longitudinal position x along the duct. The integral solution outlined below was constructed by Sparrow [2].

Based on the things learned in Chapter 2, we expect the formation of velocity boundary layers along the two walls in the region x "close enough" to the mouth of the channel: in this region the tip of each plate is surrounded by the U stream in the same manner as in Fig. 2.1. Since the thickness δ of each layer grows in the x direction, and since one layer can only grow to a thickness $D/2$ before merging with the other layer, the channel flow can be thought of as the succession of two distinct flow regions. In the first region (the entrance or developing section), distinct boundary layers coexist with *core* fluid that has not yet felt viscously the presence of the walls. In the second region, the core has disappeared and the boundary layers are no longer distinct. We obtain a quick (back-of-the-envelope) estimate of the entrance length X by writing $\delta(X) = D/2$ in Blasius' result for boundary layer thickness [eq. (85) of Chapter 2]; the result is

$$\frac{X/D}{\mathrm{Re}_D} = 0.0103 \tag{1}$$

Relative to this simple result, the Sparrow integral solution to the entrance length problem takes into account the important fact that the core fluid is

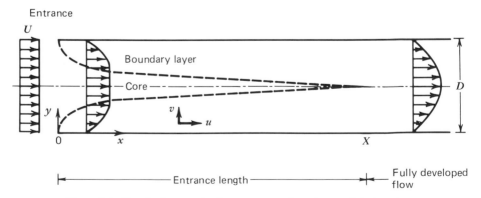

Figure 3.1 Developing flow in the entrance region of a parallel-plate duct.

squeezed, hence, accelerated in the x direction. This effect is illustrated graphically in Fig. 3.1: since more fluid stagnates near the walls as the boundary layers thicken, the core velocity U_c must increase so that the mass flowrate in each cross-section x is constant and equal to ρUD. We start with the integral momentum equation (52) of Chapter 2, where $U_\infty = U_c$ and $Y = \delta(x)$. The free-stream pressure gradient dP_∞/dx appearing in eq. (52) of Chapter 2 is now the core pressure gradient dP/dx; since the core flow is by definition inviscid, dP/dx is related to $U_c(x)$ through the Bernoulli equation $\rho U_c^2/2 + P = $ constant; hence

$$U_c \frac{dU_c}{dx} + \frac{1}{\rho} \frac{dP}{dx} = 0 \tag{2}$$

Eliminating dP/dx between eqs. (2) and (52) of Chapter 2 yields

$$\frac{d}{dx}\left[\int_0^\delta (U_c - u)u\,dy\right] + \frac{dU_c}{dx}\int_0^\delta (U_c - u)\,dy = \nu\left(\frac{\partial u}{\partial y}\right)_0 \tag{3}$$

In addition, mass conservation in the channel of half-width (from $y = 0$ to $y = D/2$) requires

$$\int_0^\delta \rho u\,dy + \int_\delta^{D/2} \rho U_c\,dy = \rho U \frac{D}{2} \tag{4}$$

Equations (3) and (4) are solved for $\delta(x)$ and $U_c(x)$, by first assuming a boundary layer profile shape. Taking $u/U_c = 2y/\delta - (y/\delta)^2$, the Sparrow solution becomes

$$\frac{x/D}{Re_D} = \frac{3}{160}\left(9\frac{U_c}{U} - 2 - 7\frac{U}{U_c} - 16\ln\frac{U_c}{U}\right) \tag{5}$$

$$\frac{\delta(x)}{D/2} = 3\left[1 - \frac{U}{U_c(x)}\right] \tag{6}$$

At the location X where the two boundary layers merge, $\delta(X) = D/2$, eqs. (5) and (6) yield $U_c(X) = \frac{3}{2}U$ and

$$\frac{X/D}{Re_D} = 0.0065 \tag{7}$$

Thus, the laminar entrance length predicted by the Sparrow solution is 37% shorter than the simplest estimate [eq. (1)]. The important contribution of both analyses is that they show analytically the extent of the laminar developing region: X scales with $D\,Re_D$, and the proportionality factor is a number of the order of 10^{-2}. Schlichting [3] solved the same problem by obtaining a series solution for the accelerated boundary layer flow in the beginning of the entrance length, and by matching this series to a second-series solution valid near the end of the laminar entrance length. Although in his solution the velocity profile varies smoothly as it asymptotically reaches the fully developed shape, the entrance region has a characteristic length perceived approximately as [3]

$$\frac{X/D}{\mathrm{Re}_D} \cong 0.04 \tag{8}$$

The fundamental difference between the entrance length and the fully developed region is illustrated further by the variation of wall frictional-shear stress as x increases. Figure 3.2 shows the skin-friction coefficient

$$C_{f,x} = \frac{\tau_{\mathrm{wall}}(x)}{\frac{1}{2}\rho U^2} \tag{9}$$

which in the case of the Sparrow solution (5)–(7) becomes

$$C_{f,x}\,\mathrm{Re}_D = \frac{8}{3}\frac{U_c}{U}\left(1 - \frac{U}{U_c}\right)^{-1} \tag{10}$$

Figure 3.2 also shows the $C_{f,x}$ calculation based on the Blasius result [eq. (92) of Chapter 2] with U_∞ replaced by U in the definition of $C_{f,x}$ and Re_x. In the entrance region, $C_{f,x}$ behaves in a manner that suggests the presence of distinct boundary layers. In the fully developed region, on the other hand, τ_{wall} and $C_{f,x}$ are no longer functions of x because the velocity profile $u(x, y)$ has

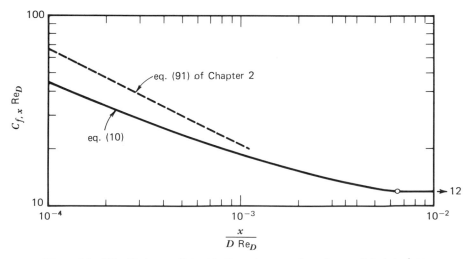

Figure 3.2   Skin-friction coefficient in the entrance region of a parallel-plate duct.

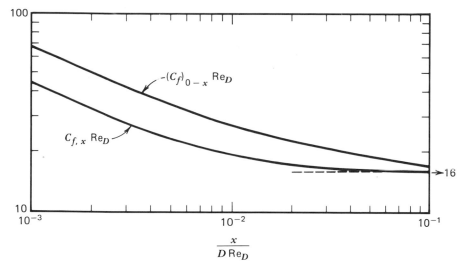

Figure 3.3 Skin-friction coefficient in the entrance region of a round tube.

become practically independent of x. We focus more closely on this effect in the next section.

The hydrodynamic entrance length problem in a round tube can be approached along similar lines. Figure 3.3 is a drawing made after Langhaar [4] that shows the variation of $C_{f,x}$ in the entrance region. Note that in most textbooks and handbooks the same drawing appears with $\mathrm{Re}_D/(x/D)$ plotted on the abscissa, which is a really strange way of plotting something $(C_{f,x})$ that varies along x. As in Fig. 3.2, the dimensionless coordinate plotted on the abscissa of Fig. 3.3 is $(x/D)/\mathrm{Re}_D$, which increases to the right as we travel deeper (to the right) into the duct of Fig. 3.1. Figure 3.3 also shows the average skin-friction coefficient, that is, the value of $C_{f,x}$ averaged from $x = 0$ to x.

FULLY DEVELOPED FLOW

The steady state mass and momentum conservation statements at any point (x, y) inside the two-dimensional channel of Fig. 3.1 are

$$\frac{\partial u}{\partial x} + \frac{\partial v}{\partial y} = 0 \tag{11}$$

$$u\frac{\partial u}{\partial x} + v\frac{\partial u}{\partial y} = -\frac{1}{\rho}\frac{\partial P}{\partial x} + \nu\left(\frac{\partial^2 u}{\partial x^2} + \frac{\partial^2 u}{\partial y^2}\right) \tag{12}$$

$$u\frac{\partial v}{\partial x} + v\frac{\partial v}{\partial y} = -\frac{1}{\rho}\frac{\partial P}{\partial y} + \nu\left(\frac{\partial^2 v}{\partial x^2} + \frac{\partial^2 v}{\partial y^2}\right) \tag{13}$$

These equations can be simplified greatly based on the following scaling argument. At any location $x \sim L$ in the fully developed region, we have $y \sim D$ and $u \sim U$. Therefore, using eq. (11), the transversal velocity in the fully developed region scales as

$$v \sim \frac{DU}{L} \tag{14}$$

We can then think of the fully developed region as the duct flow section situated far enough from the entrance, such that the scale of v is negligible. Based on this definition only, the mass continuity equation (11) requires in the fully developed flow limit

$$v = 0 \quad \text{and} \quad \frac{\partial u}{\partial x} = 0 \tag{15}$$

Traditionally, eqs. (15) are taken as *definitions* and a starting point in the analysis of the fully developed regime [1]. However, it is important to keep in mind the scaling foundations of eqs. (15). We can think of a fully developed region only because we can think of a flow region where the scale of y is D (fixed). This corresponds to a velocity variation that spreads over the entire cross-section. In the entrance region, on the other hand, the scale of y is δ (not fixed) and, as a consequence, neither v nor $\partial u/\partial x$ are negligible (reexamine the Sparrow solution (5)–(7) to see this feature of the flow in the entrance region).

Based on eqs. (15), the y momentum equation (13) reduces to

$$\frac{\partial P}{\partial y} = 0 \tag{16}$$

implying that P is a function of x only. This conclusion is similar to the one reached previously in the study of the laminar boundary layer (Chapter 2): P remains uniform in each cross-section of the fully developed region because P is only a function of x in the two entrance boundary layers merging to create the fully developed region.

Finally, the x momentum equation (12) becomes

$$\frac{dP}{dx} = \mu \frac{\partial^2 u}{\partial y^2} = \text{constant} \tag{17}$$

Each side of this equation must be *constant* because P is a function of x only and u is a function of y. Solving eq. (17) subject to the no-slip conditions

$$u = 0 \quad \text{at } y = \pm D/2 \tag{18}$$

yields the well-known Hagen–Poiseuille solution for fully developed flow between parallel plates.

$$u = \frac{3}{2} U \left[1 - \left(\frac{y}{D/2} \right)^2 \right]$$

$$U = \frac{D^2}{12\mu} \left(-\frac{dP}{dx} \right) \tag{19}$$

where y is measured away from the centerline of the channel. The velocity profile is parabolic and the velocity is proportional to the pressure drop per unit duct length in the direction of flow.

In general, for a duct of arbitrary cross-section, eq. (17) is replaced by

$$\frac{dP}{dx} = \mu \nabla^2 u = \text{constant} \tag{20}$$

where, in the Laplacian operator ∇^2, $\partial^2 u / \partial x^2 = 0$. For example, the fully developed laminar flow in a round tube of radius r_0 [eq. (20)] reads

$$\frac{dP}{dx} = \mu \left(\frac{\partial^2 u}{\partial r^2} + \frac{1}{r} \frac{\partial u}{\partial r} \right) \tag{21}$$

Solving this equation subject to $u = 0$ at the cross-section periphery $(r = r_0)$ yields

$$u = 2U \left[1 - \left(\frac{r}{r_0} \right)^2 \right]$$

$$U = \frac{r_0^2}{8\mu} \left(-\frac{dP}{dx} \right) \tag{22}$$

These results were first reported by Hagen [5] in 1839 and, independently, by Poiseuille [6] in 1840.

In eqs. (19) and (22) we see the simplest solutions available for laminar fully developed flow in the rectilinear duct. In general, as shown in the next section, the solution to the Poisson-type equation (20) is considerably more difficult.

It is common practice in the convection literature to assign to any Hagen–Poiseuille flow a *Reynolds number* defined for a round tube as $\text{Re}_D = UD/\nu$. Fluid mechanics textbooks repeat one after another the seemingly general explanation that the Reynolds number is the ratio of two scales, namely, the inertia divided by the friction force. The proliferation of unexplained terms such as the Reynolds number is perhaps the best example of the aging of fluid mechanics into a field where the new generations obediently accept as dogma the language coined by the pioneers who lived and created in a time when "anything goes" was the rule.

Although there are certain *flow regions* in which one could envision a Reynolds number as the ratio inertia/friction,[†] it is not at all clear that the Reynolds number always means the same thing for all flows. In Chapter 2, we saw that in laminar boundary layer flow, the inertia effect always balances the friction effect, and that the only physical meaning attached to the Reynolds number is the geometric feature represented by the boundary layer slenderness ratio squared [eq. (31) of Chapter 2]. In the case of Hagen–Poiseuille flow through a straight duct, it should be obvious that the inertia/friction interpretation of the Reynolds number is pure nonsense. Hagen–Poiseuille flows are flows in which the fluid inertia is zero everywhere: these flows are governed by a permanent and perfect balance between the imposed (driving) longitudinal pressure gradient $(-dP/dx)$ and the opposing effect of friction posed by the wall on the flow. So, if we must define a dimensionless group for fully developed laminar flow through a straight duct, this group can only be the ratio

$$\frac{\text{longitudinal pressure force}}{\text{friction force}}$$

According to the momentum equation for Hagen–Poiseuille flow and all subsequent solutions available for such flows, the order of magnitude of this ratio is unity

$$\frac{-dP/dx}{\mu\,\partial^2 u/\partial r^2} = O(1)$$

revealing again the force balance that *means* Hagen–Poiseuille flow.

There are other signs that Re_D has no meaning in fully developed laminar duct flow. First, it is well-known that the calculated Re_D in such flows can reach as high as 2300 (note that this number is considerably greater than unity). Such a large number, coupled with the inertia/friction interpretation of the Reynolds number, suggests that inertia overwhelms friction in Hagen–Poiseuille flow—a totally absurd conclusion. Another sign is the friction factor f defined later in eq. (24): in laminar flow both f and τ_w are very sensitive to changes in Re_D, demonstrating that the Reynolds number is an inappropriate ratio of scales, and that $\frac{1}{2}\rho U^2$ is an inappropriate pressure unit for Hagen–Poiseuille flow.[‡]

[†]As in Problem 12 at the end of Chapter 1.
[‡]The group $\frac{1}{2}\rho U^2$ is inappropriate as pressure unit here and in Chapter 2 because it is a Bernoulli equation concept, that is, a *reversible flow* concept [9]. Note that laminar boundary layer regions and fully developed flows through ducts are flows governed by friction (thermodynamic irreversibility); hence, the use of a reversible flow concept to discuss such flows is a conceptual inconsistency.

The fact that the Reynolds number is not conceptually justified in the discussion of fully developed laminar flow does not mean that it is not a *useful* dimensionless group for convective heat transfer engineering. The Reynolds number is useful, particularly in the presentation of duct friction data on a *single* plot for both the turbulent and the laminar regimes (as in the Moody chart of Fig. 7.9). For this reason, the classical Reynolds number nomenclature is retained in the present treatment. But, at the same time, the reader should be constantly aware of the fact that the alleged general meaning of the concept of Reynolds number is not at all clear, and that it would be very timely for future research to clarify this meaning. A proposal in this direction is advanced in Chapter 6, in which the concept of Reynolds number is linked to the equivalence of two characteristic time scales that rule the phenomenon of transition to turbulence.

HYDRAULIC DIAMETER AND PRESSURE DROP

The engineering objective of the preceding analysis is the calculation of the pressure drop in a duct with prescribed flowrate or the calculation of the flowrate in a duct with prescribed pressure drop. In fully developed laminar flow, $\dot{m}$ and ΔP are proportional to one another. In general (especially in turbulent flow), the $\dot{m}(\Delta P)$ relationship is not as simple and is usually the object of laboratory measurements. The historically empirical approach to determining the $\dot{m}(\Delta P)$ relationship[†] is responsible for the terminology in use today.

Consider the duct of length L and *arbitrary* flow cross-section A shown in Fig. 3.4: regarding the AL control volume as a black box, that is, without looking inside to see the actual flow, the momentum theorem in the longitudi-

---

[†] This approach was the trademark of hydraulics, the precursor of moder fluid mechanics. About the hydraulics of the nineteenth century, Tietjens [7] wrote in 1934:

> The hydraulics, which tried to answer the multitudinous problems of practice, disintegrated into a collection of unrelated problems. Each individual question was solved by assuming a formula containing some undetermined coefficients and then determining these by experiments. Each problem was treated as a separate case and there was lacking an underlying theory by which the various problems could be correlated.

In contemporary fluid mechanics research we like to think that the classical mechanics embodied in the Navier–Stokes equations offers a common theoretical basis for all fluid flow phenomena. Despite this thought, the contemporary study of turbulence has retraced the empirical course chosen by hydraulics: in fact, replace the name hydraulics with turbulence in the above quotation from Tietjens, and you will obtain a fairly close description of what has been going on in twentieth century turbulence research. Granted, the empirical constants of hydraulics have been replaced by shady (however, equally empirical) concepts such as eddy diffusivity, numerical models, and many universal constants. It is on this background of unyielding empiricism that the scaling laws of Chapter 6 offer some hope that a single idea may be used to account for many of the flow phenomena we have been calling turbulence (see Chapters 7 and 8).

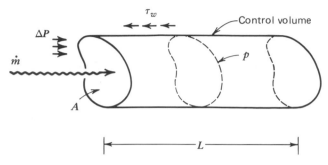

Figure 3.4   Control volume showing the force balance expressed by eq. (23).

nal direction (Chapter 1) yields

$$A\Delta P = \tau_w pL \tag{23}$$

where p is the perimeter of the cross-section (the wetted perimeter). The unknown wall shear stress τ_w is replaced by a dimensionless unknown, the *friction factor* defined as[†]

$$f = \frac{\tau_w}{\frac{1}{2}\rho U^2} \tag{24}$$

This definition is essentially the same as eq. (9) for the skin-friction coefficient in the entrance region; one important difference between f and $C_{f,x}$ is that the friction factor f is x-independent because it is a *fully developed regime* concept.

After all these definitions, the pressure drop ΔP across the black-box duct can be calculated from the formula

$$\Delta P = f\frac{pL}{A}\left(\frac{1}{2}\rho U^2\right) \tag{25}$$

Finally, it is recognized that A/p represents a characteristic linear dimension of the duct cross-section, namely,

$$r_h = \frac{A}{p}, \quad \text{the hydraulic radius}$$

[†]Definition (24) is not unique; the heat transfer literature also uses $4f$ as the friction factor, since the product $4f$ appears explicitly in eq. (27). The student can tell which f definition is used in a particular treatment (if the f definition is missing), by checking the f formula for a round tube. Thus, if $f = 16/\text{Re}$, the present definition [eq. (24)] was used.

or (26)

$$D_h = 4r_h = \frac{4A}{p}, \quad \text{the hydraulic diameter}$$

Table 3.1 shows a vertically aligned column of five different cross-section shapes and sizes, all having the same hydraulic diameter D_h. This arrangement shows the meaning of hydraulic diameter: it is a conventional length that accounts for "how close" the wall and its resistive effect are positioned relative to the stream. Thus, in the case of highly asymmetric cross-sections such as the gap between two infinite parallel plates, the hydraulic diameter scales with the *smallest* of the two dimensions of the cross-section.

Table 3.1   Scale Drawing of Five Duct Sizes Having the Same Hydraulic Diameter

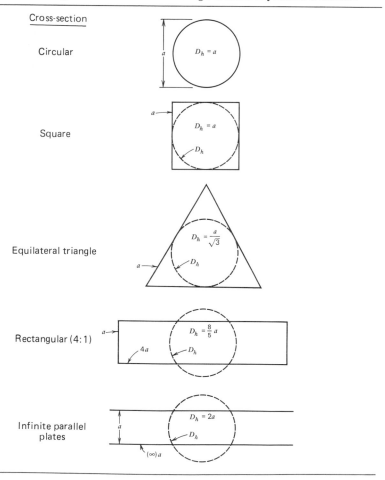

The pressure drop formula (25) can be written as

$$\Delta P = f \frac{4L}{D_h} \left(\frac{1}{2} \rho U^2 \right) \tag{27}$$

The calculation of ΔP is possible provided we know the flow parameter f. The friction factors derived from the Hagen–Poiseuille solutions (19) and (22) are, respectively:

$$f - \frac{24}{Re_{D_h}}, \quad D_h - 2D, \quad \text{parallel plates } (D = \text{gap thickness}) \tag{28}$$

$$f = \frac{16}{Re_{D_h}}, \quad D_h = D, \quad \text{round tube } (D = \text{tube diameter}) \tag{29}$$

These formulas hold as long as the flow is in the laminar regime ($Re_{D_h} < 2000$, empirically).

The literature is rich in results that are equivalent to eqs. (28) and (29) for other duct cross-sections. Most of these numbers (the numerical values of $f\,Re_{D_h}$) have been compiled in a recent reference work [8] and some are shown here in Table 3.2. The friction factor f emerges after solving the Poisson

Table 3.2   The Effect of Cross-Sectional Shape on f and Nu in Fully Developed Duct Flow

Cross-Section Geometry	$f\,Re_{D_h}$	$(\pi D_h^2/4)/A_{duct}$	Nu = hD_h/k Uniform q''	Uniform T_0
(triangle, 60°)	13.3	0.605	3	2.35
(square)	14.2	0.785	3.63	2.89
(circle)	16	1	4.364	3.66
(rectangle $a \times 4a$)	18.3	1.26	5.35	4.65
(parallel plates)	24	1.57	8.235	7.54
(parallel plates, one side insulated)	24	1.57	5.385	4.86

One side insulated

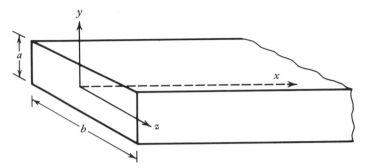

Figure 3.5   Coordinate system for a duct with a rectangular cross-section.

equation (20) in the duct cross-section of interest. To illustrate this procedure beyond the two simplest examples given in the preceding section, consider the fully developed laminar flow through a duct of rectangular cross-section (Fig. 3.5). We have to solve

$$\frac{dP}{dx} = \mu\left(\frac{\partial^2 u}{\partial y^2} + \frac{\partial^2 u}{\partial z^2}\right) = \text{constant} \tag{30}$$

in the y-z plane in which the cross-section dimensions are a and b, respectively. Equation (30) can be solved for $u(y, z)$ by Fourier series (see Problem 4). In this section we outline a more direct, albeit less accurate approach to the needed answer.

To calculate f or τ_w we need the velocity distribution $u(y, z)$: from the parallel plate and round tube solutions discussed previously, we know enough to expect $u(y, z)$ to be adequately represented by the expression

$$u(y, z) = u_0\left[1 - \left(\frac{y}{a/2}\right)^2\right]\left[1 - \left(\frac{z}{b/2}\right)^2\right] \tag{31}$$

where u_0 is the centerline (peak) velocity. The problem reduces to calculating u_0 from eq. (30): since $u(y, z)$ of eq. (31) will not satisfy eq. (30) at every point (y, z), we can select u_0 such that expression (31) satisfies eq. (30) *integrated* over the entire cross-section:

$$ab\frac{dP}{dx} = \mu\int_{-a/2}^{a/2}\int_{-b/2}^{b/2}\left(\frac{\partial^2 u}{\partial y^2} + \frac{\partial^2 u}{\partial z^2}\right) dz\, dy \tag{32}$$

The result is

$$ab\frac{dP}{dx} = -\frac{16}{3}\mu u_0\left(\frac{b}{a} + \frac{a}{b}\right) \tag{33}$$

From the definition of average velocity U,

$$abU = \int_{-a/2}^{a/2} \int_{-b/2}^{b/2} u \, dz \, dy \tag{34}$$

we also obtain

$$u_0 = \frac{9}{4} U \tag{35}$$

Substituting eqs. (33) and (35) into the pressure drop formula [eq. (27)] yields

$$f = \frac{a^2 + b^2}{(a+b)^2} \frac{24}{\mathrm{Re}_{D_h}} \tag{36}$$

where

$$D_h = \frac{4ab}{2(a+b)} \tag{37}$$

The friction factor f [eq. (36)] is invariant to the transformation $a \to b, b \to a$, because the cross-section geometry is invariant (rectangular) to the transformation $y \to z, z \to y$ (Fig. 3.5). Figure 3.6 shows this relatively simple result next to the numerical calculation of $f \, \mathrm{Re}_{D_h}$ [8]: the present $f \, \mathrm{Re}_{D_h}$ result coincides

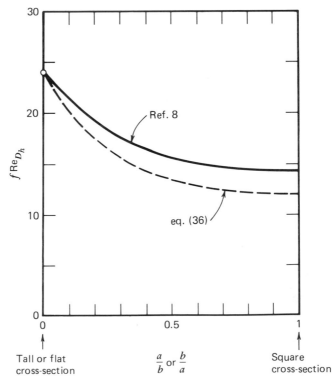

Figure 3.6 Friction factor for fully developed flow in a duct with a rectangular cross-section (Fig. 3.5).

with the numerical result in the tall and flat cross-section shape limits, because in those limits the profile shape assumption (31) is exactly the Hagen–Poiseuille profile shape. Overall, the agreement between eq. (36) and numerically derived results is better than 15 percent.

Table 3.2 shows a useful compilation of friction factors for laminar fully developed flow in some of the most common duct geometries. Regardless of cross-sectional shape, the value of $f \, \mathrm{Re}_{D_h}$ is consistently of the order of 20, thus stressing the usefulness of the hydrodynamic diameter scaling discussed right after eqs. (26). The fact that in Table 3.2 the numerical value of $f \, \mathrm{Re}_{D_h}$ varies from 13.3 to 24 is further evidence that the length scale D_h accounts for the effective distance between the walls "squeezing" the flow. It is clear from Table 3.1 that in equilateral triangles D_h underestimates this *effective distance*, whereas in parallel-plate channels the effective wall-to-wall distance is overestimated by D_h. This numerical mismatch between D_h and the wall-to-wall distance seems to explain the fact that $f \, \mathrm{Re}_{D_h}$ *increases* in Table 3.2 from the equilateral triangle to the parallel-plate channel. Indeed, as shown in Fig. 3.7,

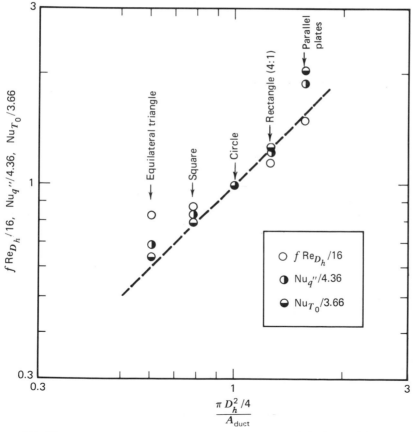

Figure 3.7 The effect of cross-sectional shape on fully developed friction and heat transfer in a straight duct.

there appears to exist an approximate proportionality between $f \, \mathrm{Re}_{D_h}$ and the degree to which D_h misjudges the wall-to-wall distance. The mismatch between D_h and the average wall-to-wall distance can be measured as the ratio $(\pi D_h^2/4)/A_{\mathrm{duct}}$ [in the case of extremely flat cross-sections, A_{duct} is equal to $(a D_h)$, where a is the actual plate-to-plate distance (see Table 3.1)]. The usefulness of the new dimensionless group $(\pi D_h^2/4)/A_{\mathrm{duct}}$ is illustrated further in Problem 6.

HEAT TRANSFER TO FULLY DEVELOPED DUCT FLOW

The basic question in connection with the heat transfer to any duct flow concerns the relationship between the wall–stream temperature difference and the wall–stream heat transfer rate (or the longitudinal temperature variation of the stream). Without loss of generality at this point, consider a tube of radius r_0 and average axial velocity U. Thus the mass flowrate through the duct is $\dot{m} = \rho \pi r_0^2 U$ (Fig. 3.8). From the engineering thermodynamics of flow systems (Chapter 1), we know that the net heat transfer from the wall to the stream $(q'' 2\pi r_0 \, dx)$ reflects in the enthalpy rise (gain) experienced by the stream; for the control volume of length dx in Fig. 3.8, the First Law of Thermodynamics in the steady state yields

$$q'' 2\pi r_0 \, dx = \dot{m}\left(h_{x+dx} - h_x \right) \tag{38}$$

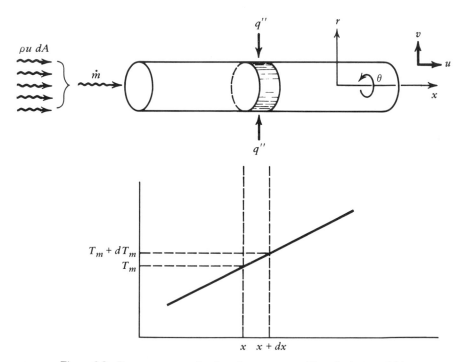

Figure 3.8 Energy conservation in a duct segment of length dx [eq. (39)].

Modeling the fluid as an ideal gas ($dh = c_P \, dT_m$) or an incompressible liquid with negligible pressure changes ($dh \cong c \, dT_m$)[†] eq. (38) yields

$$\frac{dT_m}{dx} = \frac{2}{r_0} \frac{q''}{\rho c_P U} \tag{39}$$

where c_P is replaced by c for incompressible liquids.

The temperature T_m appearing in the first law analysis of the duct as a "control volume" is the *bulk temperature* of the stream. In heat transfer engineering—an activity that developed in parallel with and independent of modern thermodynamics [9]—T_m is the *mean temperature* of the stream. Implicit in this name is the fact that the fluid temperature cannot be uniform over the duct cross-section at $x =$ fixed: for example, if the stream is to be heated by the wall, then the fluid layer or *lamina* situated closer to the wall will necessarily be warmer than a layer situated farther from the wall. Of course, a relationship must exist between the temperature at every point in the cross-section $T(x, r)$ and the mean temperature $T_m(x)$. However, T_m is not just any average, it is the mean temperature whose definition is the First Law for bulk flow [eq. (38)]. We can write the First Law again, this time for the bundle of mini-streams $\rho u \, dA$ piercing the tube cross-section,

$$q'' 2\pi r_0 \, dx = d \iint_A \rho u c_P T \, dA \tag{40}$$

Combining eq. (40) with eq. (39), we obtain the formula for calculating T_m

$$T_m \rho c_P U A = \iint_A \rho c_P u T \, dA \tag{41}$$

In constant property tube flow, eq. (41) reduces to

$$T_m = \frac{1}{\pi r_0^2 U} \int_0^{2\pi} \int_0^{r_0} u T r \, dr \, d\theta \tag{42}$$

Returning now to the basic heat transfer question in duct flow, we want to know the relationship between q'' and the wall–fluid temperature difference. Since the fluid temperature varies over the duct cross-section, $\Delta T = T_0 - T_m$ is *conventionally* selected as the wall–fluid temperature difference. We are then interested in the heat-transfer coefficient

$$h = \frac{q''}{T_0 - T_m} = \frac{k (\partial T / \partial r)_{r=r_0}}{T_0 - T_m} \tag{43}$$

[†] This approximation is correct only if in the duct length of interest, $c \Delta T_m$ is considerably greater than $\Delta P / \rho$ (see Table 1.1).

where, as in Chapter 2, q'' is defined as positive when proceeding from the wall into the fluid [compare eq. (43) with eq. (6) of Chapter 2].

Fully Developed Temperature Profile

Equation (43) outlines the path to follow: we must first determine the temperature field in the fluid $T(x, r)$ by solving the energy equation subject to appropriate wall-temperature boundary conditions. For steady, θ-symmetric flow through a round tube the energy equation (1.43b) reduces to

$$\frac{1}{\alpha}\left(u\frac{\partial T}{\partial x} + v\frac{\partial T}{\partial r}\right) = \frac{\partial^2 T}{\partial r^2} + \frac{1}{r}\frac{\partial T}{\partial r} + \frac{\partial^2 T}{\partial x^2} \tag{44}$$

In the *hydrodynamic* fully developed region, we have $v = 0$ and $u = u(r)$; hence

$$\frac{u(r)}{\alpha}\frac{\partial T}{\partial x} = \frac{\partial^2 T}{\partial r^2} + \frac{1}{r}\frac{\partial T}{\partial r} + \frac{\partial^2 T}{\partial x^2} \tag{45}$$

The energy equation (45) expresses a balance between a maximum of three possible effects: axial convection, radial conduction, and axial conduction. The respective scales of these three effects are

$$\overbrace{\frac{U}{\alpha}\left(\frac{q''}{D\rho c_p U}\right)}^{\text{Convection}}, \qquad \overbrace{\underbrace{\frac{\Delta T}{D^2}}_{\text{radial}}, \quad \underbrace{\frac{1}{x}\left(\frac{q''}{D\rho c_p U}\right)}_{\text{longitudinal}}}^{\text{Conduction}} \tag{46}$$

where we used eq. (39) to recognize $\partial T/\partial x \sim q''/(D\rho c_p U)$. Of the three scales in (46), the radial conduction effect will always be present because without it the heat transfer problem of this chapter evaporates. Multiplying scales (46) by $D^2/\Delta T$ and using the definition of heat-transfer coefficient $h = q''/\Delta T$, we obtain

$$\overbrace{\frac{hD}{k}}^{\text{Convection}}, \qquad \overbrace{\underbrace{1}_{\text{radial}}, \quad \underbrace{\left(\frac{hD}{k}\right)^2\left(\frac{\alpha}{UD}\right)^2}_{\text{longitudinal}}}^{\text{Conduction}} \tag{47}$$

We conclude that in the limit of large Peclet numbers

$$\mathrm{Pe}_D = \frac{UD}{\alpha} \gg 1 \tag{48}$$

the axial conduction effect is negligible. Most important, from the convection ~ radial conduction balance, we learn that the Nusselt number is a constant of order one.

$$\text{Nu} = \frac{hD}{k} \sim 1 \tag{49}$$

This Nu scaling is confirmed by many (less approximate) solutions. In the same domain ($\text{Pe}_D \gg 1$), the energy equation to solve for $T(x, r)$ is therefore

$$\frac{u(r)}{\alpha} \frac{\partial T}{\partial x} = \frac{\partial^2 T}{\partial r^2} + \frac{1}{r} \frac{\partial T}{\partial r} \tag{50}$$

At this point it is instructive to summarize the assumptions on which the simplified energy equation (50) is based. First, we assumed that the flow is hydrodynamically fully developed; hence, the velocity profile $u(r)$ is the same at any x along the duct. Second, we assumed that the scale of $\partial^2 T/\partial r^2$ is $\Delta T/D^2$ [eq. (46)], in other words, the effect of thermal diffusion has had time to reach the centerline of the stream. Certainly, this last assumption is not valid in a thermal entrance region X_T near the duct entrance, where the proper scale of $\partial^2 T/\partial r^2$ is $\Delta T/\delta_T^2$, with $\delta_T \ll D$. The extent of X_T and the heat-transfer coefficient in the thermal entrance region are determined later in this chapter.

On theoretical scaling grounds, the two assumptions listed above should be sufficient for regarding the temperature profile $T(x, r)$ as *fully developed*: it is simply the profile in the region situated downstream from the two entrance regions (X, X_T) where both u and T are developing. The scaling feature of the thermally developed region is $\text{Nu} = \text{constant} = O(1)$ [eq. (49)]. It is worth noting that the heat transfer literature defines the fully developed temperature as [1, 10–12]

$$\frac{T_0 - T}{T_0 - T_m} = \phi\left(\frac{r}{r_0}\right) \tag{51}$$

where, in general, T, T_0, and T_m can be functions of x. This special analytical expression for $T(x, r)$ is purely the result of the scaling law $\text{Nu} \sim 1$; to see the relationship between eqs. (51) and (49), recall that

$$\text{Nu} = \frac{hD}{k} = \frac{D}{k} \frac{q''}{T_0 - T_m} \tag{52}$$

hence,

$$\text{Nu} = D\frac{(\partial T/\partial r)_{r=r_0}}{T_0 - T_m} \sim 1 \tag{53}$$

Thus, the x variation of $(\partial T/\partial r)_{r=r_0}$ must be the same as that of $T_0(x) - T_m(x)$. Since $\partial T/\partial r$ is a function of both x and r, then, according to eq. (53),

$$\frac{\partial T/\partial(r/r_0)}{T_0(x) - T_m(x)} = f_1\left(\frac{r}{r_0}\right) = O(1) \tag{54}$$

Integrating this expression with respect to r/r_0 yields

$$T\left(x, \frac{r}{r_0}\right) = (T_0 - T_m)f_2\left(\frac{r}{r_0}\right) + f_3(x) \tag{55}$$

where f_2 and f_3 are arbitrary functions of r/r_0 and x, respectively. Note that expression (55) is essentially the same as eq. (51) used as a quick "definition" of thermal development.

Uniform Wall Heat Flux

If q'' is not a function of x, then eq. (50) can be solved analytically, because $\partial T/\partial x$ is a constant proportional to q''. In order to see this property of $\partial T/\partial x$, we rewrite the fully developed temperature profile as

$$T(x, r) = T_0(x) - \frac{q''}{h}\phi\left(\frac{r}{r_0}\right) \tag{56}$$

hence,

$$\frac{\partial T}{\partial x} = \frac{dT_0}{dx} \tag{57}$$

we also have

$$\frac{dT_0}{dx} = \frac{dT_m}{dx} \tag{58}$$

In conclusion, combining eqs. (57) and (58) and using the First Law statement (39), we obtain

$$\frac{\partial T}{\partial x} = \frac{2}{r_0}\frac{q''}{\rho c_p U} = \text{constant} \tag{59}$$

This means that the temperature everywhere in the cross-section varies linearly in x, the slope of the line being proportional to q''. The main features of this temperature field are summarized schematically in Fig. 3.9.

The radial variation of T, namely, the dimensionless profile $\phi(r/r_0)$, is obtained finally by solving the energy equation for thermally developed flow

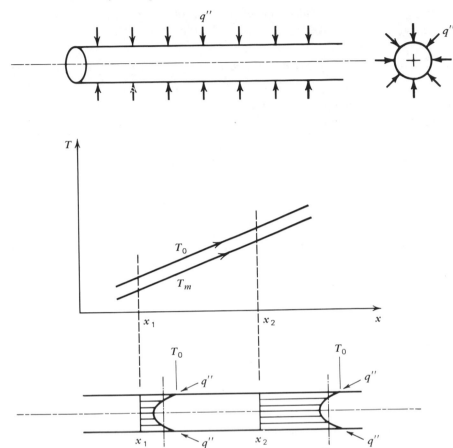

Figure 3.9 Fully developed temperature profile in a round tube with uniform heat flux.

[eq. (50)]. Substituting the temperature profile (56) and the Hagen–Poiscuille velocity profile (22) into eq. (50) leads to the following dimensionless equation for $\phi(r_*)$, where $r_* = r/r_0$,

$$-2\frac{hD}{k}(1 - r_*^2) = \frac{d^2\phi}{dr_*^2} + \frac{1}{r_*}\frac{d\phi}{dr_*} \tag{60}$$

Note that the object of this analysis, $hD/k = \mathrm{Nu}$, appears explicitly in eq. (60); integrating this equation twice and invoking one boundary condition (finite ϕ' at $r_* = 0$) yields

$$\phi = C_2 - 2\,\mathrm{Nu}\left(\frac{r_*^2}{4} - \frac{r_*^4}{16}\right) \tag{61}$$

where C_2 is the second, undetermined constant of integration. Combining eqs. (61) and (56) and setting $T = T_0$ at $r_* = 1$ to determine C_2, we obtain

$$T = T_0 - (T_0 - T_m) \mathrm{Nu} \left(\frac{3}{8} - \frac{{r_*}^2}{2} + \frac{{r_*}^4}{8} \right) \tag{62}$$

Finally, the mean temperature difference $T_0 - T_m$ follows from the definition of bulk (mean) temperature [eq. (42)],

$$T_0 - T_m = \frac{1}{\pi r_0^2 U} \int_0^{2\pi} \int_0^{r_0} (T_0 - T) u r \, dr \, d\theta$$

$$= 4 \int_0^1 (T_0 - T)(1 - r_*^2) r_* \, dr_* \tag{63}$$

Using expression (62) for $(T_0 - T)$ under the integral sign, the mean temperature difference drops out from both sides of the equal sign, leaving

$$1 = 4 \, \mathrm{Nu} \int_0^1 \left(\frac{3}{8} - \frac{r_*^2}{2} + \frac{r_*^4}{8} \right)(1 - r_*^2) r_* \, dr_* = \frac{11}{48} \mathrm{Nu} \tag{64}$$

The Nusselt number for thermally fully developed Hagen–Poiseuille flow with uniform heat flux is therefore

$$\mathrm{Nu} = \frac{48}{11} = 4.36 \tag{65}$$

which agrees in an order of magnitude sense with the scaling law (49). The Nu values corresponding to other duct cross-section shapes are listed in Table 3.2. For noncircular cross-sections, the Nusselt number is based on the hydraulic diameter

$$\mathrm{Nu} = \frac{h D_h}{k} \tag{66}$$

and, for this reason, the Nu values of Table 3.2 appear to vary in the same manner as the area ratio $(\pi D_h^2/4)/A_{\mathrm{duct}}$ (see the discussion on p. 81 and Problem 6). For noncircular cross-sections the problem is solved numerically, starting with the equation

$$\frac{u}{\alpha} \frac{\partial T}{\partial x} = \nabla^2 T \tag{67}$$

and replacing $\partial T/\partial x$ by a First Law balance of type (39), namely,

$$\frac{dT_m}{dx} = \frac{q'}{\rho c_p A U} = \text{constant} \tag{68}$$

Here q' is the heat transfer rate per unit duct length, regarded as independent of x. The wall temperature T_0 at a given x is usually assumed uniform around the noncircular periphery of the duct; consequently, the local heat flux around the periphery is nonuniform, varying from a maximum in wall regions close to the stream to a minimum in wall regions close to other wall regions (q'' drops to zero in the sharp corners of the cross-section). Thus q' is the perimeter line integral of q''. Also, since q'' varies along the perimeter, the heat-transfer coefficient varies too: the Nusselt numbers listed in Table 3.2 refer to the heat-transfer coefficient averaged over the duct perimeter. For further information regarding the Nusselt numbers of various duct geometries and the numerical method for calculating the Nu values, the reader is directed to Refs. 8, 13, and 14.

Uniform Wall Temperature

In a round tube with T_0 independent of x, we expect a temperature field of the kind sketched in Fig. 3.10. Suppose the stream bulk temperature is T_1 at some

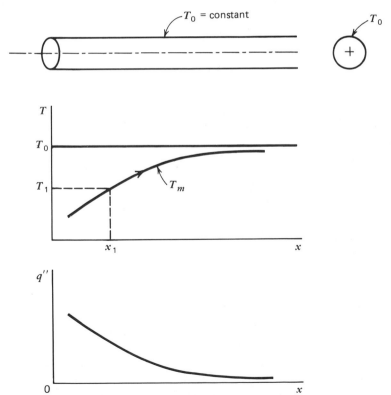

Figure 3.10 The longitudinal variation of mean temperature and wall heat flux in thermally developed flow through a tube with isothermal wall.

place $x = x_1$ in the fully developed region. Given the temperature difference $T_0 - T_1$, heat will be transfered from the wall to the stream and, as a result, the stream temperature will rise monotonically in the direction of flow. This also means that $T_0 - T_m$ (and q'') will monotonically decrease in the x direction. This discussion can easily be translated into analysis by combining the First Law of Thermodynamics [eq. (39)] with the only other thing we know at this point ($T_0 = $ constant); thus

$$q''(x) = h[T_0 - T_m(x)] \qquad (69)$$

where, on scaling grounds, h is also a constant [see eq. (49)]. Eliminating $q''(x)$ between eqs. (69) and (39), and integrating the result from $T_m = T_1$ at $x = x_1$, yields

$$T_0 - T_m(x) = (T_0 - T_1)\exp\left[-\frac{\alpha \text{Nu}}{r_0^2 U}(x - x_1)\right] \qquad (70)$$

The mean temperature difference decreases exponentially in the direction of flow and so does the heat flux [eq. (69)]. These qualities are illustrated in Fig. 3.10.

The Nusselt number appearing in eq. (70) can be calculated by again solving the energy equation (50). This time $\partial T / \partial x$ is replaced by

$$\frac{\partial T}{\partial x} = \frac{\partial}{\partial x}[T_0 - \phi(T_0 - T_m)] = \phi \frac{dT_m}{dx} \qquad (71)$$

Substituting this, the Hagen–Poiseuille profile, and $T = T_0 - \phi(T_0 - T_m)$ into the energy equation (50) yields the following dimensionless equation in the unknown $\phi(r_*)$:

$$-2\,\text{Nu}(1 - r_*^2)\phi = \frac{d^2\phi}{dr_*^2} + \frac{1}{r_*}\frac{d\phi}{dr_*} \qquad (72)$$

This equation is similar to eq. (60) for uniform heat flux, except that now the unknown $\phi(r_*)$ is present on the left-hand side of the equal sign. Furthermore, we now have two boundary conditions to impose on ϕ:

(i) $\qquad\qquad d\phi/dr_* = 0 \quad$ at $r_* = 0$, radial symmetry

(ii) $\qquad\qquad \phi = 0 \quad$ at $r_* = 1$, isothermal wall $\qquad\qquad$ (73)

In principle, eq. (72) and conditions (73) are sufficient for determining the unknown function ϕ. However, in view of the make-up of eq. (72), the radial profile ϕ will be a function of both r_* and Nu, where Nu is the real unknown in this problem. The necessary additional condition for determining Nu uniquely is the definition of the heat-transfer coefficient (i.e., Nu), [eq. (43)]; this condition can be written as

$$\text{Nu} = -2\left(\frac{d\phi}{dr_*}\right)_{r_*=1} \tag{74}$$

The problem statement is now complete: the value of Nu must be such that the $\phi(r_*, \text{Nu})$ solution of eqs. (72) and (73) satisfies the Nu definition (74). The actual solution may be pursued in a number of ways, for example, by successively approximating (guessing) and improving the ϕ solution (see Problem 7). Nowadays, however, it is more convenient to solve the problem numerically: the differential equation (72) is first approximated by finite differences, and, next, integrated from $r_* = 1$ to $r_* = 0$. To perform the integration at all, we must first guess the value of Nu, which also gives us a guess for the initial slope of the ensuing $\phi(r_*)$ curve [see eq. (74)]. The success of any Nu guess is judged by means of the first of conditions (73); the refined result is ultimately

$$\text{Nu} = 3.66 \tag{75}$$

which, again, agrees with the scaling law (49).

Table 3.2 lists other Nu values for noncircular cross-sections. The slight variation in these values appears to mimic that of $(\pi D_h^2/4)/A_{\text{duct}}$, implying that it is caused by hydraulic diameter nondimensionalization, that is, by the mismatch between hydraulic diameter and effective wall–stream distance (Fig. 3.7).

Tube Surrounded by Isothermal Fluid

Based on the $q'' = \text{constant}$ and $T_0 = \text{constant}$ analysis and the Nusselt numbers compiled in Table 3.2, we conclude that in thermally developed laminar flow Nu is influenced by the cross-section geometry *and* the way in which T_0, T_m, or q'' vary with x. For example, in a round tube Nu drops from 4.36 to 3.66 as the mean temperature variation changes from linear in x (when $q'' = \text{constant}$) to exponential in x (when $T_0 = \text{constant}$). The effect of axial variation of imposed heating or cooling conditions is illustrated further by a very instructive problem considered by Sparrow and Patankar [15].

Figure 3.11 shows the expected temperature variation along a tube heated by an isothermal external fluid. Assuming that the tube wall thickness and its thermal resistance are negligible, the local heat flux may be taken as proportional to the ambient-tube temperature difference

$$q'' = h_e\left[T_\infty - T_0(x)\right] \tag{76}$$

Here h_e is the external heat-transfer coefficient, assumed known and constant.

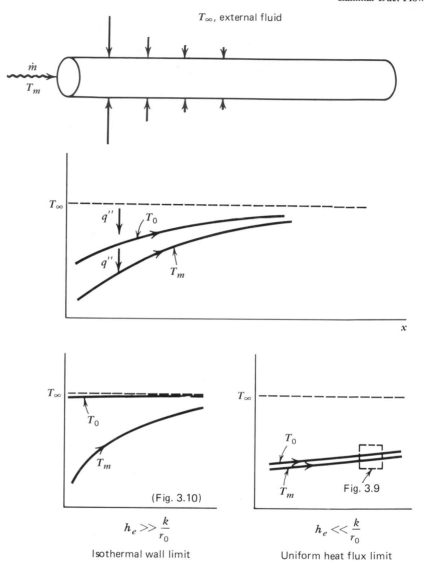

Figure 3.11 The general case of fully developed tube flow surrounded by isothermal external fluid.

If the tube wall resistance is not negligible, then h_e in eq. (76) must be replaced by an effective external coefficient

$$h_{\text{eff}} = \left(\frac{r_0}{k_w} \ln \frac{r_w}{r_0} + \frac{r_0/r_w}{h_e} \right)^{-1} \tag{77}$$

where k_w and r_w are the tube wall conductivity and outer radius. Comparing the external conductance h_e with the internal conductance $h(\sim k/r_0)$, we distinguish two limiting situations. When the external thermal contact is superior, the wall temperature T_0 approaches the isothermal condition $T_0 = T_\infty$ = constant, which was the case considered in the preceding section (Fig. 3.10). When the external contact is relatively poor, $h_e r_0/k \ll 1$, the mean temperature difference $T_0 - T_m$ is locally independent of x; both T_0 and T_m vary linearly with x, however, very slowly because the heat flux is controlled (throttled) by the external resistance. Thus, the previous two analyses are the limiting cases of the more general arrangement sketched in Fig. 3.11. When $h_e r_0/k$ is finite, T_0, T_m, and q'' vary exponentially in the flow direction.

Since the wall temperature floats between the known ambient temperature T_∞ and the unknown bulk temperature T_m, it is more useful to define the Nusselt number as [15]

$$\hat{\mathrm{Nu}} = \frac{q''}{T_\infty - T_m} \frac{D}{k} \tag{78}$$

The relationship between Nu and the duct-side Nusselt number [eq. (49)]

$$\mathrm{Nu} = \frac{q''}{T_0 - T_m} \frac{D}{k} \tag{49'}$$

is, from a resistance series argument,

$$\frac{2}{\hat{\mathrm{Nu}}} = \frac{1}{\mathrm{Bi}} + \frac{2}{\mathrm{Nu}} \tag{79}$$

where

$$\mathrm{Bi} = \frac{h_e}{k/r_0}, \quad \text{the Biot number} \tag{80}$$

Writing the following expression for the thermally developed profile,

$$T = T_\infty - [T_\infty - T_m(x)]\theta(r_*) \tag{81}$$

the energy equation (50) becomes

$$-2\hat{\mathrm{Nu}}(1 - r_*^2)\theta = \frac{d^2\theta}{dr_*^2} + \frac{1}{r_*}\frac{d\theta}{dr_*} \tag{82}$$

with the two boundary conditions

(i) $\qquad\qquad d\theta/dr_* = 0, \quad \text{at } r_* = 0$

(ii) $\qquad\qquad d\theta/dr_* = -\mathrm{Bi}\,\theta, \quad \text{at } r_* = 1 \tag{83}$

The second condition is the statement of heat flux continuity through the $r = r_0$ surface.

$$q'' = k\left(\frac{\partial T}{\partial r}\right) = h_e(T_\infty - T_0), \quad \text{at } r = r_0 \tag{84}$$

Finally, the $\hat{\text{Nu}}$ definition (78) yields

$$\frac{d\theta}{dr_*} = -\frac{\hat{\text{Nu}}}{2}, \quad \text{at } r_* = 1 \tag{85}$$

The $\hat{\text{Nu}}$ eigenvalue problem represented by eqs. (82)–(85) was solved numerically and the results are tabulated in Ref. 15. The same results have been plotted as solid lines in Fig. 3.12. The duct-side Nusselt number Nu varies smoothly from 4.36 to 3.66 as Bi increases from 0 to ∞; this behavior was anticipated qualitatively in the discussion of Fig. 3.11. The overall heat-transfer coefficient ($\hat{\text{Nu}}$) becomes proportional to Bi in the limit of very poor external thermal contact [from eqs. (76) and (78), $\hat{\text{Nu}} \to 0$]. In the isothermal wall limit (Bi $\to \infty$), Nu and $\hat{\text{Nu}}$ are identical in accordance with eq. (79).

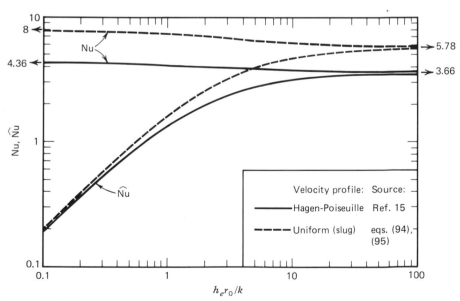

Figure 3.12 Nusselt number for thermally developed flow in a round table surrounded by isothermal fluid.

HEAT TRANSFER TO DEVELOPING FLOW

The heat transfer results listed in Table 3.2 and as solid lines in Fig. 3.12 apply exclusively to laminar duct flow regions where both the velocity and temperature profiles are fully developed. Measuring x from the actual entrance of the duct (Fig. 3.13), these results are valid in the downstream section delineated by

$$x > \max(X, X_T) \tag{86}$$

where X and X_T are the hydrodynamic and thermal entrance lengths, respectively. We know already from the discussion of hydrodynamic entrance length earlier in this chapter that the extent of X_T must be determined by the point where the entrance thermal boundary layer thickness δ_T becomes of the same order as the hydraulic diameter of the duct.

Scale Analysis

Figure 3.13 shows the first qualitative result of the $\delta_T(X_T) \sim D_h$ scaling, namely, the relative size of X and X_T as influenced by the Prandtl number. Since, according to Chapter 2, the ratio δ/δ_T increases monotonically with Pr, the ratio X/X_T must decrease monotonically as Pr increases. To determine the X_T scale, hence, the ratio X/X_T, first consider the limit of small Prandtl number fluids.

(i) Pr $\ll$ 1. According to eq. (37) of Chapter 2, δ_T develops faster than δ,

$$\delta_T(x) \sim x \, \mathrm{Pr}^{-1/2} \mathrm{Re}_x^{-1/2} \tag{87}$$

Claiming now that at the end of thermal development $x \sim X_T$ and $\delta_T \sim D_h$, we have

$$X_T \mathrm{Pr}^{-1/2} \mathrm{Re}_{X_T}^{-1/2} \sim D_h \tag{88}$$

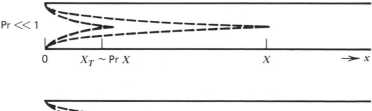

Figure 3.13 The effect of the Prandtl number on the size of hydrodynamic entrance length X relative to the size of thermal entrance length X_T.

or

$$\left(\frac{X_T/D_h}{\text{Re}_{D_h}\text{Pr}} \right)^{1/2} \sim 1 \qquad\qquad (89)$$

This is a well-known X_T result listed in other reference books as

$$\frac{X_T/D_h}{\text{Re}_{D_h}\text{Pr}} \sim \text{constant} \qquad\qquad (90)$$

where the constant is identified empirically as "approximately 0.1" [16]. Note that eq. (90) is of the same type as eq. (1) where the constant is also less than unity. In view of the apparent discrepancy between eqs. (89) and (90), the reader should keep in mind that eq. (89) is the correct way of writing $\delta_T \sim D_h$ [17]. Squaring any proportionality in which the coefficient is $O(1)$ but *numerically* less than 1.0 leads to a proportionality of type (90) where the coefficient is no longer $O(1)$: this coefficient later assumes the role of *transition constant*, adding to the long list of empirical constants the handbooks ask us to memorize.

In conclusion, the correct scaling for the transition from the developing to the thermally developed temperature profile is $\delta_T \sim D_h$, which means that the proper dimensionless group governing this transition is

$$\left(\frac{X_T/D_h}{\text{Re}_{D_h}\text{Pr}} \right)^{1/2} \qquad\qquad (89')$$

Likewise, the proper dimensionless group governing the transition from the developing to the fully developed velocity profile is

$$\left(\frac{X/D_h}{\text{Re}_{D_h}} \right)^{1/2} \sim 1 \qquad\qquad (1')$$

The reason these groups govern transition phenomena is that they become of order one during transition: they become of order one only because they represent the competition between the *two* proper scales which, after all, make the concept of *transition* possible. (This point is discussed further in Chapter 6).

(ii) $\text{Pr} \gg 1$. In the case of fluids such as water and oils (bottom-half of Fig. 3.13), it is tempting to compare D_h with the δ_T given by eq. (42) of Chapter 2. Such a comparison would be incorrect because the δ_T scale (42) of Chapter 2 is valid in boundary layers where the velocity thickness is consistently much greater than δ_T (so that the u scale is $(\delta_T/\delta)U_\infty$ inside the layer of thickness δ_T). In a duct, unlike in external flow (Chapter 2), the velocity profile spreads over D_h; hence, the u scale inside the δ_T layer is U itself. Thus, it is easy to show that $\delta_T(x) \sim x \, \mathrm{Pr}^{-1/2} \mathrm{Re}_x^{-1/2}$, which is identical to the scaling encountered in $\mathrm{Pr} \ll 1$ cases. In conclusion, criterion (89) is general and applies also in the case of $\mathrm{Pr} \gg 1$ fluids. Dividing eqs. (89) and (1') side-by-side we learn finally that

$$\frac{X_T}{X} \sim \mathrm{Pr} \tag{91}$$

This scaling is valid for all values of Pr (Fig. 3.13).

The local Nusselt number in the thermally developing section ($x \ll X_T$) scales as follows:

$$\mathrm{Nu} = \frac{hD_h}{k} \sim \frac{q''}{\Delta T} \frac{D_h}{k} \sim \frac{D_h}{\delta_T} \sim \left(\frac{x/D_h}{\mathrm{Re}_{D_h} \mathrm{Pr}} \right)^{-1/2} \tag{92}$$

Since the δ_T scale is $x \, \mathrm{Pr}^{-1/2} \mathrm{Re}_x^{-1/2}$ over the entire Pr range, the Nusselt number result (92) should be valid for all values of Pr: as is shown later in this section (Figs. 3.14–3.16), this conclusion is supported by numerical analysis.

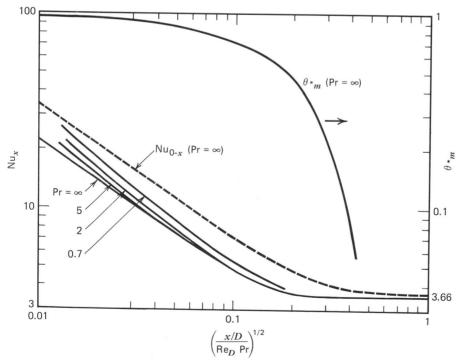

Figure 3.14 The heat transfer characteristics of the thermal entrance region of a round tube with isothermal wall (drawn based on data from Refs. 8 and 18).

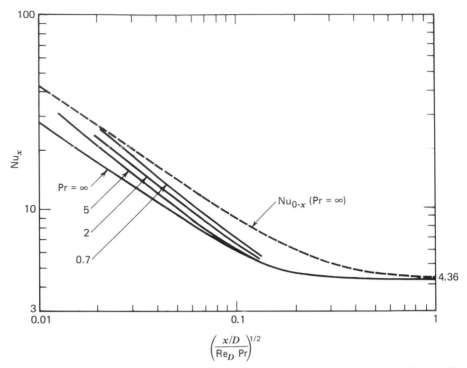

Figure 3.15 The heat transfer characteristics of the thermal entrance region of a round tube with uniform heat flux (drawn based on data from Refs. 8 and 18).

Thermally Developed Uniform (Slug) Flow

Perhaps the simplest way to demonstrate the effect of incomplete development on heat transfer is to consider the top-half of Fig. 3.13 in the limit $Pr \to 0$. In this limit the hydrodynamic length is infinitely greater than the thermal length, therefore, we can think of a tube section sufficiently far downstream from $x \sim X_T$, and, at the same time, sufficiently far upstream from $x \sim X$, in which the temperature profile is fully developed while the velocity profile is still uniform. In real life, it is hard to picture such a tube section in real fluids flowing through tubes, because the velocity profile is constantly departing from the uniform flow description. However, the thermally developed uniform flow condition can occur in the case of a solid (rod) moving with good thermal contact through a heated sleeve.

The Nusselt number can be obtained analytically in the most general case of a tube surrounded by an isothermal fluid (Fig. 3.11). The eigenvalue problem (82)–(85) is first modified by replacing the Hagen–Poiseuille flow $2(1 - r_*^2)$ in eq. (82) with the factor "1" representing uniform flow. The solution can be

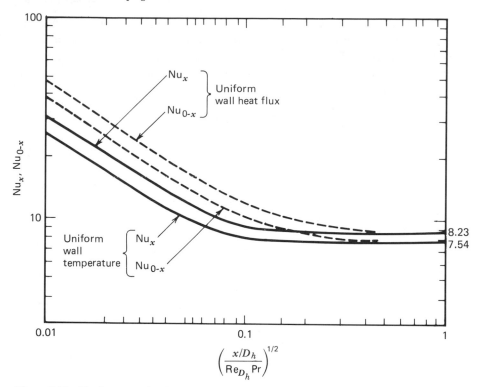

Figure 3.16 The heat transfer characteristics of thermally developing Hagen–Poiseuille flow in a parallel-plate duct (drawn based on data from Ref. 8).

expressed in terms of Bessel functions,

$$\theta = \frac{\hat{N}u^{1/2}}{2} \frac{J_0(r_*\hat{N}u^{1/2})}{J_1(\hat{N}u^{1/2})} \tag{93}$$

with the Nusselt number given implicitly by

$$\hat{N}u^{1/2}J_1(\hat{N}u^{1/2}) = Bi\, J_0(\hat{N}u^{1/2}) \tag{94}$$

This result is shown by the dashed curves on Fig. 3.12 [note that the corresponding inner-side Nusselt number Nu follows from eq. (79)]. The behavior of Nu in the two Bi extremes is

$$\lim_{Bi \to 0} \hat{N}u = 2\,Bi, \qquad \lim_{Bi \to \infty} \hat{N}u = 5.783 \tag{95}$$

Comparing the uniform flow with the fully developed flow Nusselt numbers in Fig. 3.12, we conclude that even when the temperature profile is fully developed, the underdevelopment of the velocity profile has the effect of increasing the value of Nu, that is, the effect of enhancing heat transfer. Physically, this conclusion makes sense because in uniform (slug) flow the fluid does not stick to the wall and, in this way, it does not insulate the wall as effectively as in Hagen–Poiseuille flow.

Thermally Developing Hagen–Poiseuille Flow

The next step on the ladder of difficulty is to consider the high Pr limit (bottom-half of Fig 3.13) and focus on the tube section described by $X < x < X_T$. In such a flow region the velocity profile is fully developed while the temperature profile is just being developed. Neglecting the effect of axial conduction ($Pe_x \gg 1$), we must solve eq. (50) subject to the conditions of

Uniform wall temperature, $T_0 =$ constant.

Symmetry about the centerline, $\partial T / \partial r = 0$ at $r = 0$.

Isothermal entering fluid, $T = T_{IN}$ for $x < 0$, where x is measured (positive) downstream from the location X in the bottom-half of Fig. 3.13.

This problem was treated for the first time by Graetz [19] in 1883 and is recognized in the heat transfer literature as the *Graetz problem*. In dimensionless form, the problem statement is

$$\frac{1}{2}(1 - r_*^2)\frac{\partial \theta_*}{\partial x_*} = \frac{\partial^2 \theta_*}{\partial r_*^2} + \frac{1}{r_*}\frac{\partial \theta_*}{\partial r_*}$$

$$\theta_* = 0 \quad \text{at } r_* = 1$$

$$\frac{\partial \theta_*}{\partial r_*} = 0 \quad \text{at } r_* = 0 \tag{96}$$

$$\theta_* = 1 \quad \text{at } x_* = 0$$

with the following notation

$$\theta_* = \frac{T - T_0}{T_{IN} - T_0}, \quad r_* = \frac{r}{r_0}, \quad x_* = \frac{x/D}{Re_D Pr} \tag{97}$$

The energy equation is clearly linear and homogeneous. Separation of variables is achieved by assuming a product solution for $\theta_*(r_*, x_*)$,

$$\theta_* = R(r_*)\Xi(x_*) \tag{98}$$

that yields two linear and homogeneous equations for R and Ξ,

$$R'' + \frac{1}{r_*}R' + \lambda^2(1 - r_*^2)R = 0 \qquad (99)$$

$$\Xi' + 2\lambda^2\Xi = 0 \qquad (100)$$

The Ξ equation admits solutions of the type $\Xi = C\exp(-2\lambda^2 x_*)$ where both C and λ^2 are arbitrary constants. The R equation is of the Sturm–Liouville type and its solution is obtainable as infinite series [20]; the θ_* solution satisfying the $r_* = 0, 1$ boundary conditions in (96) becomes the *Graetz series*

$$\theta_* = \sum_{n=0}^{\infty} C_n R_n(r_*)\exp(-2\lambda_n^2 x_*) \qquad (101)$$

where R_n and λ_n are eigenfunctions and eigenvalues, while C_n are constants determined by the $x_* = 0$ condition in the problem statement (96). For the analytical details concerning the derivation of the Graetz series, the reader is directed to Refs. 21 and 22; the heat transfer conclusions of interest are

$$\theta_{*m} = \frac{T_m - T_0}{T_{IN} - T_0} = 8\sum_{n=0}^{\infty} \frac{G_n}{\lambda_n^2}\exp(-2\lambda_n^2 x_*)$$

$$\mathrm{Nu}_x = \frac{\sum_{n=0}^{\infty} G_n \exp(-2\lambda_n^2 x_*)}{2\sum_{n=0}^{\infty}(G_n/\lambda_n^2)\exp(-2\lambda_n^2 x_*)} \qquad (102)$$

$$\mathrm{Nu}_{0-x} = \frac{1}{4x_*}\ln\left(\frac{1}{\theta_{*m}}\right)$$

θ_{*m}, Nu_x, and Nu_{0-x} are the bulk dimensionless temperature, the local Nusselt number, and the average Nusselt number for the entrance of section of length x. The eigenvalues λ_n and the constants $G_n = -(C_n/2)R_n'(1)$ for the first five terms of the infinite series are tabulated in Table 3.3.

Table 3.3   Graetz Series Solution Eigenvalues and Constants
(Round Tube, Isothermal Wall, Hagen–Poiseuille Flow) [8]

n	λ_n	G_n
0	2.704	0.7488
1	6.679	0.5438
2	10.673	0.4629
3	14.671	0.4154
4	18.67	0.3829
5	22.67	0.3587

Figure 3.14 shows the above results [expressions (102)] as the curves labeled "Pr $= \infty$" for θ_{*m}, Nu_x, and Nu_{0-x}. The group $x_*^{1/2} = (x/D/Re_D Pr)^{1/2}$ is used on the abscissa in order to illustrate the scaling law (89) that rules the transition from thermally developing to thermally fully developed duct flow. The Nusselt number curves show a knee at $x_*^{1/2} = O(1)$, in agreement with eq. (89): the validity of this transition criterion is dramatized further by the sudden drop of the bulk temperature from the inlet value ($\theta_{*m} = 1$) to the wall value ($\theta_{*m} = 0$). Finally, all the finite Pr curves show that in the entrance region ($x_*^{1/2} \ll 1$) the Nusselt number obeys a relationship of the type

$$Nu_x = (\text{constant})\left(\frac{x/D}{Re_D Pr}\right)^{-1/2} \tag{103}$$

where (constant) $= O(1)$. This proves the validity of the scaling law (92) and, considering the century-long effort of obtaining and perfecting the curves of Fig. 3.14, illustrates the power of proper scale analysis. The log–log presentation of Fig. 3.14 (and Figs. 3.15 and 3.16) is intentional, in order to illustrate the existence of the scaling law (92) in the form of *lines of slope* -1.

The Pr $= \infty$ curves on Fig. 3.15 show the corresponding Nu information for thermally developing Hagen–Poiseuille flow in the entrance to a tube with uniform heat flux. Figure 3.16 shows the same type of information for parallel-plate channels with either $T_0 = $ constant or $q = $ constant wall heating.

Thermally and Hydraulically Developing Flow

The most realistic (and most difficult) version of the problem consists of solving eqs. (96) with the Hagen–Poiseuille profile $2(1 - r_*^2)$ replaced by the actual x-dependent velocity profile present in the hydrodynamic entry region. This problem has been solved numerically by a number of investigators who used finite-difference formulations of problem (96): the history of this numerical work is recounted in Ref. 8. Figures 3.14 and 3.15 show a sample of the voluminous finite Pr data available in the literature.

SYMBOLS

a, b	dimensions of rectangular duct cross-section (Fig. 3.5)
A	duct cross-sectional area (Fig. 3.4)
Bi	Biot number [eq. (80)]
c_P	specific heat at constant pressure
$C_{f,x}$	local skin friction coefficient [eq. (9)]
D	plate-to-plate spacing (Fig. 3.1)
D_h	hydraulic diameter [eq. (26)]
f	friction factor [eq. (24)]
h	heat transfer coefficient; specific enthalpy in eq. (38)
L	duct length

Nu	Nusselt number in the fully developed region [eq. (52)]
Nu_x	local Nusselt number in the developing (entrance) region [eq. (102)]
p	wetted perimeter (Fig. 3.4)
P	pressure
ΔP	pressure drop
Pr	Prandtl number
q'	heat transfer rate per unit duct length [W/m]
q''	heat flux [W/m^2]
r	radial coordinate
r_0	tube radius
r_h	hydraulic radius [eq. (26)]
Re_D	Reynolds number (UD/ν)
Re_{D_h}	Reynolds number based on hydraulic diameter (UD_h/ν)
T	temperature
T_0	wall temperature
T_m	bulk temperature [eq. (42)]
T_∞	external fluid temperature
T_{IN}	inlet temperature
u	longitudinal velocity component
U	duct-averaged velocity
u_0, U_c	centerline velocity
v	transversal velocity component
x	longitudinal coordinate
X	hydrodynamic entrance length
X_T	thermal entrance length
y	transversal coordinate
α	thermal diffusivity
δ	velocity boundary layer thickness
δ_T	thermal boundary layer thickness
μ	viscosity
ν	kinematic viscosity
ρ	density
τ_w	wall shear stress
ϕ	fully developed temperature profile function [eq. (51)]

REFERENCES

1. W. M. Kays and M. E. Crawford, *Convective Heat and Mass Transfer*, 2nd ed., McGraw-Hill, New York, 1980, pp. 59, 90.

2. E. M. Sparrow, Analysis of laminar forced convection heat transfer in the entrance region of flat rectangular ducts, NACA TN 3331, 1955.

3. H. Schlichting, *Boundary Layer Theory*, 4th ed., McGraw-Hill, New York, 1960, p. 169.

4. H. L. Langhaar, Steady flow in the transition length of a straight tube, *J. Appl. Mech.*, Vol. 9, 1942, pp. A55–A58.

5. G. Hagen, *Über die Bewegung des Wassers in engen zylindrischen Röhren*, Pogg. Ann., Vol. 46, 1839, p. 423.

6. J. Poiseuille, *Récherches expérimentales sur le mouvement des liquides dans les tubes de très petits diamètres*, Comptes Rendus, Vol. 11, 1840, pp. 961, 1041.

7. O. G. Tietjens, Preface to *Fundamentals of Hydro and Aeromechanics*, L. Prandtl and O. G. Tietjens, Dover, New York, 1957, p. VII.

8. R. K. Shah and A. L. London, *Laminar Flow Forced Convection in Ducts*, Supplement 1 to *Advances in Heat Transfer*, Academic Press, New York, 1978.

9. A. Bejan, *Entropy Generation through Heat and Fluid Flow*, Wiley, New York, 1982, Preface.

10. F. P. Incropera and D. P. Dewitt, *Fundamentals of Heat Transfer*, Wiley, New York, 1981, p. 385.

11. W. M. Rohsenow and H. Y. Choi, *Heat, Mass and Momentum Transfer*, Prentice-Hall, Englewood Cliffs, NJ, 1961, p. 139.

12. J. H. Lienhard, *A Heat Transfer Textbook*, Prentice-Hall, Englewood Cliffs, NJ, 1981, p. 311.

13. R. K. Shah, Laminar flow friction and forced convection heat transfer in ducts of arbitrary geometry, *Int. J. Heat Mass Transfer*, Vol. 18, 1975, pp. 849–862.

14. R. K. Shah and A. L. London, Thermal boundary conditions and some solutions for laminar duct flow forced convection, *J. Heat Transfer*, Vol. 96, 1974, pp. 159–165.

15. E. M. Sparrow and S. V. Patankar, Relationship among boundary conditions and Nusselt numbers for thermally developed duct flows, *J. Heat Transfer*, Vol. 99, 1977, pp. 483–485.

16. W. M. Kays and H. C. Perkins, Forced convection, internal flow in ducts, in W. M. Rohsenow and J. P. Hartnett, eds., *Handbook of Heat Transfer*, McGraw-Hill, New York, 1973, p. 7-22.

17. D. Poulikakos, private discussions.

18. R. W. Hornbeck, An all-numerical method for heat transfer in the inlet of a tube, *Am. Soc. Mech. Eng.*, Paper 65-WA/HT-36, 1965.

19. L. Graetz, *Über die Warmeleitungsfähigkeit von Flüssigkeiten* (On the thermal conductivity of liquids) Part 1. *Ann. Phys. Chem.*, Vol. 18, pp. 79–94, 1883, Part 2; *Ann. Phys. Chem.*, Vol. 25, pp. 337–357, 1885.

20. F. B. Hildebrand, *Advanced Calculus for Applications*, Prentice-Hall, Englewood Cliffs, NJ, 1962, p. 208.

21. T. B. Drew, Mathematical attacks on forced convection problems: A review, *Trans. Am. Inst. Chem. Eng.*, Vol. 26, 1931, pp. 26–80.

22. M. Jakob, *Heat Transfer*, Vol. 1, Wiley, New York, 1949.

PROBLEMS

1. Determine the skin-friction coefficient $C_{f,x}$ for hydrodynamically developing flow in a parallel-plate duct by using the Sparrow integral solution for the velocity distribution [eqs. (5) and (6)].

2. Determine the velocity distribution corresponding to fully developed (Hagen–Poiseuille) flow through the annular space formed between two concentric tubes. Let r_i and r_o be the inner and outer radii, respectively. Show that the inner wall shear stress τ_{w,r_i} differs from the value along the outer wall, τ_{w,r_o}. Calculate the friction factor for this flow by using instead of τ_w in eq. (24) the *average* τ_w value defined based on a force balance of type (23):

$$\pi(r_o^2 - r_i^2)\,\Delta P = \tau_{w,\text{avg}}2\pi(r_o + r_i)L = 2\pi L(r_o\tau_{w,r_o} + r_i\tau_{w,r_i}).$$

3. Determine the velocity distribution and friction factor for Hagen–Poiseuille flow through a duct whose cross-section has the shape of an extremely slender wedge (a triangle with tip angle $\varepsilon \ll 1$, and long sides equal to b). Neglect the friction effect introduced by the short wall opposing the tip angle, whose length is εb. Start with eq. (30) where y is along b and z along εb. Check the relative order of magnitude of the two terms on the right-hand side of eq. (30) and neglect the insignificant one. To calculate the friction factor, use the *perimeter-averaged* wall shear stress $\tau_{w,\text{avg}}$ defined in the preceding problem.

4. Determine the Hagen–Poiseuille flow through a duct of rectangular cross-section (Fig. 3.5) by solving eq. (30) for $u(y, z)$ as a Fourier series. This problem is analytically identical to determining the temperature distribution inside a rectangular object with internal heat generation and isothermal walls. With reference to Fig. 3.5, the problem statement is

$$\nabla^2 u = \frac{1}{\mu}\frac{dP}{dx} = \text{constant}$$

$$u = 0 \quad \text{at } y = \pm a/2 \qquad\qquad\text{(A)}$$

$$u = 0 \quad \text{at } z = \pm b/2$$

To solve it, assume

$$u(y, z) = u_1(y) + u_2(y, z)$$

where $u_1(y)$ is the Hagen–Poiseuille flow through the infinite parallel-plate channel of width $2a$,

$$\frac{d^2 u_1}{dy^2} = \frac{1}{\mu}\frac{dP}{dx} \qquad\qquad\qquad\text{(B)}$$

$$u_1 = 0 \quad \text{at } y = \pm a/2$$

and where u_2 is the necessary *correction*,

$$\nabla^2 u_2 = 0$$

$$u_2 = 0, \quad \text{at } y = \pm a/2 \qquad\qquad\text{(C)}$$

$$u_2 = -u_1(y), \quad \text{at } z = \pm b/2$$

Note that adding problems B and C equation-by-equation yields the original problem A. The advantage of decomposing the problem as A = B + C is that problem C can be solved by Fourier series expansion, while problem A cannot. (Problem C is solvable because the equation $\nabla^2 u_2 = 0$ is homogeneous *and* one set of boundary conditions (y) is homogeneous.)

5. Consider the approximate solution to Hagen–Poiseuille flow through a duct with rectangular cross-section [eqs. (31)–(37)]. Retrace the analytical steps of this solution by starting with an equally "reasonable" velocity profile instead of eq. (31), for example,

$$u(y, z) = u_0 \cos\frac{\pi y}{a} \cos\frac{\pi z}{b}$$

6. Table 3.2 and Fig. 3.7 suggest that a very direct engineering approximation for f and Nu in fully developed duct flow is

$$f = \frac{16}{\mathrm{Re}_{D_h}} \frac{\pi D_h^2/4}{A_{\mathrm{duct}}}$$

$$\mathrm{Nu} = 3.66 \frac{\pi D_h^2/4}{A_{\mathrm{duct}}}, \quad T_0 = \text{constant}$$

$$\mathrm{Nu} = 4.36 \frac{\pi D_h^2/4}{A_{\mathrm{duct}}}, \quad q'' = \text{constant}$$

Check the accuracy of these geometric correlations by completing the following table for friction and heat transfer through a duct with regular *hexagonal* cross-section:

	Numerical Results [8]	Approximate Results
$f\,\mathrm{Re}_{D_h}$	15.05	
Nu (T_0 = constant)	3.34	
Nu (q'' = constant)	4.00	

7. Determine the fully developed temperature profile in a tube with constant wall temperature by solving eqs. (72)–(74) using the method of *successive approximations*. The technique consists of guessing a particular polynomial for $\phi(r_*)$, substituting this guess into the left-hand side of eq. (72), and, finally, integrating eq. (72) to obtain a better guess (approximation) for $\phi(r_*)$. The procedure can be repeated until the change in the Nu value from one approximation to the next is below a cost-determined percentage. To start the procedure, a good initial guess is $\phi_0(r_*) = 1$.

8. Verify that in cases where the wall thermal resistance is not negligible, the effective external heat-transfer coefficient for a tube immersed in an isothermal fluid is given by eq. (77).

9. Use Fig. 3.11 and the limiting values of the Biot number $h_e r_0/k$ to show that T_0 = constant and q_0 = constant are two special cases of the heat transfer problem involving fully developed flow through a tube surrounded by isothermal fluid.

10. Consider the problem of thermal development in the entrance region X_T of a tube with uniform (slug) flow throughout the X_T length. This assumption amounts to imagining a $\mathrm{Pr} = 0$ fluid; in addition, the energy equation in eqs. (96) is simplified as $(1 - r_*^2)$ is being replaced by the constant $\frac{1}{2}$. Solve this simplified version of the problem using the separation of variables indicated in eq. (98). As a guide, use a conduction heat transfer textbook and the observation that the simplified problem is analytically identical to the problem of transient heat conduction in an initially isothermal ($\theta_* = 1$) cylindrical object with isothermal boundary ($\theta_* = 0$), where x_* assumes the role of dimensionless time.

11. Consider the Graetz series solution for Nu_x in a thermally developing Hagen–Poiseuille flow in a tube [Table 3.3 and eqs. (102)]. Show that in the range $x_* > O(1)$, the series expressions for Nu_x and Nu_{0-x} tend to the fully developed value of 3.66 (Table 3.2).

12. Evaluate the hydraulic diameter of a tube of internal diameter D, which has a slowly twisting tape insert (dividing wall) positioned right through the middle (see figure).

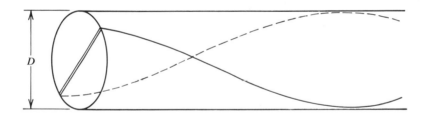

13. A water stream is heated in fully developed flow through a pipe with uniform heat flux at the wall. The flowrate is $\dot{m} = 10$ g/s, the heat flux $q'' = 0.1$ W/cm^2, and the pipe radius $r_0 = 1$ cm. The properties of water are $\mu = 0.01$ g/(cm s) and $k = 0.006$ W/(cm K). Calculate

(a) The Reynolds number based on pipe diameter and mean fluid velocity.

(b) The heat-transfer coefficient.

(c) The difference between the wall temperature and the mean (bulk) fluid temperature.

14. A water chiller passes a stream of 0.1 kg/s through a pipe immersed in a bath containing a mixture of crushed ice and water. Thus, the pipe wall temperature may be assumed to be $T_w = 0°C$. The original (inlet) temperature of the stream is $T_1 = 40°C$, and the specific heat of water is $c = 4.182$ J/(g K).

(a) If the effectiveness of this heat exchanger is $\varepsilon = 0.85$, calculate the final (outlet) temperature of the water stream.

(b) Under the same conditions, what is the overall heat transfer rate between the stream and ice–water bath?

(c) Assume that the pipe length is L, the diameter D, and the flow regime fully developed laminar. If a new pipe is to be used ($D_1 = D/2$) and if the effectiveness is to remain unchanged, then what should be the length of the new pipe (relative to the old length L), $L_1/L = ?$

15. Consider the Hagen–Poiseuille flow through a tube of radius r_0. The flow is extremely viscous so that the energy equation reduces to

$$0 = k\frac{1}{r}\frac{d}{dr}\left(r\frac{dT}{dr}\right) + \mu\Phi$$

where Φ is the viscous dissipation term $\Phi = (du/dr)^2$. Determine the temperature distribution inside the pipe, subject to $T = T_o$ (constant) at $r = r_0$. Let Q be the total heat transfer rate through the pipe wall, over a pipe length L. Prove that: $Q = \dot{m}\,\Delta P/\rho$, where $\dot{m}$ and ΔP are the mass flowrate and the pressure drop over the length L. Comment on the thermodynamic (lost-work) significance of this result; show that it is the same as eq. (48) of Chapter 1.

4

Laminar Natural Convection

From a thermodynamic standpoint, the basic feature that links the external flow problems of Chapter 2 with the internal flow problems of Chapter 3 is the implied presence of an additional entity (a mechanism) that, in all cases, is responsible for creating the flow. Fluid flow is not merely a body of fluid such as the stagnant water pool in a glass: it is the relative motion of one fluid layer past an adjacent fluid layer or a solid surface. Since it is an intrinsic property of fluid flow to destroy (dissipate) available work [1], any fluid flow requires a driving mechanism in order to exist. For example, in order to witness the boundary layer forming along a smooth wall in parallel uniform flow (Fig. 2.1), somebody must steadily sink mechanical power into the task of dragging the solid wall through the fluid: in the case of motorized sea-transport, that "somebody" is the ship's power plant. Likewise, in order to witness the duct flows addressed in Chapter 3, the fluid must be first pumped (forced to flow) through the duct: the operation of a pump always requires the expense of mechanical power [2].

For the thermodynamic reason outlined above, the convective heat transfer problems of Chapters 2 and 3 can be regarded as examples of forced convection, in other words, examples of heat transport by fluid motion which is forced to happen. The creation and maintainence of the flow requires a consistent sacrifice of mechanical power (available work). Indeed, the destruction of available work through fluid flow is the thermodynamic reason for the existence of a *fluid friction question* in heat transfer engineering [eq. (1) of Chapter 2].

From a mathematical standpoint, the flow problem in forced convection appears to be decoupled from the heat transfer problem. Historically, this has clearly been the case as the Blasius flow solution preceded the Pohlhausen heat transfer solution, and as the Hagen–Poiseuille flow emerged almost 100 years before the fully developed heat transfer results summarized in Table 3.2. Of course, the flow field must be known or, at least, assumed (postulated) before proceeding with an analysis of heat transfer in forced convection. However, unless model-polishing, second-order effects such as temperature-dependent physical properties are taken into account, the flow field can be analyzed independently of the temperature field.

In the present chapter we focus on a class of convective heat transfer problems that differ fundamentally from the forced convection class considered so far. The difference is thermodynamic and mathematical as well. Thermodynamically, the flows of this chapter are not forced but happen naturally: these flows are driven by the buoyancy effect due to the presence of gravitational acceleration and density variations from one fluid layer to another. Mathematically, the flow field is intimately coupled to the temperature field, as temperature variations within the fluid can induce density variations; hence, a buoyancy-driven flow.

NATURAL CONVECTION VERSUS FORCED CONVECTION

The simplest configuration for the study of natural convection is shown in Fig. 4.1. Think of a body of temperature T_0 and height H immersed in a fluid of temperature T_∞. For a more meaningful discussion, we think of a heated body immersed in a cold fluid reservoir, such as an old-fashioned stove in the middle

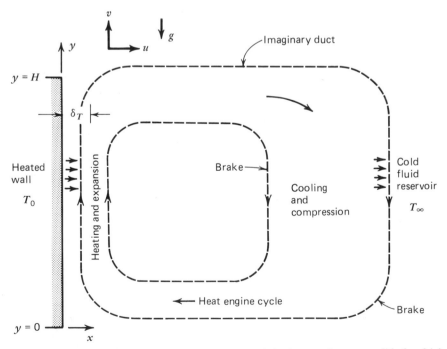

Figure 4.1 Natural convection along a vertical wall and the heat engine responsible for driving the flow.

of a room. Since air at constant pressure expands upon being heated, the air layer adjacent to the wall expands (becomes lighter, less dense) and rises. At the same time, the cold reservoir fluid is displaced *en masse* downward. Thus, the wall–reservoir temperature difference drives the all-familiar "natural circulation" or free convection cell sketched in Fig. 4.1. In view of the thermodynamics of forced convection discussed in the preceding segment, it is appropriate to ask the question: Who or what power plant is responsible for the steady cyclic flow encountered in natural convection?

To answer this question we follow the evolution of a fluid packet through the imaginary closed duct that holds the cellular flow. Starting from the bottom of the heated wall, the packet is heated by the wall and expands as it rises to lower pressures in the hydrostatic pressure field maintained by the reservoir. Later on, along the downflowing branch of the cycle, the fluid packet is cooled by the reservoir and compressed as it reaches the depths of the reservoir. From the circuit executed by each fluid packet we learn that the cellular flow is the succession of four processes,

$$\text{heating} \rightarrow \text{expansion} \rightarrow \text{cooling} \rightarrow \text{compression}$$

In other words, the convection loop of Fig. 4.1 is equivalent to the cycle executed by the working fluid in a heat engine. In principle, this heat engine cycle should be capable of delivering useful work if we insert a properly designed propeller in the stream: this is the origin of the "wind power" discussed nowadays in connection with the harnessing of solar work indirectly from the atmospheric heat engine loop. In the absence of work-collecting devices (e.g., windmill propellers), the heat engine cycle drives its working fluid fast enough so that its entire work output potential is dissipated by friction in the brake at the interface between what moves and what does not move. The real circulation pattern of Fig. 4.1 can be regarded as an infinity of such heat engine cycles—one nested inside the next.

The thermodynamics of Fig. 4.1 illustrates the fundamental difference between forced convection and natural convection. As was shown earlier, in forced convection the heat engine driving the flow is external, whereas in natural convection the heat engine is built into the flow itself.

The natural convection loop of Fig. 4.1 will be analyzed in this chapter and the next by focusing on two possible extremes. In this chapter, we will discuss mainly the interaction between a vertical heated object and a much larger fluid reservoir, so large that the downward motion sketched in Fig. 4.1 is negligible. Thus, the vertical velocity of the reservoir fluid situated sufficiently far from the heated wall will be taken as equal to zero. In Chapter 5, on the other hand, we will study the circulation present in finite-size fluid layers heated from the side (Fig. 5.1): in that case the downward portion of the natural convection loop of Fig. 4.1 is as strong as the upward portion, because both portions are driven by two vertical walls maintained at different temperatures.

VERTICAL BOUNDARY LAYER EQUATIONS

The chief heat transfer problem in connection with Fig. 4.1 is to predict the heat transfer rate Q when the wall reservoir temperature difference is known,

$$Q = (HW)h_{0-H}(T_0 - T_\infty) \tag{1}$$

in other words, to calculate the wall-averaged heat-transfer coefficient h_{0-H}. Note that HW is the wall area and W the wall dimension in the direction perpendicular to the x–y plane. In this section we focus on the boundary layer regime, where the scale of h_{0-H} is κ/δ_T and the thermal boundary layer thickness δ_T is negligibly small in comparison with H.

Proceeding as in Chapter 2, the complete Navier–Stokes equations for the steady, constant property, two-dimensional flow of Fig. 4.1 are

$$\frac{\partial u}{\partial x} + \frac{\partial v}{\partial y} = 0 \tag{2}$$

$$\rho\left(u\frac{\partial u}{\partial x} + v\frac{\partial u}{\partial y}\right) = -\frac{\partial P}{\partial x} + \mu\nabla^2 u \tag{3}$$

$$\rho\left(u\frac{\partial v}{\partial x} + v\frac{\partial v}{\partial y}\right) = -\frac{\partial P}{\partial y} + \mu\nabla^2 v - \rho g \tag{4}$$

$$u\frac{\partial T}{\partial x} + v\frac{\partial T}{\partial y} = \alpha\nabla^2 T \tag{5}$$

Compared with the equations of Chapter 2, note the presence of the body force term $-\rho g$ in the vertical momentum equation (4). The governing equations (2)–(5) reduce to simpler forms if the focus of the analysis is the boundary layer region ($x \sim \delta_T$, $y \sim H$, and $\delta_T \ll H$). Thus, only the $\partial^2/\partial x^2$ term survives in the ∇^2 operator, and, as demonstrated in Chapter 2, the transversal momentum equation (3) reduces to the statement that in the boundary layer, the pressure is a function of longitudinal position only,

$$\frac{\partial P}{\partial y} = \frac{dP}{dy} = \frac{dP_\infty}{dy} \tag{6}$$

The boundary layer equations for momentum and energy are then

$$\rho\left(u\frac{\partial v}{\partial x} + v\frac{\partial v}{\partial y}\right) = -\frac{dP_\infty}{dy} + \mu\frac{\partial^2 v}{\partial x^2} - \rho g \tag{7}$$

$$u\frac{\partial T}{\partial x} + v\frac{\partial T}{\partial y} = \alpha\frac{\partial^2 T}{\partial x^2} \tag{8}$$

Noting further that dP_∞/dy is the hydrostatic pressure gradient dictated by the reservoir fluid of density ρ_∞, $dP_\infty/dy = -\rho_\infty g$, the momentum equation (7) becomes

$$\rho\left(u\frac{\partial v}{\partial x} + v\frac{\partial v}{\partial y}\right) = \mu\frac{\partial^2 v}{\partial x^2} + (\rho_\infty - \rho)g \tag{9}$$

Equations (2), (8), and (9) must be solved in order to determine u, v, and T in the boundary layer. Through the body force term $(\rho_\infty - \rho)g$ in the momentum equation (9), the flow is driven by the density field $\rho(x, y)$ generated by the temperature field $T(x, y)$. Equations (8) and (9) are coupled via the *equation of state* of the fluid, for example,

$$P = \rho RT \tag{10}$$

if the fluid behaves according to the ideal gas model [3]. At any level y we have

$$\rho = \frac{P_\infty/R}{T} \quad \text{and} \quad \rho_\infty = \frac{P_\infty/R}{T_\infty} \tag{11}$$

hence,

$$\rho - \rho_\infty = \rho\left(1 - \frac{T}{T_\infty}\right) \tag{12}$$

This last expression can be rearranged as

$$\frac{\rho_\infty - \rho}{\rho_\infty}\left(1 - \frac{\rho_\infty - \rho}{\rho_\infty}\right)^{-1} = \frac{T - T_\infty}{T_\infty} \tag{13}$$

which in the limit $(T - T_\infty) \ll T_\infty$ yields

$$\rho \simeq \rho_\infty\left[1 - \frac{1}{T_\infty}(T - T_\infty) + \cdots\right] \tag{14}$$

This result states that the density decreases slightly below ρ_∞ as the local *absolute* temperature increases slightly above the reservoir absolute temperature T_∞. In general, expression (14) is written as

$$\rho \simeq \rho_\infty\left[1 - \beta(T - T_\infty) + \cdots\right] \tag{15}$$

where β is the volume expansion coefficient at constant pressure [4]

$$\beta = -\frac{1}{\rho}\left(\frac{\partial\rho}{\partial T}\right)_P \tag{16}$$

Implicit in expression (15) is the assumption that the dimensionless product $\beta(T - T_\infty)$ is considerably smaller than unity.

The *Boussinesq approximation* of the boundary layer equations amounts to substituting eq. (15) into eqs. (8) and (9) and, in each case, retaining the dominant term. For example, in the momentum equation (9), ρ appears in the inertia terms as well as in the body force term; using the Boussinesq approximation (15), the inertia terms will be multiplied by the dominant term ρ_∞ = constant, whereas the leading body force term becomes $\rho_\infty \beta g(T - T_\infty)$. Therefore the momentum equation (9) can be written as

$$u\frac{\partial v}{\partial x} + v\frac{\partial v}{\partial y} = \nu\frac{\partial^2 v}{\partial x^2} + g\beta(T - T_\infty) \qquad (17)$$

where g, β, T_∞, and $\nu = \mu/\rho_\infty$ are constants. Likewise, the thermal diffusivity appearing in the energy equation (8), $\alpha = k/(\rho_\infty c_P)$ is a constant.

The Boussinesq-approximated momentum equation (17) illustrates the coupling between the temperature field and the flow field. If the fluid is isothermal $(T = T_\infty)$, then the driving force is zero everywhere and eqs. (17) and (2) yield the "no-flow" solution $u = v = 0$. When the fluid is heated by the wall, the body force term is finite $[\sim g\beta(T_0 - T_\infty)]$ and so are the velocity components u, v. In what follows we focus in more detail on the solution to the Boussinesq-boundary layer equations (2), (8), and (17) subject to the following conditions:

(i) Impermeable, solid, isothermal wall

$$u = v = 0 \quad \text{and} \quad T = T_0 \quad \text{at } x = 0$$

(ii) Stagnant, isothermal infinite reservoir

$$v = 0 \quad \text{and} \quad T = T_\infty \quad \text{as } x \to \infty \qquad (18)$$

SCALE ANALYSIS

Consider the conservation of mass, momentum, and energy in the *thermal* boundary layer region ($x \sim \delta_T$, $y \sim H$), where the heating effect of the vertical wall is felt. In the steady state, the heat conducted from the wall horizontally into the fluid is swept and carried upwards as an enthalpy stream. The energy equation (8) expresses a balance between convection and conduction,

$$\underbrace{u\frac{\Delta T}{\delta_T}, \; v\frac{\Delta T}{H}}_{\text{Convection}} \sim \underbrace{\alpha\frac{\Delta T}{\delta_T^2}}_{\text{Conduction}} \qquad (19)$$

where $\Delta T = T_0 - T_\infty$ is the scale of $T - T_\infty$. From mass conservation in the same layer, that is,

$$\frac{u}{\delta_T} \sim \frac{v}{H} \tag{20}$$

we learn that two convection terms in eq. (19) are of order $v\Delta T/H$. Thus, the energy balance

$$v\frac{\Delta T}{H} \sim \alpha\frac{\Delta T}{\delta_T^2} \tag{21}$$

yields

$$v \sim \frac{\alpha H}{\delta_T^2} \tag{22}$$

where the thermal thickness δ_T is still unknown.

Turning our attention to the momentum equation (17) and focusing still on the $\delta_T \times H$ region, we recognize the interplay between three forces

$$\underbrace{u\frac{v}{\delta_T}, \ v\frac{v}{H}}_{\text{Inertia}} \quad \text{or} \quad \underbrace{\frac{\nu v}{\delta_T^2}}_{\text{Friction}} \sim \underbrace{g\beta\Delta T}_{\text{Buoyancy}} \tag{23}$$

The mass conservation scaling (20) indicates that the two inertia terms are of order v^2/H. It remains to establish under what conditions the δ_T layer is ruled by an inertia $\sim$ buoyancy balance, as opposed to a friction $\sim$ buoyancy balance (note that the buoyancy force is not negligible since, without it, there would be no flow). Dividing expression (23) through the buoyancy scale $g\beta\Delta T$ and using eq. (22) to eliminate the vertical velocity scale v, we obtain

$$\underbrace{\left(\frac{H}{\delta_T}\right)^4 \mathrm{Ra}_H^{-1}\mathrm{Pr}^{-1}}_{\text{Inertia}} \quad \underbrace{\left(\frac{H}{\delta_T}\right)^4 \mathrm{Ra}_H^{-1}}_{\text{Friction}} \quad \underbrace{1}_{\text{Buoyancy}} \tag{24}$$

where the Rayleigh number is defined as

$$\mathrm{Ra}_H = \frac{g\beta\Delta T H^3}{\alpha\nu} \tag{25}$$

From expression (24) it is clear that the competition between inertia and friction is decided by a *fluid property*, the Prandtl number: high-Pr fluids will form a δ_T layer ruled by the friction $\sim$ buoyancy balance, while low Pr fluids will form a δ_T layer with buoyancy balanced by inertia. Below, we examine these two possibilities in more detail.

High-Pr Fluids

When Pr $\gg$ 1, the friction $\sim$ buoyancy balance of eq. (24) yields

$$\delta_T \sim H \, \mathrm{Ra}_H^{-1/4} \tag{26}$$

and, using eq. (22),

$$v \sim \frac{\alpha}{H} \mathrm{Ra}_H^{1/2} \tag{27}$$

Since the heat-transfer coefficient scales as k/δ_T, the Nusselt number varies as

$$\mathrm{Nu} = \frac{hH}{k} \sim \mathrm{Ra}_H^{1/4} \tag{28}$$

It will be shown later that the Nu $\sim \mathrm{Ra}_H^{1/4}$ proportionality for Pr $\gg$ 1 fluids is confirmed by more precise analyses and numerous laboratory measurements; therefore, the δ_T and v scales derived above are the correct scales for the thermal boundary layer region.

The top-half of Fig. 4.2 shows qualitatively the conclusions reached so far: the δ_T-thick layer effects the transition from T_0 to T_∞, and at the same time, drives fluid upwards with a velocity given by eq. (27). Since we are dealing with high-Pr fluids, we must expect from a similar analysis in Chapter 2 that the fluid motion is not restricted to a layer of thickness δ_T. It is possible for the heated δ_T layer to viscously drag along a layer of outer (unheated) fluid. Let δ be the thickness of this outer layer, and let us also assume that $\delta \gg \delta_T$. Consider now the conservation of momentum in the boundary layer of thickness δ [eq. (17)]. Since the outer fluid is isothermal, the buoyancy effect is absent. The δ layer is driven (entrained) viscously by the much thinner δ_T layer, and it is restrained by its own inertia. Thus, eq. (17) dictates a balance inertia $\sim$ friction in a layer of thickness δ,

$$v\frac{v}{H} \sim \nu \frac{v}{\delta^2} \tag{29}$$

where the vertical velocity scale v is imposed by the driving instrument (the δ_T layer); eliminating v between eqs. (29) and (27) yields

$$\delta \sim H \, \mathrm{Ra}_H^{-1/4} \mathrm{Pr}^{1/2} \tag{30}$$

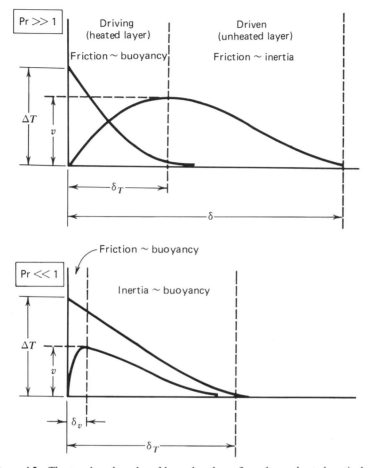

Figure 4.2   The two length scales of boundary layer flow along a heated vertical wall.

in other words,

$$\frac{\delta}{\delta_T} \sim \mathrm{Pr}^{1/2} > 1 \qquad (31)$$

In conclusion, the higher the Prandtl number, the thicker the layer of unheated fluid driven upwards by the heated layer. Influenced by the language of the older subfield of forced convection (Chapter 2), the present natural convection literature refers to δ as the *velocity boundary layer thickness* [5]. This terminology is conceptually inappropriate, because it negates the fundamental difference between forced convection boundary layers and natural convection boundary layers. This fundamental difference is illustrated in Fig. 4.2, where the velocity profile is described by *two* length scales (δ_T and δ), not by a single length (δ) as in forced convection. The velocity scale [eq. (27)] is reached within a thin layer δ_T, while the velocity decays to zero within a thick layer δ.

Low-Pr Fluids

Looking back at eq. (24), if $Pr \ll 1$, we see a balance between inertia and buoyancy in a layer of thickness δ_T. Combining this balance with the v scale of eq. (22) yields, in order,

$$\delta_T \sim H(Ra_H Pr)^{-1/4} \tag{32}$$

$$v \sim \frac{\alpha}{H}(Ra_H Pr)^{1/2} \tag{33}$$

$$Nu = \frac{hH}{k} \sim (Ra_H Pr)^{1/4} \tag{34}$$

We notice here the emergence of a new dimensionless group, $Ra_H Pr$, which plays the same role for low-Pr fluids as Ra_H for high-Pr fluids. Not long ago [6], the name of *Boussinesq number* was proposed for this group

$$Bo_H = Ra_H Pr = \frac{g\beta\Delta T H^3}{\alpha^2} \tag{35}$$

The bottom-half of Fig. 4.2 shows the meaning of scales (32) and (33): The δ_T layer is driven upwards by buoyancy and restrained by inertia. This means that outside the δ_T layer, where the fluid is isothermal and the buoyancy effect is absent, the fluid is motionless. The velocity profile must then be as wide as the temperature profile. However, since the no-slip condition still applies at the wall, the location of the velocity peak is an important second-length scale in the description of the velocity profile. Let δ_v be the thickness of a very thin layer right near the wall, a layer in which the buoyancy-driven fluid is restrained viscously by the wall. The buoyancy $\sim$ friction balance in the layer of thickness δ_v yields

$$v\frac{v}{\delta_v^2} \sim g\beta\Delta T \tag{36}$$

where the v scale is dictated by the δ_T layer scale [eq. (33)]. Combining, we find

$$\delta_v \sim H\,Gr_H^{-1/4} \tag{37}$$

where the *Grashof number* is defined as

$$Gr_H = \frac{g\beta\Delta T H^3}{\nu^2} = \frac{Ra_H}{Pr} \tag{38}$$

Dividing eqs. (37) and (32) side-by-side yields

$$\frac{\delta_v}{\delta_T} \sim \mathrm{Pr}^{1/2} < 1 \tag{39}$$

This relationship is shown qualitatively in Fig. 4.2; however, it should not be confused with eq. (31), as δ_v should not be confused with δ.

Observations

Table 4.1 provides a bird's-eye view of the conclusions reached based on scale analysis. The first three columns contain the length scales governing the thermal layer and the buoyancy-driven wall jet. It is apparent that the length scale $H\,\mathrm{Ra}_H^{-1/4}$ plays the role of primary length unit and that the remaining length scales follow from $H\,\mathrm{Ra}_H^{-1/4}$ through an appropriate stretching/compression factor depending solely on Pr. The relative order of magnitude of all length scales is shown in Fig. 4.3 using $H\,\mathrm{Ra}_H^{-1/4}$ as the length unit on the ordinate: it is clear that the boundary layer geometry of Pr < 1 fluids differs fundamentally from the geometry of Pr > 1 fluids.

Although the scales of Table 4.1 are the result of remarkably simple and brief analysis, they appear to be generally unknown in contemporary natural convection research. For example, in a deservedly influential natural convection chapter, Gebhart [7] relies on scaling arguments to conclude that for Pr > 1 fluids $\delta_T \sim H\,\mathrm{Ra}_H^{-1/4}\mathrm{Pr}^{-1/4}$ and $\delta \sim \mathrm{Gr}_H^{-1/4}$; these scales contradict those of Table 4.1 and are clearly incorrect (as a test, Gebhart's scales would imply Nu $\sim (\mathrm{Ra}_H\mathrm{Pr})^{1/4}$ for Pr > 1: this heat transfer scaling is not supported either by measurements [8] or more exact analyses [9]).

The erroneous view that for Pr > 1 fluids the wall jet thickness varies as $\mathrm{Gr}_H^{-1/4}$ is responsible for the widespread and incorrect use of the Grashof number in the nondimensional presentation of natural convection results in external (boundary layer) flow. The scales derived in this section demonstrate that Gr_H appears as a relevant dimensionless group only in the δ_v scale for Pr < 1 fluids [eq. (37)]: as such, Gr_H would be relevant to estimating the shear force along a vertical wall immersed in a liquid metal—not a pressing engineering problem! From a heat transfer standpoint, the important groups in external natural convection are the Rayleigh number for Pr > 1 fluids and the Boussinesq number for Pr < 1 fluids.

Another worthwhile observation concerns the very *meaning* of dimensionless numbers such as Ra_H, Bo_H, and Gr_H. For example, we are often told that the Grashof number can be interpreted as the parameter describing the ratio of buoyancy to viscous forces in the natural convection boundary layer. To see the error in this interpretation, consider the natural convection of air along the cold vertical wall in a room, where $\mathrm{Gr}_H \sim 10^8$–10^{10}: according to the above interpretation, the viscous forces must be negligible in comparison with the body force, because the Grashof number is enormous *vis-a-vis* unity. This is

Table 4.1   Summary of Flow and Heat Transfer Scales in a Natural Convection Boundary Layer Along a Vertical Wall

Prandtl Number Range	Thermal Boundary Layer Thickness	Wall Jet Velocity Profile			Nusselt Number $\mathrm{Nu} = \dfrac{hH}{k}$
		Distance from Wall to Velocity Peak	Thickness of Wall Jet	Velocity Scale	
$\mathrm{Pr} > 1$	$H\,\mathrm{Ra}_H^{-1/4}$	$H\,\mathrm{Ra}_H^{-1/4}$	$\mathrm{Pr}^{1/2}(H\,\mathrm{Ra}_H^{-1/4})$	$\dfrac{\alpha}{H}\,\mathrm{Ra}_H^{1/2}$	$\mathrm{Ra}_H^{1/4}$
$\mathrm{Pr} < 1$	$\mathrm{Pr}^{-1/4}(H\,\mathrm{Ra}_H^{-1/4})$	$\mathrm{Pr}^{1/4}(H\,\mathrm{Ra}_H^{-1/4})$	$\mathrm{Pr}^{-1/4}(H\,\mathrm{Ra}_H^{-1/4})$	$\dfrac{\alpha}{H}(\mathrm{Pr}\,\mathrm{Ra}_H)^{1/2}$	$(\mathrm{Pr}\,\mathrm{Ra}_H)^{1/4}$

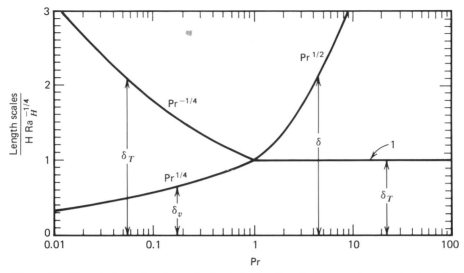

Figure 4.3 The relative order of magnitude of the length scales in natural convection boundary layer flow.

certainly not true, since in the case of air (Pr ~ 1) there always exists a balance between friction and buoyancy (or between inertia and buoyancy): without a balance of forces, the wall jet cannot exist. As shown in Problem 18 at the end of this chapter, the notion that the inertia/friction ratio scales as Gr_H comes from incorrect scale analysis performed in a flow region distinct from the boundary layer region.

By themselves, dimensionless numbers such as Ra_H, Bo_H, and Gr_H have no meaning. What has meaning is the 1/4th power of these numbers:

$$Ra_H^{1/4} \sim \frac{\text{wall height}}{\text{thermal boundary layer thickness}}, \qquad \text{if } Pr > 1$$

$$Bo_H^{1/4} \sim \frac{\text{wall height}}{\text{thermal boundary layer thickness}}, \qquad \text{if } Pr < 1$$

$$Gr_H^{1/4} \sim \frac{\text{wall height}}{\text{wall shear layer thickness}}, \qquad \text{if } Pr < 1$$

The meaning of $Ra_H^{1/4}$, $Bo_H^{1/4}$, and $Gr_H^{1/4}$ is purely *geometric*: these numerical values account for the *slenderness* of the boundary layer region occupied by the buoyancy-induced flow. When, in an actual problem, the calculated value of Ra_H or Gr_H is enormous compared with unity, Nature is trying to tell us something: it tells us that the real-life slenderness ratio $Ra_H^{1/4}$ (already greater than one) was unnecessarily raised to the fourth power.

INTEGRAL SOLUTION

As shown in Chapter 2, an integral solution to the governing equations may be used to determine the actual y variation of features such as local heat flux (q''), thermal boundary layer thickness (δ_T), and wall jet velocity profiles. So far, we know only the order of magnitude of the relevant flow and heat transfer parameters (Table 4.1). Integrating the momentum equation (17) and the energy equation (8) from the wall $(x = 0)$ to a far enough plane $x = X$ in the motionless isothermal cold reservoir, we obtain the integral boundary layer equations for momentum and energy.

$$\frac{d}{dy} \int_0^X v^2 dx = -\nu \left(\frac{\partial v}{\partial x} \right)_{x=0} + g\beta \int_0^X (T - T_\infty) \, dx \qquad (40)$$

$$\frac{d}{dy} \int_0^X v(T_\infty - T) \, dx = \alpha \left(\frac{\partial T}{\partial x} \right)_{x=0} \qquad (41)$$

The length scales of Table 4.1 and Fig. 4.3 are very useful in selecting the proper shapes of v and T profiles to be substituted into the integral equations (40) and (41). To begin with, we must carry out the integral analysis in two parts, for $Pr > 1$ and $Pr < 1$, as the boundary layer constitution changes dramatically across $Pr \sim 1$. The other lesson learned from Fig. 4.3 is that the velocity profile shape is governed by two length scales, one for the wall shear layer and another for the overall thickness of the moving layer of fluid.

High-Pr Fluids

A suitable set of profiles for $Pr > 1$ fluids, compatible with the top-half of Fig. 4.2, is

$$T - T_\infty = \Delta T e^{-x/\delta_T} \qquad (42)$$

$$v = Ve^{-x/\delta}(1 - e^{-x/\delta_T}) \qquad (43)$$

where V, δ_T, and δ are unknown functions of altitude (y), and $\Delta T = T_0 - T_\infty$ = constant. Substituting profiles (42) and (43) into the momentum and energy integrals, and setting $X \to \infty$ yields

$$\frac{d}{dy} \left[\frac{V^2 \delta q^2}{2(2 + q)(1 + q)} \right] = -\frac{\nu V q}{\delta} + g\beta \Delta T \frac{\delta}{q} \qquad (44)$$

$$\frac{d}{dy} \left[\frac{V\delta}{(1 + q)(1 + 2q)} \right] = \frac{\alpha}{\delta} \qquad (45)$$

where q is the Pr function (Fig. 4.3).

$$q(\text{Pr}) = \frac{\delta}{\delta_T} \tag{46}$$

In eqs. (44) and (45) we have two equations for three unknowns: $V(y)$, $\delta(y)$, and $q(\text{Pr})$. The third equation, necessary for determining V, δ, and q uniquely, is a challenging proposition. Historically, older integral analyses such as Squire's [10] avoided this problem altogether by assuming [11] $\delta_T = \delta$ from the outset (i.e., $q = 1$). However, since a great deal of the information relating to boundary layer geometry is buried in the δ/δ_T function (Table 4.1, Fig. 4.3), it is instructive to meet the challenge of an integral analysis with $\delta \neq \delta_T$. It is up to the individual researcher to come up with a third equation that, next to eqs. (44) and (45), determines V, δ, and q uniquely. First, we must keep in mind that eqs. (44) and (45) are approximate and, yes, arbitrary substitutes for the real equations to be satisfied [eqs. (17) and (8)]. So, we have the freedom to bring along into the analysis any other condition (equation) that spells approximately conservation of momentum or conservation of energy. Since the energy equation is, in a scaling sense, less ambiguous that the momentum equation,[†] it makes sense to select as a third equation a force balance statement: one which is both clear and analytically brief is the fact that in the no-slip layer $0 < x < 0^+$ the inertia terms of eq. (17) are identically zero.

$$0 = \nu \frac{\partial^2 v}{\partial x^2} + g\beta(T_0 - T_\infty) \tag{47}$$

This is to say that right next to the wall there is no ambiguity associated with whether inertia is indeed negligible compared with both friction and buoyancy, regardless of the Prandtl number.

Equations (44), (45), and (47) are solved for V, δ, and q by first noting that $\delta \sim y^{1/4}$ and $V \sim y^{1/2}$; the chief results of this solution are

$$\text{Pr} = \frac{5}{6} q^2 \frac{q + \frac{1}{2}}{q + 2} \tag{48}$$

which is an implicit result for $q(\text{Pr})$, and

$$\text{Nu} = \frac{q''}{T_0 - T_\infty} \frac{y}{k} = \left[\frac{3}{8} \frac{q^3}{(q + 1)(q + \frac{1}{2})(q + 2)} \right]^{1/4} \text{Ra}_y^{1/4} \tag{49}$$

In the limit $\text{Pr} \to \infty$, this solution reduces to

$$\frac{\delta}{\delta_T} = \left(\frac{6}{5} \text{Pr} \right)^{1/2} \quad \text{and} \quad \text{Nu} = 0.783 \, \text{Ra}_y^{1/4} \tag{50}$$

thus confirming the scaling laws summarized in Table 4.1.

[†] Because in natural boundary layer flow the energy equation spells conduction $\sim$ convection, while the momentum equation spells either friction $\sim$ buoyancy or inertia $\sim$ buoyancy.

Low-Pr Fluids

According to the bottom-half of Fig. 4.2, for a $Pr < 1$ fluid, we combine the exponential decay temperature profile (42) with a new velocity profile

$$v = V_1 e^{-x/\delta_T}(1 - e^{-x/\delta_v}) \tag{51}$$

where V_1, δ_T, and δ_v are unknown functions of y. Using again eqs. (40), (41), and (47), and noticing from Table 4.1 that $\delta_T \sim y^{1/4}$, $\delta_v \sim y^{1/4}$, and $V_1 \sim y^{1/2}$, the solution boils down to

$$Pr = \frac{5}{3}\left(\frac{q_1}{1+q_1}\right)^2, \qquad q_1 = \frac{\delta_v}{\delta_T} \tag{52}$$

$$Nu = \frac{q''}{T_0 - T_\infty}\frac{y}{k} = \left(\frac{3}{8}\right)^{1/4}\left(\frac{q_1}{2q_1+1}\right)^{1/2} Ra_y^{1/4} \tag{53}$$

In the limit $Pr \rightarrow 0$, these results reduce to

$$\frac{\delta_v}{\delta_T} = \left(\frac{3}{5}Pr\right)^{1/2} \quad \text{and} \quad Nu = 0.689(Pr\,Ra_y)^{1/4} \tag{54}$$

Once again, this limiting behavior confirms within a numerical factor of order one the scaling laws discovered in the preceding section.

The integral heat transfer results are summarized in Fig. 4.4 next to the similarity solution outlined in the following section. The Nu expressions (49)

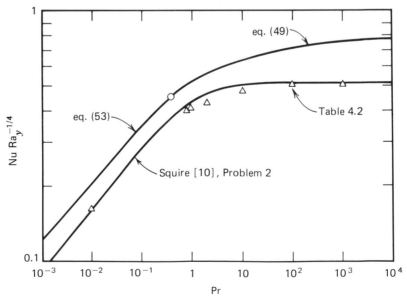

Figure 4.4 Local Nusselt number for laminar natural convection boundary layer flow along a vertical wall: integral versus similarity results.

and (53) match at $Pr = 5/12$ when the assumed velocity profiles are identical $(q = 1, q_1 = 1)$.

As shown earlier in Table 2.1, the Nusselt number calculations depend to some extent on the choice of analytical expressions for velocity and temperature profile. This choice is always a trade-off between what constitutes a reasonable profile shape and the function that leads to least analytical complications. In the preceding example, the choice of exponentials in the make-up of temperature and velocity profiles led to a relatively simple analysis. Figure 4.4 shows also the Nusselt number predicted by Squire's integral analysis [10], which assumes polynomial temperature and velocity profiles with $\delta_T = \delta$: this analysis is outlined in Problem 2 at the end of this chapter. Although the $\delta_T = \delta$ assumption is justified only for fluids with $Pr \sim 1$, the Squire analysis predicts the correct Nusselt number in a wide Pr range.

SIMILARITY SOLUTION

Following the argument centered around Fig. 2.5, we can think of temperature and wall jet profiles whose shape remains unchanged as both profiles occupy wider areas as y increases. From Table 4.1 and the integral solution we know that any length scale of the boundary layer region is proportional to $y^{1/4}$. The dimensionless similarity variable $\eta(x, y)$ can then be constructed as x divided by any of the length scales summarized in Table 4.1; selecting the $Pr > 1$ thermal boundary layer thickness $y\,Ra_y^{-1/4}$ as the most appropriate length scale (Fig. 4.3), the similarity variable emerges as

$$\eta = \frac{x}{y} Ra_y^{1/4} \tag{55}$$

Introducing the streamfunction $u = \partial\psi/\partial y, v = -\partial\psi/\partial x$ in place of the continuity equation (2), the boundary layer equations (8) and (17) become

$$\frac{\partial\psi}{\partial y}\frac{\partial T}{\partial x} - \frac{\partial\psi}{\partial x}\frac{\partial T}{\partial y} = \alpha\frac{\partial^2 T}{\partial x^2} \tag{56}$$

$$-\frac{\partial\psi}{\partial y}\frac{\partial^2\psi}{\partial x^2} + \frac{\partial\psi}{\partial x}\frac{\partial^2\psi}{\partial x\partial y} = -\nu\frac{\partial^3\psi}{\partial x^3} + g\beta(T - T_\infty) \tag{57}$$

Now, from the first column of Table 4.1 we note that, in general, the dimensionless temperature profile will be a function of both $\eta(x, y)$ and Pr; let this unknown function be $\theta(\eta, Pr)$, defined as

$$\frac{T - T_\infty}{T_0 - T_\infty} = \theta(\eta, Pr) \tag{58}$$

For the vertical velocity profile v, from the fourth column of Table 4.1 (Pr > 1) we select the expression

$$v = \frac{\alpha}{y} \mathrm{Ra}_y^{1/2} G(\eta, \mathrm{Pr}) \tag{59}$$

where the front terms represent the scale of v, and $G(\eta, \mathrm{Pr})$ is the $O(1)$ dimensionless similarity profile of the wall jet. From the definition $v = -\partial \psi / \partial x$, we conclude that the streamfunction expression must be

$$\psi = \alpha \mathrm{Ra}_y^{1/4} F(\eta, \mathrm{Pr}) \tag{60}$$

where $G = -\partial F / \partial \eta$. Substituting eqs. (58) and (60) into the boundary layer equations for energy and momentum [eqs. (56) and (57)], we obtain the following system of dimensionless equations

$$\tfrac{3}{4} F\theta' = \theta'' \tag{61}$$

$$\frac{1}{\mathrm{Pr}} \left(\tfrac{1}{2} F'^2 - \tfrac{3}{4} FF'' \right) = -F''' + \theta \tag{62}$$

where $(\ )'$ is shorthand notation for $\partial (\ )/\partial \eta$. These equations show once again the meaning of the Pr > 1 scaling adopted in the definition of η and G (both of order one) [eqs. (55) and (59)]. The energy equation (61) is a balance between convection and conduction, while the momentum equation (62) reduces to a balance between friction and buoyancy as Pr $\to \infty$, that is, as the inertia effect vanishes.

Equations (61) and (62) must be solved subject to the similarity formulation of the appropriate boundary conditions [eqs. (18)]:

(i) At $x = 0$	$u = 0$	$F = 0$	
$(\eta = 0)$	$v = 0$	$F' = 0$	
	$T = T_0$	$\theta = 1$	
(ii) As $x \to \infty$	$v = 0$	$F' = 0$	
$(\eta \to \infty)$	$T = T_\infty$	$\theta = 0$	(63)

Figures 4.5a and b present the solution as temperature profiles and velocity profiles in the thermal boundary layer region $\eta = O(1)$. As anticipated by Fig. 4.2, in the limit Pr $\to \infty$, the temperature profiles collapse onto a unique curve.

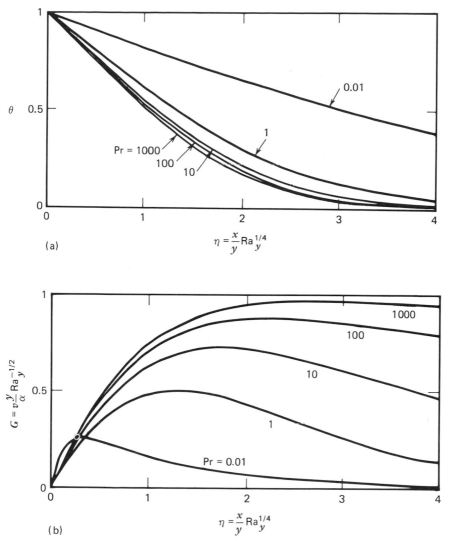

Figure 4.5 Similarity solution for laminar natural convection boundary layer flow along a vertical wall: (*a*) temperature profiles; (*b*) vertical velocity profiles. These drawings are based on the correct scales of the δ_T-thick region of Pr > 1 fluids.

Also, in the same limit, the $\eta \sim 1$ portions of the velocity profiles approach a unique curve, while the dimensionless velocity peak is consistently a number of order one (the velocity peak falls in the region occupied by the thermal boundary layer). As Pr increases, the velocity profile extends farther and farther into isothermal fluid. All these observations support very strongly the scale analysis whose results have been summarized in Table 4.1.

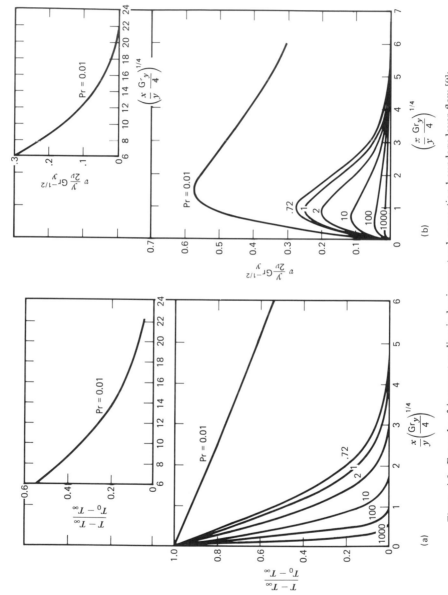

Figure 4.6 Example of incorrect scaling in laminar natural convection boundary layer flow [9]: (a) temperature profiles; (b) vertical velocity profiles.

The numerical solution plotted in Fig. 4.5 was obtained by modifying the numerical results already published by Ostrach as a solution to a different formulation of the same problem [9]. Ostrach's results and original plots are reproduced here as Figs. 4.6a and b. It is instructive to compare Figs. 4.5 and 4.6 side-by-side, in order to recognize what *proper scaling* does to the presentation of numerical results. Note first that for the boundary layer length scale Ostrach chose $y \, Gr_y^{-1/4}$ which, as observed on p. 119 of this chapter is a *fictitious* scale (this scale had been used earlier by Schmidt and Beckmann [12]). The best indication that $y \, Gr_y^{-1/4}$ is not appropriate as boundary layer thickness is that in both Pr limits (Pr → 0 and Pr → ∞), the temperature and velocity profiles of Fig. 4.6 shift as Pr changes. Note also that the vertical velocity scale chosen by Ostrach $[(v/y)(Gr_y)^{1/2}]$ is also fictitious, since it does not characterize either the Pr → 0 limit or the Pr → ∞ limit (Table 4.1): the best indicator that Ostrach's vertical velocity scale is incorrect is that the velocity profiles of Fig. 4.6b constantly move as Pr changes, and the peak dimensionless velocity is not $O(1)$.

Another fascinating aspect of Ostrach's results is that, improperly scaled as they are, they have been reproduced identically in virtually every heat transfer textbook published during the past three decades [e.g., Refs. 5, 7, 11, and 13–16]. This state of affairs illustrates the timeliness of not only clarifying the scaling laws of natural convection (the mission of this chapter), but also the importance of stressing scaling analysis in heat transfer in general.

The local heat-transfer coefficient predicted by the similarity solution is

$$Nu = \frac{hy}{k} = -(\theta')_{\eta=0} Ra_y^{1/4} \tag{64}$$

which is the expected scaling law in the Pr ≫ 1 range (Table 4.1). The numerical coefficient $-(\theta')_{\eta=0}$ is, in general, a function of the Prandtl number, as is shown in Table 4.2 and Fig. 4.4. In the two Pr limits of interest, the Nusselt number approaches the following asymptotes [17].

$$Nu = 0.503 \, Ra_y^{1/4}, \quad \text{as Pr} \to \infty \tag{65}$$

$$Nu = 0.6(Ra_y Pr)^{1/4}, \quad \text{as Pr} \to 0 \tag{66}$$

Noting that since $h \sim y^{-1/4}$ the average heat-transfer coefficient for a wall of height H is $h_{0-H} = 4/3h(y = H)$, the average Nusselt number $Nu_{0-H} = h_{0-H}H/k$ is equal to $4/3 \, Nu(y = H)$. Therefore, the wall-averaged heat transfer results corresponding to the two Pr limits are

$$Nu_{0-H} = 0.671 \, Ra_H^{1/4}, \quad \text{as Pr} \to \infty \tag{65'}$$

$$Nu_{0-H} = 0.8(Ra_H Pr)^{1/4}, \quad \text{as Pr} \to 0 \tag{66'}$$

Table 4.2   Similarity Solution Heat Transfer Results for Natural Convection Boundary Layer Along a Vertical Isothermal Wall[a]

Pr	0.01	0.72	1	2	10	100	1000
Nu $\mathrm{Ra}_y^{-1/4}$	0.162	0.387	0.401	0.426	0.465	0.490	0.499

[a] Numerical values calculated from Ostrach's solution [9]

These conclusions are anticipated within 30 percent by the scaling laws of Table 4.1: such agreement is not uncommon in cases where the scale analysis is correct.

Figure 4.4 shows that despite the factor of 10 increase in the Prandtl number from air (Pr = 0.72) to water (Pr $\simeq$ 5–7), the Nusselt number varies by only 15 percent if the Rayleigh number is held constant. This observation is the basis for the simulation of air natural convection heat transfer of room-size systems in small-scale laboratory systems using water as working fluid (see Problem 4).

UNIFORM WALL HEAT FLUX

The analyses presented so far are based on the assumption that the vertical wall is isothermal. This would be a good approximation in cases where the vertical wall is massive and highly conducting in the vertical y direction: indeed the object of Problem 5 is to estimate the needed vertical conductance through the wall so that the T_0 = constant description is valid.

From a practical standpoint, however, an equally important wall model is the uniform heat flux condition $q'' = $ constant. In many applications, the wall heating effect is the result of radiation heating[†] from the other side or, as in the case of electronic components, the result of resistive heating. The heat transfer problem in such cases consists of predicting the wall-ambient temperature difference $T_0(y) - T_\infty$ when the uniform heat flux q'' is given. The solution to this heat transfer problem can be pursued step-by-step according to the methodology outlined in the preceding three sections. To avoid repetition, however, we outline only the scale analysis, leaving the integral and similarity solutions as journal-supervised homework for the student (Problems 6 and 7).

Regardless of how q'', ΔT, and δ_T vary with altitude y, the definition of wall heat flux requires

$$q'' \sim k \frac{\Delta T}{\delta_T} \tag{67}$$

[†] The constant heat flux condition applies to nuclear radiation heating, and only under special conditions to thermal radiation heating (in general, in thermal radiation heat transfer the wall heat flux depends on the wall temperature).

Figure 4.7a illustrates this scaling law in the case of an isothermal wall, where both ΔT and the product $q''\delta_T$ are independent of y. The bottom-half of the figure (Fig. 4.7b) shows what to expect in the case of constant q'', namely, identical ΔT and δ_T functions of y. To determine these y functions, we make the observation that the scaling analysis starting with eq. (19) is general, in other words, in that analysis δ_T and ΔT represent the respective order of magnitudes of thermal layer thickness and wall-ambient temperature difference along a wall of height H.

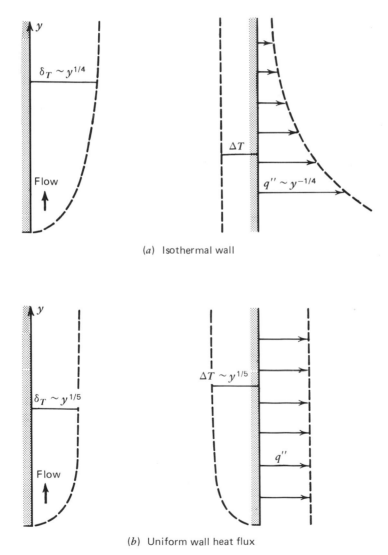

(a) Isothermal wall

(b) Uniform wall heat flux

Figure 4.7   The effect of wall heating conditions on the natural convection boundary layer.

For $\mathrm{Pr} \gg 1$ fluids, eq. (26) recommends

$$\delta_T \sim H\left(\frac{g\beta\Delta T H^3}{\alpha\nu}\right)^{-1/4} \tag{68}$$

Recognizing that in the present problem ΔT is not given (q'' is), we use eq. (67) to eliminate ΔT and solve for δ_T,

$$\delta_T \sim H\,\mathrm{Ra}_{*H}^{-1/5} \tag{69}$$

where Ra_* is a Rayleigh number based on q'',

$$\mathrm{Ra}_{*H} = \frac{g\beta H^4 q''}{\alpha\nu k} \tag{70}$$

From eq. (67), the corresponding ($\mathrm{Pr} \gg 1$) scale of the wall-ambient temperature difference is

$$\Delta T \sim \frac{q''}{k} H\,\mathrm{Ra}_{*H}^{-1/5} \tag{71}$$

Note that both δ_T and ΔT are proportional to $H^{1/5}$; since the H-averaged quantities are proportional to $H^{1/5}$, we draw the conclusion that the local values of δ_T and ΔT are proportional to $y^{1/5}$. This conclusion is illustrated in the bottom-half of Fig. 4.7.

The local Nusselt number for a constant heat flux wall may be defined as

$$\mathrm{Nu} = \frac{q''}{T_0(y) - T_\infty}\frac{y}{k} \tag{72}$$

Therefore, in the range $\mathrm{Pr} > 1$, the Nusselt number must scale as

$$\mathrm{Nu} \sim \frac{H}{\delta_T} \sim \mathrm{Ra}_{*H}^{1/5} \tag{73}$$

For the low Prandtl number fluids we start with eq. (32) and, using the same analysis as above, we obtain

$$\delta_T \sim H(\mathrm{Ra}_{*H}\mathrm{Pr})^{-1/5}$$

$$\Delta T \sim \frac{q''}{k} H(\mathrm{Ra}_{*H}\mathrm{Pr})^{-1/5}$$

$$\mathrm{Nu} \sim (\mathrm{Ra}_{*H}\mathrm{Pr})^{1/5} \tag{74}$$

The validity of these scaling results can be tested by referring to more exact analyses published on the same topic. Sparrow [18] carried out an integral analysis of the same type as Squire's [10] (i.e., assuming only one length scale δ_T for the velocity profile) and arrived at the following local Nusselt number

$$\mathrm{Nu} = \frac{2}{360^{1/5}} \left(\frac{\mathrm{Pr}}{\frac{4}{5} + \mathrm{Pr}} \right)^{1/5} \mathrm{Ra}_{*y}^{1/5} \qquad (75)$$

The similarity solution was reported by Sparrow and Gregg [19], who found that eq. (75) is in fact an adequate curve fit for the similarity Nu results in the range $0.01 < \mathrm{Pr} < 100$. Thus, in the two Pr limits, eq. (75) yields

$$\mathrm{Nu} = 0.616 \, \mathrm{Ra}_{*y}^{1/5}, \quad \mathrm{Pr} \to \infty$$

$$\mathrm{Nu} = 0.644 \, \mathrm{Ra}_{*y}^{1/5} \mathrm{Pr}^{1/5}, \quad \mathrm{Pr} \to 0 \qquad (76)$$

These limiting expressions are anticipated correctly by the scale laws [eqs. (73) and (74)].

The similarity temperature and velocity profiles for $q'' = $ constant are available in graphic form in Ref. 19. It is worth noting that the Sparrow and Gregg formulation [19] uses the fictitious scale $y(\mathrm{Ra}_{*y}/\mathrm{Pr})^{-1/5}$ as horizontal length unit in the construction of the similarity variable. Consequently, the plotted profiles and their dependence on Pr are very similar to Ostrach's graphs for the isothermal wall (Fig. 4.6).

Similarity solutions can be developed for an infinity of wall temperature conditions, provided they obey either the power law $T_0 - T_\infty = Ay^m$ [20], or the exponential law $T_0 - T_\infty = Ae^{my}$ [20], or the line $T_0 - T_\infty = A + By$ [21], where A, B, and m are all constants. Thus, the $T_0 = $ constant and $q'' = $ constant problems discussed so far are only two special cases of the vast analytically accessible class of problems. From an engineering standpoint, however, the $T_0 = $ constant and $q'' = $ constant results are by far the most useful.

THE EFFECT OF THERMAL STRATIFICATION

In the hope of shedding some light on the basics of natural convection along a vertical wall, we considered so far the simplest model possible, that is, the heat transfer interaction between a vertical wall and an isothermal semiinfinite fluid reservoir (Fig. 4.1). Proceeding now on the road from the simple to the complex, we take a closer look at the problem of modeling a heat transfer situation involving natural convection. Vertical walls are rarely in communication with semiinfinite isothermal pools of fluid: more often, their height is finite and the heated boundary layer eventually hits the ceiling. At that point, the heated stream has no choice but to discharge horizontally into the fluid reservoir (to the right in Fig. 4.1): the direction of this discharge is horizontal because the discharge contains fluid warmer than the rest of the reservoir. The long-time effect of this discharge process is the thermal stratification characterized by warm fluid layers floating on top of increasingly colder layers. Indeed, thermal stratification is a characteristic of all fluid bodies surrounded by differentially heated side walls lined by boundary layers, as is demonstrated in Chapter 5. At this point it is sufficient to recognize that the air in any room with the doors closed is thermally stratified in such a way that the lowest layers assume the temperature of the coldest wall and the highest layers near the ceiling approach the temperature of the warmest wall.

In view of this discussion, Fig. 4.8 is a more general model for the buoyancy-driven flow near a vertical wall. The fluid reservoir is now linearly stratified,

$$T_\infty(y) = T_{\infty,0} + \gamma y \tag{77}$$

$T_{\infty,0}$ being the lowest temperature in the arrangement and γ the constant temperature gradient (assumed known). The dashed line in Fig. 4.8 shows the location of the isothermal reservoir model employed so far ($\gamma = 0°C/m$). From the outset, if the bottom temperature difference $T_0 - T_{\infty,0}$ remains constant, we expect the overall heat transfer rate to *decrease* as γ increases: the reason for this expectation is that the effective (mean) temperature difference between wall and fluid decreases as γ increases.

A similarity formulation for the problem of Fig. 4.8 is not possible [22]: this is unfortunate because the temperature boundary conditions of Fig. 4.8 are by far the simplest and easiest to relate to thermally stratified fluids in actuality. An integral solution can be constructed in a relatively straightforward manner by combining the integral equations (40) and (41) with Squire-type temperature and velocity profiles.

$$T - T_\infty = (T_0 - T_\infty)\left(1 - \frac{x}{\delta_T}\right)^2$$

$$v = V\frac{x}{\delta}\left(1 - \frac{x}{\delta}\right)^2 \tag{78}$$

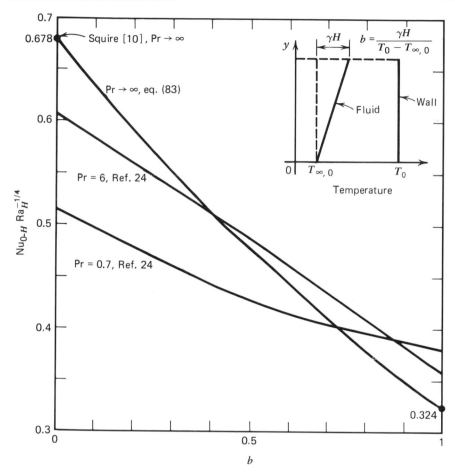

Figure 4.8 The effect of reservoir stratification on the heat transfer from an isothermal vertical wall.

where $\delta_T = \delta$ and, according to eq. (77), $T_0 - T_\infty(y) = T_0 - T_{\infty,0} - \gamma y$. In the dimensionless form required by numerical integration, the momentum and energy equations reduce to

$$\frac{1}{(105)\text{Pr}} \frac{d}{dy_*}(V_*^2 \delta_*) = -\frac{V_*}{\delta_*} + \frac{\delta_*}{3}(1 - by_*) \tag{79}$$

$$\frac{d}{dy_*}[V_* \delta_*(1 - by_*)] = \frac{60}{\delta_*}(1 - by_*) \tag{80}$$

with the following notation (Table 4.1):

$$y_* = \frac{y}{H}, \qquad \delta_* = \frac{\delta}{H\,\mathrm{Ra}_H^{-1/4}}$$

$$V_* = \frac{V}{\frac{\alpha}{H}\,\mathrm{Ra}_H^{1/2}}, \qquad \mathrm{Ra}_H = \frac{g\beta H^3(T_0 - T_{\infty,0})}{\alpha\nu}$$

$$b = \frac{\gamma H}{T_0 - T_{\infty,0}}, \qquad \text{(the stratification parameter)} \tag{81}$$

Equations (79) and (80) can be integrated numerically from $y_* = 0$ to $y_* = 1$ to determine $\delta_*(y_*, b)$ and $V_*(y_*, b)$; the local heat flux is then

$$q'' = -k\left(\frac{\partial T}{\partial x}\right)_{x=0} = \frac{k(T_0 - T_{\infty,0})}{H\,\mathrm{Ra}_H^{-1/4}}\,\frac{2}{\delta_*}(1 - by_*) \tag{82}$$

Integrating q'' over the wall height H yields q'; hence, the overall Nusselt number

$$\mathrm{Nu}_{0-H} = \frac{q'}{k(T_0 - T_{\infty,0})} = \mathrm{Ra}_H^{1/4}\int_0^1 \frac{2}{\delta_*}(1 - by_*)\,dy_* \tag{83}$$

The result of this calculation for $\mathrm{Pr} \to \infty$ is shown in Fig. 4.8 as $\mathrm{Nu}_{0-H}\mathrm{Ra}_H^{-1/4}$ versus the stratification parameter b, where it should be kept in mind that both Nu_{0-H} and Ra_H are based on the *maximum* temperature difference $T_0 - T_{\infty,0}$. Of special interest is the heat transfer rate in the fully stratified limit ($b = 1$)

$$\mathrm{Nu}_{0-H} = 0.324\,\mathrm{Ra}_H^{1/4}, \qquad \mathrm{Pr} \to \infty \tag{84}$$

that matches within 11 percent the conclusion of an Oseen-linearized analysis of boundary layer convection in a stratified enclosure [23]. (The details of this analysis are given in Chapter 5.)

Figure 4.8 also shows the $\mathrm{Pr} = 6$ and $\mathrm{Pr} = 0.7$ results predicted for the same problem by Chen and Eichhorn [24], based on the *local nonsimilarity technique* [25–27]. The general trend is the same as in the integral solution, namely, a gradual decrease in Nu_{0-H} as the stratification degree b increases. In the isothermal reservoir limit, the Chen and Eichhorn results match the similarity solution (Table 4.2); hence, they fall slightly below the corresponding results based on the Squire-type integral analysis (Fig. 4.4).

CONJUGATE BOUNDARY LAYERS

There are many engineering situations in which the vertical wall heating a certain buoyant boundary layer is itself heated on the other (back) side by a sinking boundary layer. Such is the case in walls, partitions, and baffles encountered regularly in the thermal design of living quarters and insulation systems. As is shown in Fig. 4.9, this heat transfer arrangement describes the common *single-pane window* where an insulating impermeable wall of finite thickness separates two fluid reservoirs at different temperatures. Boundary layers form on both sides of the wall, however, the wall temperature or heat flux are not known *a priori* as in the simpler models considered earlier: the condition of the wall is the result of the heat transfer interaction between the two boundary layers. It is often said that, depending on the layer-to-layer interaction, the wall temperature "floats" to an equilibrium distribution between the two extreme temperatures maintained by the two reservoirs. Since

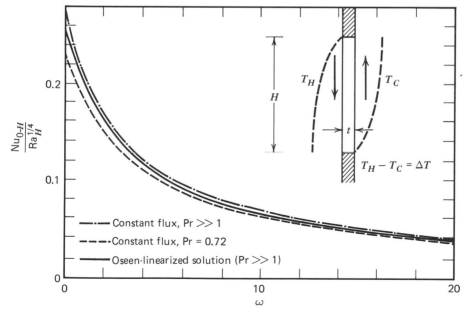

Figure 4.9 The heat transfer between two fluid reservoirs separated by a vertical wall with natural convection boundary layers on both sides [28]. Note that both Nu_{0-H} and Ra_H are based on $\Delta T = T_H - T_C$.

one boundary drives the other, the boundary layers are termed *conjugate* (as two oxen engaged in the same yoke: note the Latin verb *conjungĕre* = to yoke).

Despite its common occurrence in real life, the conjugate boundary layer configuration has only recently become the subject of heat transfer research [28–31]. Figure 4.9 shows the Nusselt number predicted analytically in the $Pr \rightarrow \infty$ limit based on the Oseen-linearization method (Chapter 5) [28]. This approach consists of writing integral conservation equations analogous to eqs. (40) and (41) for both sides of the wall, with the additional complication that the wall temperature $T_0(y)$ is unknown. The additional equation necessary for determining T_0 is the condition of heat flux continuity in the x direction, from one face of the wall to the other. In Fig. 4.9 both the overall Nusselt number and the Rayleigh number are based on the overall temperature difference imposed by the two fluid reservoirs. The heat transfer rate (hence, the ratio $Nu_{0-H}/Ra_H^{1/4}$) decreases as the wall thickness resistance parameter ω increases. The wall parameter is defined as [28]

$$\omega = \frac{t}{H}\frac{k}{k_W}Ra_H^{1/4} \qquad (85)$$

where t, H, k, and k_W are the wall thickness, wall height, fluid conductivity, and wall conductivity, respectively. Note that the dimensionless parameter ω is the ratio of wall thermal resistance $t/(Hk_W)$ divided by the thermal resistance scale of one boundary layer $(H\,Ra_H^{-1/4})/(Hk)$.

One of the chief contributions of the analysis described in Ref. 28 is to show that the wall heat flux distribution negotiated between two conjugate natural convection boundary layers is satisfactorily approximated by the $q'' = $ constant model discussed earlier. Therefore, as is shown in Problem 9 and in Ref. 28, an estimate of the $Nu_{0-H}(\omega, Ra_H)$ relationship can be obtained by adding in series the three resistances constituted by the two $q'' = $ constant boundary layers sandwiching the wall. Figure 4.9 shows good agreement between this more direct approach and the Oseen-linearized analysis.

The effect of thermal stratification on the conjugate boundary layer configuration has been documented in Ref. 31. Figure 4.10 shows that the coefficient in the $Nu_{0-H} \sim Ra_H^{1/4}$ proportionality increases as the degree of thermal stratification on either side increases (the dimensionless temperature gradients a and b are defined graphically in Fig. 4.10.) This behavior appears to contradict the effect shown in Fig. 4.8 for a single isothermal wall. The apparent contradiction is explained by the fact that in Fig. 4.8 both Nu_{0-H} and Ra_H are based on the maximum temperature difference, whereas in Fig. 4.10 (and in Fig. 4.9) the same numbers are based on the reservoir-to-reservoir temperature difference evaluated at midheight. Thus, the ΔT sketched in Fig. 4.10 is the arithmetically averaged temperature difference between the two stratified reservoirs.

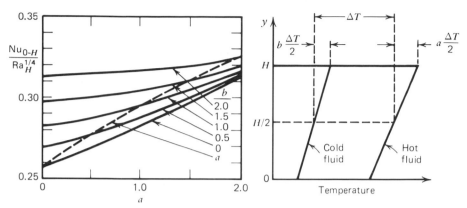

Figure 4.10 The effect of thermal stratification on the heat transfer between two fluid reservoirs separated by a vertical wall [31] ($\omega = 0$, Pr > 1).

VERTICAL CHANNEL FLOW

Consider now the interaction between the natural convection boundary layers formed along two parallel walls facing each other (Fig. 4.11). If the boundary layer thickness scales are much smaller than the wall-to-wall spacing D, then the flow driven along one wall may be regarded (approximately) as a wall jet unaffected by the presence of another wall. On the other hand, if the boundary layer grows to the point that its thickness becomes comparable to D, then the two wall jets merge into a single buoyant stream rising through the chimney formed by the two walls. The latter case is characteristic of vertical fin-to-fin cooling channels provided in certain pieces of electronic equipment. It is clear from Fig. 4.11 that the channel flow departs from the wall jet description in the same way that the duct flows of Chapter 3 depart from the pure boundary layer flows of Chapter 2. Needless to say, due to this fundamental departure and to a sufficient number of engineering applications, the buoyant channel-flow configuration constitutes an important segment in natural convection research (see, e.g., Refs. 32–34). As an exercise, we focus below on the simplest analysis of the channel flow, with the final objective of predicting the capability of this flow to cool or heat the walls of the channel.

The flow part of the problem may be solved by considering the momentum equation in the y direction (for notation, see Fig. 4.11).

$$\rho\left(u\frac{\partial v}{\partial x} + v\frac{\partial v}{\partial y}\right) = -\frac{\partial P}{\partial y} + \mu\nabla^2 v - \rho g \tag{86}$$

The mass continuity equation, in conjunction with the idea that the channel is long enough so that the u scale becomes sufficiently small, leads to the concept

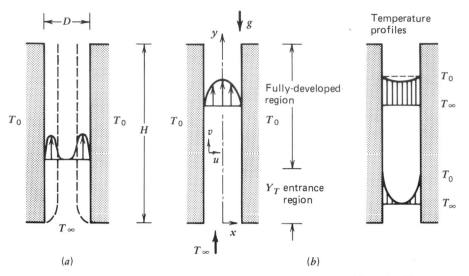

Figure 4.11   Natural convection in the channel between two vertical heated walls.

of *fully developed flow*, for which we have

$$u = 0 \quad \text{and} \quad \frac{\partial v}{\partial x} = 0 \tag{87}$$

Furthermore, the momentum equation in the lateral direction x can be used to show that the pressure in the fully developed region is a function of y only, and, since both ends of the channel are open to the ambient of density ρ_∞,

$$\frac{\partial P}{\partial y} = \frac{dP}{dy} = -\rho_\infty g \tag{88}$$

Combining eqs. (86)–(88) and using once more the Boussinesq approximation yields the much simpler equation

$$\frac{d^2 v}{dx^2} = -\frac{g\beta}{\nu}(T - T_\infty) \tag{89}$$

which is the natural convection equivalent of the Hagen–Poiseuille equation encountered in Chapter 3.

In order to solve eq. (89) it is necessary to first derive the temperature profile $T - T_\infty$; the student can verify that in order to derive the temperature profile from the energy equation, one must know the velocity profile. The two profiles, velocity and temperature, are coupled; hence eq. (89) and the energy equation must be solved simultaneously. However, a much simpler solution approach is possible if we observe that in the fully developed region between two *isothermal* walls, the temperature difference can be approximated by $(T_0 - T_\infty)$; in other words,

$$(T_0 - T) \ll (T_0 - T_\infty). \tag{90}$$

The range of validity of this approximation will be determined later in this section. Based on this approximation, the right-hand side of eq. (89) becomes a constant, and the heat transfer solution follows without difficulty in a few algebraic steps. The main features of this solution are:

Velocity profile

$$v = \frac{g\beta D^2 (T_0 - T_\infty)}{8\nu} \left[1 - \left(\frac{x}{D/2} \right)^2 \right]$$

Mass flowrate per unit length normal to the plane of Fig. 4.11

$$\dot{m} = \frac{\rho g \beta D^3 (T_0 - T_\infty)}{(12)\nu}$$

Total heat transfer rate between stream and channel walls

$$Q = \dot{m} \text{ (outlet enthalpy–inlet enthalpy)}$$

$$= \dot{m} c_P (T_0 - T_\infty)$$

Average heat flux

$$q''_{0-H} = Q/(2H)$$

Average Nusselt number

$$\frac{q''_{0-H} H}{(T_0 - T_\infty)k} = \frac{\mathrm{Ra}_D}{24} \tag{91}$$

Note that the dimensionless group emerging from this analysis is the Rayleigh number based on wall-to-wall spacing,

$$\mathrm{Ra}_D = \frac{g\beta D^3 (T_0 - T_\infty)}{\alpha \nu} \tag{92}$$

and that the Grashof number is once again absent from the discussion (this observation raises again the question of whether the Grashof number is a relevant dimensionless group in natural convection).

The fully developed flow and heat transfer solution (91) is valid for all Prandtl numbers. The Rayleigh number range of its validity follows from the requirement that the thermal entrance length Y_T be much smaller than the channel height H,

$$Y_T < H \tag{93}$$

The order of magnitude of Y_T follows from the observation that the thermal boundary layer thickness δ_T becomes of order $D/2$ when y is of order Y_T, that is,

$$Y_T \operatorname{Ra}_{Y_T}^{-1/4} \sim \frac{D}{2}, \qquad \Pr > 1$$

$$Y_T \operatorname{Bo}_{Y_T}^{-1/4} \sim \frac{D}{2}, \qquad \Pr < 1 \tag{94}$$

Evaluating Y_T from above, criterion (93) becomes

$$\operatorname{Ra}_D^{1/4} < 2\left(\frac{H}{D}\right)^{1/4}, \qquad \Pr > 1$$

$$\operatorname{Bo}_D^{1/4} < 2\left(\frac{H}{D}\right)^{1/4}, \qquad \Pr < 1 \tag{95}$$

This criterion states that the fully developed flow and the temperature profile assumption breaks down if the Rayleigh number exceeds a certain order of magnitude dictated by the geometric aspect ratio of the channel, H/D. (See also Problem 13 at the end of this chapter).

COMBINED NATURAL AND FORCED CONVECTION

If we think of the basic heat-transfer configuration examined in this chapter (Figs. 4.1 and 4.2) as a model of the flow near a heated wall in a room, then a major limitation of this model is the assumption that the fluid reservoir is motionless sufficiently far from the wall. Look around any modern building and you will see that the air inventory of each room is replenished continuously or intermittently by, in most cases, a central air-conditioning system. This means that relative to each heated wall or cooled window, the room air-reservoir is actually in motion: the reservoir is *forced* into and out of the room by an external agent (the fan in the ventilation system). Depending on the strength of this forced circulation, the heat transfer from the wall to the room air may be ruled by either natural convection or forced convection or a combination of natural and forced convection.

To understand the interaction between natural convection and reservoir-driven forced convection is important, because many engineering systems rely on a combination of these two mechanisms. There are many ways in which these two mechanisms can interact and fight one another for dominance, as there are many ways in which the reservoir fluid may move relative to the direction of buoyant flow near the wall. (Just think of the heated wall jet rising on the outer surface of a flat solar collector in winter time, and how this wall jet will be affected by the changing wind direction and changing wind velocity). Due to the diversity of the natural–forced convection interaction, it is impossible to treat this subject fully; however, it is instructive to study one simple example of such an interaction and to experience once again the power and cost-effectiveness of pure scaling arguments.

As is shown in the insert of Fig. 4.12, let us consider the heat transfer from a vertical heated wall (T_0) to an isothermal fluid reservoir moving upward (T_∞, U_∞), that is, in the same direction as the natural wall jet present when $U_\infty = 0$. This basic problem has been solved by a number of authors using approximate analytical methods, for example, in Refs. 35–37. From a heat transfer standpoint, the key question is: under what conditions is the combined natural–forced phenomenon characterized (approximately) by the scales of pure natural convection, and, conversely, under what conditions by the scales of pure forced convection? In other words, what is the criterion for the *transition* from one convection mechanism to another?

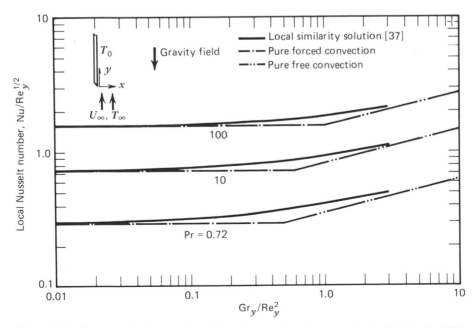

Figure 4.12   Heat transfer by natural and forced convection along a vertical wall (after Ref. 37).

Regardless of the type of convection, the heat transfer is always between the wall and the fluid reservoir. If the mechanism is natural convection, then the thermal distance between the heat-exchanging entities is of order

$$(\delta_T)_{\text{NC}} \sim y\,\text{Ra}_y^{-1/4}, \qquad \text{Pr} > 1 \qquad (96)$$

as reservoir fluid supplies the buoyant wall jet of thermal boundary layer thickness $(\delta_T)_{\text{NC}}$. On the other hand, if the mechanism is forced convection, then the wall and the reservoir are separated by a thermal length of order (Chapter 2)

$$(\delta_T)_{\text{FC}} \sim y\,\text{Re}_y^{-1/2}\text{Pr}^{-1/3}, \qquad \text{Pr} > 1 \qquad (97)$$

The type of convection mechanism is decided by the smaller of the two distances, $(\delta_T)_{\text{NC}}$ or $(\delta_T)_{\text{FC}}$, since the wall will leak heat to the nearest heat sink. Thus, the scale criterion for transition from natural to forced convection is

$$(\delta_T)_{\text{NC}} < (\delta_T)_{\text{FC}}, \quad \text{Natural convection}$$

$$(\delta_T)_{\text{NC}} > (\delta_T)_{\text{FC}}, \quad \text{Forced convection} \qquad (98)$$

in other words, for $\text{Pr} > 1$ fluids,

$$\frac{\text{Ra}_y^{1/4}}{\text{Re}_y^{1/2}\text{Pr}^{1/3}} \begin{cases} > O(1), & \text{Natural convection} \\ < O(1), & \text{Forced convection} \end{cases} \qquad (99)$$

To verify the validity of this criterion, we examine the Lloyd and Sparrow [37] local similarity solution to the combined heat transfer problem (Fig. 4.12). This solution shows that forced convection dominates at small values of $\text{Gr}_y/\text{Re}_y^2$, while natural convection takes over at large values of the same parameter. Note, however, that the knee in each Nusselt number curve shifts to the right as Pr increases: this effect is due to the fact that the abscissa parameter $\text{Gr}_y/\text{Re}_y^2$ is not the dimensionless group that properly serves as transition parameter in eq. (99).

$$\frac{\text{Gr}_y}{\text{Re}_y^2} = \left(\frac{\text{Ra}_y^{1/4}}{\text{Re}_y^{1/2}\text{Pr}^{1/3}} \right)^4 \text{Pr}^{1/3} \qquad (100)$$

Figure 4.13 shows the replotting of Lloyd and Sparrow's results using the proper transition parameter on the abscissa and the proper forced convection Nusselt number scaling on the ordinate. The sign of correct scaling analysis [hence, the validity of criterion (99)], is that: (1) the $\text{Pr} > 1$ curves fall on top of each other, and (2) the knee of all the curves is at $O(1)$ on the abscissa.

Figure 4.13 shows that an engineer can estimate with sufficient accuracy and efficiency the Nusselt number by first using criterion (99) and by relying on the appropriate asymptotic scaling (i.e., pure natural convection or pure forced convection).

Repeating the geometric argument of eq. (98), this time for $Pr < 1$ fluids, we find the following transition criterion:

$$\frac{Bo_y^{1/4}}{Pe_y^{1/2}} \begin{cases} > O(1), & \text{Natural convection} \\ < O(1), & \text{Forced convection} \end{cases} \tag{101}$$

Note that the dimensionless group $Bo_y^{1/4} Pe_y^{-1/2}$ is equal to Lloyd and Sparrow's abscissa parameter Gr_y/Re_y^2 raised to the 1/4th power (Fig. 4.12). Criteria

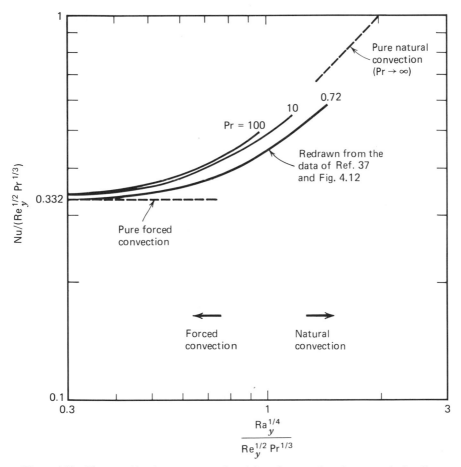

Figure 4.13 The transition between natural and forced convection along a vertical wall.

(99) and (101) suggest one more time that the Grashof number is not an appropriate dimensionless group in natural convection.

FILM CONDENSATION

Two-phase convection phenomena such as condensation, vaporization, and melting occupy a distinct place in the classical treatment of heat transfer (see, e.g., Refs. 11, 14, and 38). The objective of this last section is to demonstrate that the principles behind *thin-film* convection processes with phase change are exactly the same as those learned in the beginning of this chapter for natural convection boundary layers. In other words, the mission of this section is to unify conceptually two important segments of heat transfer education— boundary-layer natural convection and film condensation, vaporization, and melting.

Consider Nusselt's classical problem [39] of thin-film condensation along a vertical cold wall (T_0) bathed by a saturated vapor (T_{sat}) such that $T_0 < T_{sat}$ (see Fig. 4.14). As an exercise, let us use scale analysis to predict the heat transfer rate from T_{sat} to T_0; in other words, let us estimate the heat flux scale $q'' \sim k(T_{sat} - T_0)/\delta$, where δ is the liquid film thickness. Of course, this heat transfer problem reduces to evaluating the thickness scale δ: the analysis begins with writing the governing equations for *boundary layer* flow in the liquid film.

Mass:
$$\frac{\partial u}{\partial x} + \frac{\partial v}{\partial y} = 0 \tag{102}$$

Energy:
$$u\frac{\partial T}{\partial x} + v\frac{\partial T}{\partial y} = \alpha\frac{\partial^2 T}{\partial x^2} \tag{103}$$

Momentum:
$$\rho\left(u\frac{\partial v}{\partial x} + v\frac{\partial v}{\partial y}\right) = -\frac{dP}{dy} + \mu\frac{\partial^2 v}{\partial x^2} + \rho g \tag{104}$$

Note that eq. (104) is the boundary layer theory result of accounting for force balances in both x and y directions (Chapter 2). Furthermore, since $dP/dy = \rho_{vap}g$, where ρ_{vap} is the density of saturated vapor residing outside the film, the momentum equation (104) reduces to

$$u\frac{\partial v}{\partial x} + v\frac{\partial v}{\partial y} = \nu\frac{\partial^2 v}{\partial x^2} + \frac{\rho - \rho_{vap}}{\rho}g \tag{105}$$

Now, by examining equations (102), (103), and (105), we make the important observation that the equations governing the liquid film are nearly

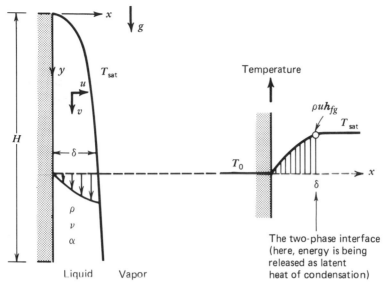

Figure 4.14 Schematic of laminar film condensation on a cold vertical wall bathed by saturated vapor.

identical to those encountered in the case of a bouyant boundary layer [eqs. (2), (5), and (17)], the only difference being that the dimensionless ratio $(\rho - \rho_{vap})/\rho$ now replaces the $\beta(T - T_\infty)$ group in the bouyancy term of the momentum equation. Such similarities between governing equations suggest the existence of similar scaling laws; indeed, from the mass and momentum equations (102) and (105) we have

$$\frac{u}{\delta} \sim \frac{v}{H} \tag{106}$$

$$v \frac{v}{\delta^2} \sim \frac{\rho - \rho_{vap}}{\rho} g \tag{107}$$

Note that eq. (107) is actually a balance between friction and the sinking effect: the limitations of this assumption are discussed at the end of this section (these limitations are not usually discussed in the classical presentation of Nusselt's integral analysis of the film flow of Fig. 4.14). The scaling law implied by the energy equation (103) is more visible if we integrate this equation from $x = 0$ to $x = \delta$.

$$\rho c_P \left[(uT)_{x=\delta} + \frac{d}{dy} \int_0^\delta vT \, dx \right] = \left(k \frac{\partial T}{\partial x} \right)_{x=\delta} - \left(k \frac{\partial T}{\partial x} \right)_{x=0} \tag{108}$$

There are four terms in this equation, and their respective scales are

$$\underbrace{\rho c_P u \Delta T, \ \rho c_P \frac{v \Delta T \delta}{H}} \qquad \underbrace{\rho u h_{fg}} \qquad \underbrace{k \frac{\Delta T}{\delta}}$$

Convection scales, both of order $\rho c_P u \Delta T$ according to eq. (106)	Rate of energy release at the liquid–vapor interface	Rate of heat transfer into the wall (the object of this entire analysis) (109)

where $\Delta T = T_{sat} - T_0$. Of special effect is the $(\rho u h_{fg})$ scale associated with the heating of the interface through the condensation of saturated vapor, which is entrained with velocity u into the liquid film: the release of the latent heat of condensation at the liquid–vapor interface is responsible for the sharp break (the two-valued slope) in the temperature profile at $x = \delta$ (Fig. 4.14). In conclusion, the energy balance (108) requires

$$k \frac{\Delta T}{\delta} \sim \left(\rho u h_{fg} \right) \quad \text{or} \quad \left(\rho c_P u \Delta T \right) \tag{110}$$

According to the principles of scale analysis put forth in Chapter 1, the energy scaling (110) can also be written as

$$k \frac{\Delta T}{\delta} \sim \rho u \left[h_{fg} + (r) c_P \Delta T \right] \tag{111}$$

where (r) is an unknown dimensionless coefficient of order $O(1)$.

To summarize the scaling arguments constructed thus far, we now have three equations [(106), (107), and (111)] for determining three unknowns (u, v, δ). Of primary interest to heat transfer engineers is the film thickness: the resulting expression for δ is

$$\frac{H}{\delta} = \left[\frac{g(\rho - \rho_{vap}) H^3 h'_{fg}}{k v (T_{sat} - T_0)} \right]^{1/4} \tag{112}$$

where

$$h'_{fg} = h_{fg} + (r) c_P \Delta T \tag{113}$$

Since $h \sim k/\delta$, the Nusselt number scale corresponding to result (112) is

$$\frac{hH}{k} \sim \left[\frac{g(\rho - \rho_{vap}) H^3 h'_{fg}}{k v (T_{sat} - T_0)} \right]^{1/4} \tag{114}$$

These conclusions are analytically similar to those reached for bouyant boundary layers, therefore, it makes sense to refer to the quantity appearing in square brackets in eqs. (112) and (114) as the *film Rayleigh number.*

$$\text{Ra}_{\text{film}, H} = \frac{g(\rho - \rho_{\text{vap}})H^3 h'_{fg}}{k\nu(T_{\text{sat}} - T_0)} \tag{115}$$

How accurate are the results of (112) and (114)? Rohsenow's integral solution of the Nusselt film condensation problem yielded the following formulas for the local and the *H*-averaged Nusselt numbers [40].

$$\text{Nu} = 0.707\,\text{Ra}_{\text{film}, y}^{1/4} \tag{116}$$

$$\text{Nu}_{0-H} = 0.943\,\text{Ra}_{\text{film}, H}^{1/4} \tag{117}$$

where $h'_{fg} = h_{fg} + 0.68 c_P\,(T_{\text{sat}} - T_0)$, or $(r) = 0.68$ as argued from scaling considerations. The agreement between scaling predictions and integral analysis is remarkable. The same scaling laws anticipate correctly the average heat-transfer coefficient for a horizontal tube covered by a thin film of condensate [41, 42]

$$\text{Nu}_D = 0.728\,\text{Ra}_{\text{film}, D}^{1/4} \tag{118}$$

where D is the outer diameter of the tube (i.e., the vertical length scale of the cold wall).

The scale analysis reported above is valid only when the sinking force on the film is balanced by the wall friction force [eq. (107)]. In general, the momentum equation (105) is ruled by three scales

$$
\begin{array}{ccc}
\text{Inertia} & \text{Friction} & \text{Bouyancy} \\[4pt]
\dfrac{v^2}{H} & \nu\dfrac{v}{\delta^2} & \dfrac{\Delta\rho}{\rho}g
\end{array}
\tag{119}
$$

Dividing each of the above scales by $\Delta\rho g/\rho$, and using the scales that led to eq. (112), we obtain

$$
\begin{array}{ccc}
\text{Inertia} & \text{Friction} & \text{Bouyancy} \\[4pt]
\left(\text{Pr}\dfrac{h'_{fg}}{c_P \Delta T}\right)^{-1} & 1 & 1
\end{array}
\tag{120}
$$

This means that the scaling results (112)–(118) are valid only when a new group, *the film Prandtl number*, is greater than unity.

$$\text{Pr}_{\text{film}} = \text{Pr}\frac{h'_{fg}}{c_P \Delta T} > 1 \tag{121}$$

It is interesting that both Pr and $h'_{fg}/(c_P \Delta T)$ are being used *separately* in the film condensation literature, and that the fundamental importance of the new group $\mathrm{Pr}_{film} = \mathrm{Pr}\, h'_{fg}/(c_P \Delta T)$ has been overlooked. The film Prandtl number Pr_{film} rules the film force balance in exactly the same way that the Prandtl number influences the force balance in a natural convection boundary layer. Consequently, in the case of $\mathrm{Pr}_{film} < 1$, the momentum balance is between inertia and sinking effects,

$$\frac{v^2}{H} \sim \frac{\Delta \rho}{\rho} g \tag{122}$$

which, combined with the mass and energy conservation laws (106) and (111), yields

$$\frac{H}{\delta} \sim \mathrm{Bo}_{film,\,H}^{1/4}, \qquad (\mathrm{Pr}_{film} < 1) \tag{123}$$

$$\frac{hH}{k} \sim \mathrm{Bo}_{film,\,H}^{1/4}, \qquad (\mathrm{Pr}_{film} < 1) \tag{124}$$

The new dimensionless group $\mathrm{Bo}_{film,\,H}$ is *the film Boussinesq* number.

$$\mathrm{Bo}_{film,\,H} = \frac{g\rho(\rho - \rho_{vap})H^3 h'_{fg}}{k^2(T_{sat} - T_0)^2} = \mathrm{Pr}_{film}\mathrm{Ra}_{film,\,H} \tag{125}$$

Figure 4.15 shows that the film Prandtl number is the correct dimensionless group for establishing the transition from a film restrained by inertia ($\mathrm{Pr}_{film} < 1$)

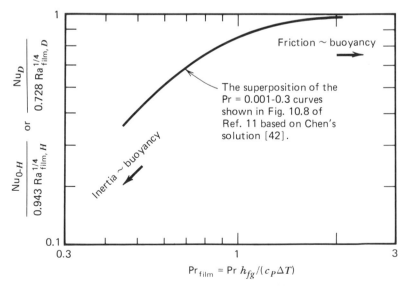

Figure 4.15 The Pr_{film} transition from inertia-restrained to friction-restrained film condensation flow on a vertical wall or on a single horizontal cylinder.

to one restrained by wall friction ($Pr_{film} > 1$). The lone curve shown in Fig. 4.15 is actually the superposition of the family of curves traditionally shown in the literature as one curve for each Pr, with $Nu_{0-H}/(0.943\,Ra^{1/4}_{film,\,H})$ on the ordinate and $c_P\,\Delta T/h'_{fg}$ on the abscissa (see Ref. 11, p. 249). Note further that a liquid film does not necessarily have to contain $Pr < 1$ fluid in order to constitute a low Pr_{film} flow.

The scale analysis built around Fig. 4.14 can be applied with very few changes to other two-phase convection phenomena, such as film vaporization and film melting. In the latter, the latent heat factor in the film Rayleigh number (115) is replaced by the latent heat of melting h_{sf}, as the wall surface becomes the locus of two-phase quasiequilibrium states. The state of the art in the field of phase-change heat transfer has been reviewed by Viskanta in a recent monograph [43]: this study is recommended to the reader.

SYMBOLS

b	thermal stratification parameter [eq. (81)]
Bo_H	Boussinesq number [eq. (35)]
$Bo_{film,\,H}$	film Boussinesq number [eq. (125)]
D	plate-to-plate spacing (Fig. 4.11)
F	streamfunction similarity profile [eq. (60)]
g	gravitational acceleration
Gr_H	Grashof number [eq. (38)]
h	heat transfer coefficient
h_{fg}	latent heat of condensation
h'_{fg}	effective h_{fg} defined by eq. (113)
H	wall height
k	fluid thermal conductivity
k_W	wall thermal conductivity
Nu	local Nusselt number
P	pressure
Pe_y	local Peclet number ($U_\infty y/\alpha$)
Pr	Prandtl number
Pr_{film}	film Prandtl number [eq. (121)]
q	function of Pr [eq. (48)]
q''	heat flux [W/m^2]
Q	heat transfer rate [W]
(r)	dimensionless coefficient of order $O(1)$ [eq. (111)]
R	ideal gas constant
Ra_H	Rayleigh number [eq. (25)]
Ra_{*H}	Rayleigh number based on heat flux [eq. (70)]
$Ra_{film,\,H}$	film Rayleigh number [eq. (115)]
Re_y	Reynolds number ($U_\infty y/\nu$)

t	wall thickness (Fig. 4.9)
T	temperature
T_0	wall temperature
ΔT	temperature difference $(T_0 - T_\infty)$
$T_{\infty,0}$	bottom temperature of a thermally stratified fluid reservoir (Fig. 4.8)
u	horizontal velocity
v	vertical velocity
W	width
x	horizontal coordinate
X	the transversal coordinate of a point situated outside the boundary layer
y	vertical coordinate
Y_T	thermal entrance length
α	thermal diffusivity
β	coefficient of thermal expansion
γ	vertical temperature gradient [K/m] [eq. (77)]
δ	outer thickness of a Pr > 1 natural convection boundary layer (Figure 4.2)
δ_T	thermal boundary layer thickness
δ_v	inner viscous layer thickness of a Pr < 1 natural convection boundary layer (Fig. 4.2)
η	similarity variable [eq. (55)]
θ	similarity temperature profile [eq. (58)]
μ	viscosity
ν	kinematic viscosity
ρ	density
ρ_{vap}	density of saturated vapor
ψ	streamfunction
ω	wall parameter [eq. (85)]
$(\ )_{FC}$	forced convection
$(\ )_{NC}$	natural convection
$(\ )_{0-H}$	averaged from $y = 0$ to $y = H$
$(\ )_\infty$	property of reservoir fluid

REFERENCES

1. A. Bejan, *Entropy Generation through Heat and Fluid Flow*, Wiley, New York, 1982, Chapter 3.

2. E. G. Cravalho and J. L. Smith, Jr., *Engineering Thermodynamics*, Pitman, Boston, MA, 1981, Chapter 11.

3. E. G. Cravalho and J. L. Smith, Jr., *op. Cit.*, Chapters 4 and 9.

4. E. G. Cravalho and J. L. Smith, Jr., *op. Cit.*, p. 288.

5. Y. Jaluria, *Natural Convection Heat and Mass Transfer*, Pergamon, Oxford, 1980, p. 25.

6. D. Grand and Ph. Vernier, Combined Convection in Liquid Metals, Proceedings of the NATO Advanced Study Institute on Turbulent Forced Convection in Channels and Rod Bundles, Istanbul, 1978.

7. B. Gebhart, *Heat Transfer*, 2nd ed., McGraw-Hill, New York, 1971, pp. 327–330.

8. H. H. Lorenz, *Die Wärmeübertragung von einer eben, senkrechten Platte an Öl bei naturlicher Konvektion*, Z. Tech. Phys., Vol. 15, No. 9, pp. 362–366.

9. S. Ostrach, An analysis of laminar free-convection flow and heat transfer about a flat plate parallel to the direction of the generating body force, NACA TN 2635, 1952.

10. H. B. Squire, Integral solution published in S. Goldstein, ed., *Modern Developments in Fluid Dynamics*, Vol. II, Dover, New York, 1965, pp. 641–643.

11. W. M. Rohsenow and H. Y. Choi, *Heat, Mass and Momentum Transfer*, Prentice-Hall, Englewood Cliffs, NJ, 1961, p. 161.

12. E. Schmidt and W. Beckmann. *Das Temperatur-und Geschwindigkeitsfeld von einer wärme abgebenden senkrechten Platte bei natürlicher Konvektion*, Forsch. Ingenieurwes., Vol. 1, 1930, p. 391.

13. A. J. Chapman, *Heat Transfer*, 3rd Ed., Macmillan, New York, 1974, p. 372.

14. F. P. Incropera and D. P. Dewitt, *Fundamentals of Heat Transfer*, Wiley, New York, 1981, p. 436.

15. H. Schlichting, *Boundary Layer Theory*, 4th ed., McGraw-Hill, New York, 1960, p. 333.

16. E. R. G. Eckert and R. M. Drake, Jr., *Analysis of Heat and Mass Transfer*, McGraw-Hill, New York, 1972, pp. 528, 529.

17. E. J. LeFevre, Laminar free convection from a vertical plane surface, *Ninth International Congress of Applied Mechanics*, Brussels, 1956, Paper 1–168.

18. E. M. Sparrow, Laminar free convection on a vertical plate with prescribed nonuniform wall heat flux or prescribed nonuniform wall temperature, NACA TN 3508, July 1955.

19. E. M. Sparrow and J. L. Gregg, Laminar free convection from a vertical plate with uniform surface heat flux, *Trans. ASME*, Vol. 78, 1956, pp. 435–440.

20. E. M. Sparrow and J. L. Gregg, Similar solutions for free convection from a nonisothermal vertical plate, *Trans. ASME*, Vol. 80, 1958, pp. 379–386.

21. K. T. Yang, Possible similarity solutions for laminar free convection on vertical plates and cylinders, *J. Appl. Mech.*, vol. 27, 1960, pp. 230–236.

22. K. T. Yang, J. L. Novotny, and Y. S. Cheng, Laminar free convection from a nonisothermal plate immersed in a temperature stratified medium, *Int. J. Heat Mass Transfer*, Vol. 15, 1972, pp. 1097–1109.

23. A. Bejan, Note on Gill's Solution for free convection in a vertical enclosure, *J. Fluid Mech.*, Vol. 90, 1979, pp. 561–568.

24. C. C. Chen and R. Eichhorn, Natural convection from a vertical surface to a thermally stratified medium, *J. Heat Transfer*, Vol. 98, 1976, pp. 446–451.

25. E. M. Sparrow, H. Quack, and C. J. Boerner, Local nonsimilarity boundary layer solutions, *AIAA J.*, Vol. 8, 1970, pp. 1936–1942.

26. E. M. Sparrow and H. S. Yu, Local nonsimilarity thermal boundary layer solutions, *J. Heat Transfer*, Vol. 93, 1971, pp. 328–334.

27. W. J. Minkowycz and E. M. Sparrow, Local nonsimilar solution for natural convection on a vertical cylinder, *J. Heat Transfer*, Vol. 96, 1974, pp. 178–183.

28. R. Anderson and A. Bejan, Natural convection on both sides of a vertical wall separating fluids at different temperatures, *J. Heat Transfer*, Vol. 102, 1980, pp. 630–635.

29. E. M. Sparrow and C. Prakash, Interaction between internal natural convection in an enclosure and an external natural convection boundary layer flow, *Int. J. Heat Mass Transfer*, Vol. 24, 1981, pp. 845–907.

30. R. Viskanta and D. W. Lankford, Coupling of heat transfer between two natural convection systems separated by a wall, *Int. J. Heat Mass Transfer*, Vol. 24, 1981, pp. 1171–1177.

31. R. Anderson and A. Bejan, Heat transfer through single and double vertical walls in natural convection: theory and experiment, *Int. J. Heat Mass Transfer*, Vol. 24, 1981, pp. 1611–1620.

32. S. Ostrach, Combined natural convection laminar flow and heat transfer of fluids with and without heat sources in channels with linearly varying wall temperatures, NACA TN 3141, 1954.

33. J. R. Bodoia and J. F. Osterle, The development of free convection between heated vertical plates, *Trans. ASME, J. Heat Transfer*, Vol. 84, 1962, pp. 40–44.

34. W. Aung, Fully developed laminar free convection between vertical plates heated asymmetrically, *Int. J. Heat Mass Transfer*, Vol. 15, 1972, pp. 1577–1580.

35. A. A. Szewczyk, Combined forced and free-convection laminar flow, *J. Heat Transfer*, Vol. 86, 1964, pp. 501–507.

36. E. M. Sparrow and J. L. Gregg, Buoyancy effects in forced-convection flow and heat transfer, *J. Appl. Mech.*, Vol. 26, 1959, pp. 133–134.

37. J. R. Lloyd and E. M. Sparrow, Combined forced and free convection flow on vertical surfaces, *Int. J. Heat Mass Transfer*, Vol. 13, 1970, pp. 434–438.

38. E. R. G. Eckert and R. M. Drake, Jr., *Analysis of Heat and Mass Transfer*, McGraw-Hill, New York, 1972, Chapter 13.

39. W. Z. Nusselt, Die Oberflächenkondensation des Wasserdampfes, *Z. Ver. Dtsch Ing.*, Vol. 60, 1916, pp. 541–569.

40. W. M. Rohsenow, Heat transfer and temperature distribution in laminar-film condensation, *Trans. ASME, J. Heat Transfer*, Vol. 78, 1956, pp. 1645–1648.

41. E. M. Sparrow and J. L. Gregg, A boundary layer treatment of laminar film condensation, *Trans. ASME, J. Heat Transfer*, Vol. 81, 1959, pp. 13–18.

42. M. M. Chen, An analytical study of laminar film condensation. Part 1. Flat plates; Part 2. Single and multiple horizontal tubes, *Trans. ASME, J. Heat Transfer*, Vol. 83, 1961, pp. 48–60.

43. R. Viskanta, Phase-change heat transfer, in G. A. Lane, ed., *Solar Heat Storage: Latent Heat Materials*, CRC Press, Boca Raton, Florida, 1983.

PROBLEMS

1. Derive the integral forms of the momentum and energy equations [eqs. (40) and (41)] by integrating eqs. (17) and (8) from the wall ($x = 0$) to a plane $x = X$ outside the vertical boundary layer.

2. Perform an integral analysis of the natural convection boundary layer by assuming the following temperature and velocity profiles [10]

$$T - T_\infty = \Delta T\left(1 - \frac{x}{\delta_T}\right)^2$$

$$v = V\frac{x}{\delta}\left(1 - \frac{x}{\delta}\right)^2$$

For brevity, assume $\delta = \delta_T$. Follow the steps between eqs. (42) and (49) to show that the local Nusselt number is given by

$$
\text{Nu} = \frac{q''}{T_0 - T_\infty} \frac{y}{k} = 0.508 \left(1 + \frac{20}{21\,\text{Pr}}\right)^{-1/4} \text{Ra}_y^{1/4}
$$

Note that, since $\delta_T = \delta$, eq. (47) cannot be satisfied by the solution. Comment on the Pr range of validity of this $\delta_T = \delta$ integral solution, and explain why the predicted Nusselt number agrees with more exact calculations over a surprisingly wide range (Fig. 4.4).

3. Derive the equations for energy and momentum in the similarity solution formulation [eqs. (61) and (62)]; start with eqs. (56) and (57) and exploit the similarity variable transformation (55) and (58–60).

4. Consider the natural convection heat leak from a life-size room with one 3 m tall wall exposed to the cold ambient. The room-air temperature is 25°C, while the room-side surface of the cold wall has an average temperature of 10°C. If the room circulation is to be simulated in a small laboratory apparatus filled with water, how tall should the water cavity of the apparatus be? In the laboratory water experiment, the temperature difference between the water body and the inner surface of the cooled wall is 10°C.

5. In many analyses of natural convection heat transfer problems, the vertical wall heating a fluid or dividing two differentially heated fluids is usually modeled as isothermal. This, of course, is an approximation valid in some cases and invalid in others. To be isothermal, while bathed by natural convection in boundary layer flow, a vertical solid wall must be thick enough. Comparing the thermal conductance to vertical conduction through the wall ($k_w W/H$) with the thermal conductance to lateral heat transfer through the same wall (and the fluid boundary layer, kH/δ_T), determine below what range of wall widths W the "isothermal wall" assumption becomes inadequate (k_w, H, k, and δ_T are the wall thermal conductivity, wall height, fluid thermal conductivity, and thermal boundary layer thickness, respectively). For a wall of fixed geometry, (W, H) is the isothermal wall assumption getting better or worse as Ra_H increases?

6. Determine the local Nusselt number for the boundary layer natural convection along a $q'' = $ constant vertical wall by performing an integral analysis using the profiles [18]

$$
T - T_\infty = \Delta T \left(1 - \frac{x}{\delta_T}\right)^2
$$

$$
v = V \frac{x}{\delta}\left(1 - \frac{x}{\delta}\right)^2
$$

and taking $\delta_T = \delta$. Show that the wall temperature distribution is given by

$$T_0(y) - T_\infty = 1.622\frac{q''y}{k}\left(\frac{\frac{4}{3} + \text{Pr}}{\text{Pr}}\right)^{1/5}\text{Ra}_{*\,y}^{-1/5}$$

and that the local Nusselt number is given by eq. (75).

7. Construct the similarity formulation of the boundary layer flow problem along a vertical wall with uniform heat flux. Start with eqs. (56) and (57) and strive to obtain the equivalent of eqs. (61) and (62) by first noting the similarity variable dictated by the scaling law (69), $\eta = (x/y)\text{Ra}_{*\,y}^{1/5}$. Compare your final expressions for momentum and energy conservation with the equations of Ref. 19: keep in mind that Ref. 19 uses $(x/y)(\text{Ra}_{*\,y}/\text{Pr})^{1/5}$ as the similarity variable, not $(x/y)\text{Ra}_{*\,y}^{1/5}$.

8. Consider the integral analysis of laminar natural convection along a vertical wall bathed by a linearly stratified fluid (Fig. 4.8). Starting with eqs. (78), show that the integral momentum and energy equations can be expressed as in eqs. (79)–(81). Solve these equations in the limit of negligible inertia, $\text{Pr} \to \infty$; show that in this limit the differential equation of $\delta_*(y_*)$ is

$$\frac{d\Delta}{dy_*} = \frac{240 + \frac{8}{3}b\Delta}{1 - by_*}$$

where $\Delta = \delta_*^4$. Approximating this equation via finite differences (Chapter 12), integrate it setting $b = 1$. Compare your estimate of the average Nusselt number [eq. (83)] with Fig. 4.8 and eq. (84).

9. Estimate the $\text{Nu}_{0-H}(\omega, \text{Ra}_H)$ function displayed in Fig. 4.9 by matching in series the thermal resistance of one $q'' = $ constant boundary layer [eq. (75)], the thermal resistance of the wall, and, finally, the resistance of another $q'' = $ constant boundary layer. For the sake of this matching procedure, assume that the temperature along each face of the wall is y-independent and equal to the actual temperature "averaged" over the wall height. Keep in mind that eq. (75) describes the *local* heat transfer (hence, the local temperature difference), and that the end-result of this analysis, Nu_{0-H}, describes the *overall* heat transfer. For more details on how to approach this solution, consult pp. 634–635 of Ref. 28.

10. An electrical conductor in a certain piece of electronic equipment may be modeled as an isothermal plate (T_0) oriented vertically. The heat transfer rate generated in the plate and released via laminar natural convection to the ambient (T_∞) is *fixed* and equal to Q. The height of the plate (H) may vary.

(a) Neglecting numerical factors of order one, what is the relationship between the Nusselt number and the Rayleigh number for this arrangement?

(b) How will the temperature difference $(T_0 - T_\infty)$ vary with the height of the system? In other words, if H increases by a factor of 2, what happens to $(T_0 - T_\infty)$?

11. The heat-exchanger tube mounted horizontally in the back of a refrigerator is to be fitted with a number of vertical rectangular plate fins of height H, width W, and fin-to-fin spacing D. The fin surface may be modeled as isothermal at temperature T_0; the ambient fluid is air at temperature T_∞. The purpose of the fins is to maximize the total heat transfer rate Q between the tube of length L and the ambient air. The only design variable is the number of fins, that is, fin-to-fin spacing D (parameters H, W, T_0, T_∞, and L are assumed given, and the thickness of the fin plate is considered negligible). Also negligible is the direct heat transfer between the unfinned portions of the tube and the ambient.

Find the relationship between Q and D in the following two extremes: (1) the fully developed channel flow regime, and (2) the boundary layer regime with the boundary layer thickness much smaller than the fin-to-fin spacing. Based on these asymptotic results, show that there exists an optimum D for which Q is maximum. Develop your best engineering estimate for the optimum fin-to-fin spacing D_{opt}.

12. The fully developed natural convection channel flow analysis presented in Chapter 4 refers to only one simple geometry (the parallel-plate channel). Using this analysis as a guideline, develop the corresponding Nusselt number formula for other vertical channel crosssections, for example,

(a) circular

(b) square

(c) equilateral triangle

Express your result in the form

$$\frac{q''_{0-H} H}{(T_0 - T_\infty)k} = (?)\mathrm{Ra}_{D_h}$$

where (?) is a numerical coefficient to be determined for each channel cross-sectional shape, and Ra_{D_h} is the Rayleigh number based on hydraulic diameter.

Hint: Note that, in general, eq. (89) is replaced by

$$\nabla^2 v = -\frac{g\beta}{\nu}(T_0 - T_\infty)$$

which is of the same type as eq. (20) of Chapter 3. The mean velocity solution to this equation may be deduced from the friction factor data assembled in Table 3.2.

13. Find the Rayleigh number range in which the fully developed regime formulas (91) are valid by translating the inequality (90) in the language of

Ra_D and H/D. Demonstrate that the $\text{Ra}_D(H/D)$ criterion determined in this manner is essentially the same as criterion (95).

14. Prove that for $\text{Pr} < 1$ fluids, the scale criterion for transition from natural convection to forced convection in boundary layer flow along a vertical wall (Fig. 4.12) is given by eq. (101).

15. Develop the integral solution for heat transfer across the laminar thin film of condensate shown in Fig. 4.14. Neglect the effect of inertia in the momentum equation. For temperature, assume first a linear profile and, second, a parabolic profile; show that the resulting Nusselt number expression is eq. (117) with $(r) = 3/8$ in the first case, and $(r) = 0.68$ in the second case.

16. The laminar film condensation problem studied in connection with Fig. 4.14 is in fact a multifaceted idealization of the way in which a vapor might condense on a cold enough wall. Following Nusselt's original investigation of this problem, the vapor was assumed saturated and its possible movement along the liquid film was neglected. Discuss qualitatively the effect the removal of each simplifying assumption would have on the scaling laws developed in the text. Quantitatively, what would be the $T_{\text{vap}} \to T_0$ heat flux scale, if the vapor is superheated ($T_{\text{vap}} > T_{\text{sat}}$)? At what film Rayleigh number will turbulence set in? (Hint: read Chapter 6 and, in particular, Table 6.1.)

17. You have a bottle of beer at room-temperature. It is a hot day, and you would like to drink this beer cold and as soon as possible. The beer bottle has a height/diameter ratio of about 5. You place the bottle in the refrigerator, however, you have the option of positioning the bottle (1) vertically or (2) horizontally. The refrigerator cools by natural convection (i.e., it does not employ forced circulation). Which way should you position the bottle? Describe the goodness of your decision by calculating the ratio t_1/t_2, where t represents the order of magnitude of the time needed for the bottle to reach thermal equilibrium with the refrigeration chamber (base this calculation on scale analysis).

18. Consider a vertical wall of height H in contact with an isothermal fluid reservoir, as is shown in Fig. 4.1. For the purpose of scale analysis, select the square flow region of height H and artificial horizontal length H. Show that if in the momentum equation (17) you invoke a balance between friction and buoyancy, then the ratio inertia/friction comes out to be of order $\text{Gr}_H = g\beta\Delta T H^3/\nu^2$ (note that the $H \times H$ region is not the boundary layer region; hence, the conclusion "inertia/friction $\sim \text{Gr}_H$" does not apply to the boundary layer region). Is the vertical velocity scale derived above compatible with the v scale recommended by the energy equation (8) for the same $H \times H$ region? In other words, is the invoked balance friction $\sim$ buoyancy in the $H \times H$ region realistic?

5

Natural Convection in Enclosures

Natural convection in enclosures is a topic of contemporary importance, because enclosures filled with fluid are central components in a long list of engineering and geophysical systems. The flow and heat transfer induced, for example, in the inner air space of a double-pane window system differs fundamentally from the external natural convection boundary layer considered earlier. In this chapter we focus on natural convection as an *internal flow*: unlike the external free convection boundary layer that is caused by the heat transfer interaction between a single wall and a very large fluid reservoir, natural convection in an enclosure is the result of the complex interaction between a finite-size fluid system in thermal communication with all the walls that confine it. The complexity of this internal interaction is responsible for the diversity of flows that can exist inside enclosures. At the same time, the complexity of the phenomenon is responsible for the engineers' relative inability to predict both the flow and the heat transferred by it across the enclosure.

Natural convection in enclosures is a challenging subject: this fact is amply reflected by the size of the research effort during the past two decades dedicated to this topic [1–3]. However, instead of compiling and reviewing the multitude of individual heat transfer results reported so far,[†] we focus on the fundamentals of the mechanism responsible for heat transfer through a fluid-filled enclosure. We will identify the correct scales that, depending on circumstances, characterize the internal flow. Armed with this information, we will be able to both anticipate the flow and to devise simplified analyses to describe the flow and heat transfer analytically.

The phenomenon of natural convection in an enclosure is as varied as the geometry and orientation of the enclosure. Judging from the number of

---

[†] To a certain extent, this has already been done in Refs. 1–3.

potential engineering applications, the enclosure phenomena can loosely be organized into two large classes:

1. Enclosures heated from the side.
2. Enclosures heated from below.

The first class is representative of applications such as solar-collectors, double-wall insulations, and air circulation through the rooms in a building. In addition, we find enclosures heated from the side in cooling systems of industrial-scale rotating electric machinery (conventional, as well as superconducting). The second class refers to the functioning of thermal insulations oriented horizontally, for example, the heat transfer through a flat-roof attic space. The study of both flow classes is also relevant to our understanding of natural circulation in the atmosphere, the hydrosphere, and the molten core of the Earth.

We focus primarily on enclosures heated from the side. The reason for this choice is twofold. First, current applications in thermal insulation engineering, solar technology, rotating fluid machinery, and energy management in architectural design demand an emphasis on this particular class of flows. The second reason is historical: natural convection in fluid layers heated from below (the Bénard flow) is already a classical subject [4, 5]. By comparison, enclosures heated from the side represent a much younger subfield in convective heat transfer research; the mission of this chapter is to bring this new subfield into perspective, and to provide future researchers and engineers with a solid methodology for attacking new problems.

TRANSIENT HEATING FROM THE SIDE

Consider a two-dimensional enclosure of height H and horizontal length L, as is shown in Fig. 5.1. The enclosure is filled with a Newtonian fluid such as air or water. We are interested in the transient behavior of the cavity fluid as the side walls are instantaneously heated and, respectively, cooled to temperatures $+\Delta T/2$ and $-\Delta T/2$. The top and bottom walls ($y = 0, H$) remain insulated throughout this experiment. Initially, that is, before the establishment of the temperature difference T across the cavity, the fluid is isothermal ($T = 0$) and motionless ($u = v = 0$) everywhere inside the cavity.

The equations governing the conservation of mass, momentum, and energy at every point in the cavity are:

$$\frac{\partial u}{\partial x} + \frac{\partial v}{\partial y} = 0 \tag{1}$$

$$\frac{\partial u}{\partial t} + u\frac{\partial u}{\partial x} + v\frac{\partial u}{\partial y} = -\frac{1}{\rho}\frac{\partial P}{\partial x} + \nu\left(\frac{\partial^2 u}{\partial x^2} + \frac{\partial^2 u}{\partial y^2}\right) \tag{2}$$

$$\frac{\partial v}{\partial t} + u\frac{\partial v}{\partial x} + v\frac{\partial v}{\partial y} = -\frac{1}{\rho}\frac{\partial P}{\partial y} + \nu\left(\frac{\partial^2 v}{\partial x^2} + \frac{\partial^2 v}{\partial y^2}\right) - g[1 - \beta(T - T_0)] \tag{3}$$

$$\frac{\partial T}{\partial t} + u\frac{\partial T}{\partial x} + v\frac{\partial T}{\partial y} = \alpha\left(\frac{\partial^2 T}{\partial x^2} + \frac{\partial^2 T}{\partial y^2}\right) \tag{4}$$

The symbols appearing in eqs. (1)–(4) are defined in Fig. 5.1 and the list of symbols at the end of this chapter. Note that in writing these equations we modeled the fluid as Boussinesq—incompressible, in other words, ρ = constant everywhere except in the body force term of the y momentum equation where it is replaced by $\rho[1 - \beta(T - T_0)]$.

Instead of solving eqs. (1)–(4) numerically and displaying the experimental results, we will rely on pure scaling arguments to predict theoretically the types of flow and heat transfer patterns that can develop in the enclosure. This scaling analysis is due to Patterson and Imberger [6]. Immediately after $t = 0$,

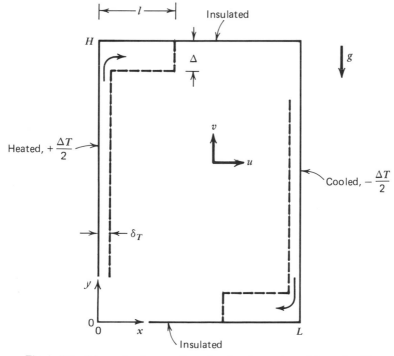

Figure 5.1   Schematic of a two-dimensional enclosure heated from the side.

the fluid bordering each side wall is motionless: this means that near the side wall the energy equation (4) expresses a balance between thermal inertia and conduction normal to the wall,

$$\frac{\Delta T}{t} \sim \alpha \frac{\Delta T}{\delta_T^2} \tag{5}$$

This equality of scales follows from recognizing ΔT, t, and δ_T as the scales of changes in T, t, and x in eq. (4). In the same equation, we took $u = v = 0$; we also recognized that $\partial^2 T/\partial y^2 \ll \partial^2 T/\partial x^2$ because near $t = 0^+$ the thermal boundary layer thickness δ_T is much smaller than the enclosure height (note that $y \sim H$ and $x \sim \delta_T$). Equation (5) dictates that in the time immediately following $t = 0$, each side wall is coated with a conduction layer whose thickness increases steadily as

$$\delta_T \sim (\alpha t)^{1/2} \tag{6}$$

The heated layer δ_T will naturally tend to rise along the heated wall: the velocity scale of this upward motion v is easier to see if we first eliminate the pressure P between the two momentum equations (2) and (3):

$$\frac{\partial}{\partial x}\left(\frac{\partial v}{\partial t} + u\frac{\partial v}{\partial x} + v\frac{\partial v}{\partial y}\right) - \frac{\partial}{\partial y}\left(\frac{\partial u}{\partial t} + u\frac{\partial u}{\partial x} + v\frac{\partial u}{\partial y}\right)$$

$$= \nu\left[\frac{\partial}{\partial x}\left(\frac{\partial^2 v}{\partial x^2} + \frac{\partial^2 v}{\partial y^2}\right) - \frac{\partial}{\partial y}\left(\frac{\partial^2 u}{\partial x^2} + \frac{\partial^2 u}{\partial y^2}\right)\right] + g\beta\frac{\partial T}{\partial x} \tag{7}$$

This new equation contains three basic groups of terms—inertia terms on the left-hand side and four viscous diffusion terms plus the buoyancy term on the right-hand side. It is easy to show that the three terms that dominate each basic group are (Problem 1):

$$
\begin{array}{ccc}
\text{Inertia} & \text{Friction} & \text{Buoyancy} \\
\dfrac{\partial^2 v}{\partial x\,\partial t}, & \nu\dfrac{\partial^3 v}{\partial x^3}, & g\beta\dfrac{\partial T}{\partial x}
\end{array} \tag{8}
$$

In terms of representative scales, the momentum balance (8) reads

$$
\frac{v}{\delta_T t}, \qquad \nu\frac{v}{\delta_T^3} \quad \sim \quad \frac{g\beta\Delta T}{\delta_T} \tag{9}
$$

The driving force in this balance is the buoyancy effect $g\beta\Delta T/\delta_T$, which is *finite*. It is important to find out whether the buoyancy effect is balanced by friction *or* inertia. Dividing eq. (9) through the friction scale, and recalling that

$\delta_T^2 \sim \alpha t$ yields

$$
\begin{array}{ccc}
\text{Inertia} & \text{Friction} & \text{Buoyancy} \\
\dfrac{1}{Pr} & 1 & \sim \dfrac{g\beta\Delta T \delta_T^2}{\nu \upsilon}
\end{array} \tag{10}
$$

Therefore, for fluids with Prandtl number of order one or greater, the correct momentum balance at $t = 0^+$ is between buoyancy and friction.

$$
1 \sim \frac{g\beta\Delta T \delta_T^2}{\nu \upsilon} \tag{11}
$$

We conclude that the initial vertical velocity scale is

$$
\upsilon \sim \frac{g\beta\Delta T \alpha t}{\nu} \tag{11'}
$$

This velocity scale is valid for fluids such as water and oils (Pr > 1), and marginally valid for gases (Pr ~ 1).

Having identified the velocity scale of the first fluid movement, we turn our attention back to the energy equation. The heat conducted from the side wall into the fluid layer δ_T is no longer spent solely on thickening the layer: part of this heat input is carried away by the layer δ_T rising with velocity υ. Thus, in the energy equation, we see a competition between three distinct effects.

$$
\begin{array}{ccc}
\text{Inertia} & \text{Convection} & \text{Conduction} \\
\dfrac{\Delta T}{t} & \upsilon\dfrac{\Delta T}{H} & \sim \alpha\dfrac{\Delta T}{\delta_T^2}
\end{array} \tag{12}
$$

As t increases, the convection effect increases [$\upsilon \sim t$, eq. (11')] while the effect of inertia decreases in importance. There comes a final time t_f when the energy equation expresses a balance between the heat conducted from the wall and the enthalpy carried away vertically by the buoyant layer,

$$
\upsilon\frac{\Delta T}{H} \sim \alpha\frac{\Delta T}{\delta_T^2} \tag{13}
$$

which yields

$$
t_f \sim \left(\frac{\nu H}{g\beta\Delta T \alpha} \right)^{1/2} \tag{13'}
$$

At such a time, the layer thickness is

$$\delta_{T,f} \sim \left(\alpha t_f\right)^{1/2} \sim H\,\mathrm{Ra}_H^{-1/4} \tag{14}$$

where Ra_H is the Rayleigh number based on the enclosure height,[†]

$$\mathrm{Ra}_H = \frac{g\beta\Delta T H^3}{\alpha\nu} \tag{15}$$

Note that the time t_f when the δ_T layer becomes convective could have been determined by setting inertia $\sim$ convection in eq. (12). Note further that beyond this time $t > t_f$, the convection $\sim$ conduction balance in eq. (12) is preserved *only* if both v and δ_T no longer increase in time. Therefore, beyond $t \sim t_f$ the thermal layers along each side wall reach a *steady state* characterized by an energy balance between conduction and convection, and a momentum balance between buoyancy and viscous diffusion (see Chapter 4).

In addition to thermal layers of thickness $\delta_{T,f}$, the side walls develop viscous (velocity) wall jets. The thickness of these jets δ_v follows from the momentum balance (7) for the region of thickness $x \sim \delta_v$ *outside* the thermal layer. In this region the buoyancy effect is minor, and we have a balance between inertia and viscous diffusion,

$$\frac{v}{\delta_v t} \sim \nu\frac{v}{\delta_v^3} \tag{16}$$

hence,

$$\delta_v \sim \left(\nu t\right)^{1/2} \sim \mathrm{Pr}^{1/2}\delta_T \tag{17}$$

In the steady state, $t > t_f$, the fluid near each side wall is characterized by a *two-layer structure*: a thermal boundary layer of thickness $\delta_{T,f}$ and a thicker wall jet $\delta_{v,f} \sim \mathrm{Pr}^{1/2}\delta_{T,f}$. The development of this structure is presented in Fig. 5.2.

Criterion for Distinct Vertical Layers

If the final thermal boundary layer thickness $\delta_{T,f}$ is smaller than the transversal extent of the enclosure (L), then the thermal layers will be distinct: using eq. (14), this criterion reads

$$\frac{H}{L} < \mathrm{Ra}_H^{1/4} \tag{18}$$

[†]The reader should be alerted to the fact that the majority of the experimental studies on natural convection in enclosures, particularly those on tall enclosures, report their results in terms of a Rayleigh number based on the *horizontal dimension L*; at least from a theoretical scaling viewpoint, this choice is without foundation. To warn the reader of the potential for confusion, in the present treatment the Rayleigh number defined by eq. (15) carries the subscript H.

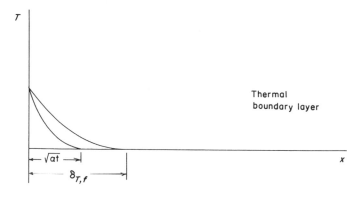

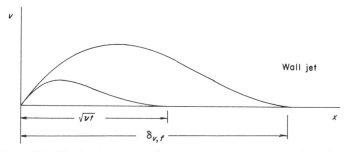

Figure 5.2 The development of the two-layer structure near the heated wall.

We could have arrived at the same inequality by stating that the vertical layers become convective in a time shorter than the thermal diffusion time between the two vertical walls ($t_f < L^2/\alpha$). The $H/L - \mathrm{Ra}_H$ subdomain in which we should expect distinct wall layers in the horizontal temperature profile is shown in Fig. 5.3. Note that this subdomain is bounded on the left by the conduction heat transfer regime (the proof that $\mathrm{Ra} < 1$ is the criterion for conduction-dominated heat transfer is proposed as an exercise to the reader; Problem 3). Note further that the corresponding criterion for distinct velocity boundary layers (wall jets) is $\delta_{v,f} < L$; hence,

$$\frac{H}{L} < \mathrm{Ra}_H^{1/4}\mathrm{Pr}^{-1/2} \tag{19}$$

Criterion for Distinct Horizontal Jets

The scaling analysis presented so far can be extended to cover events near the two horizontal adiabatic walls in an attempt to predict the shape of the temperature and velocity profiles vertically across the cavity. A detailed

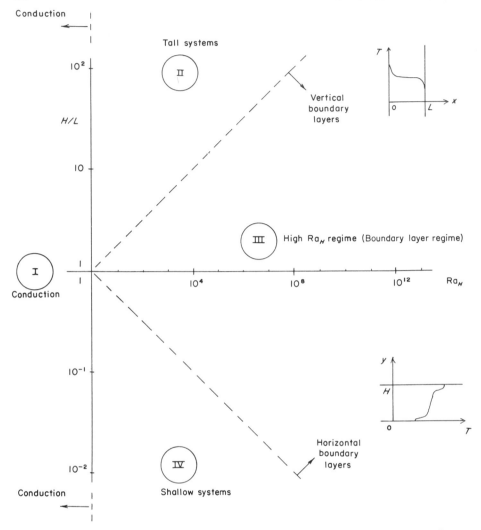

Figure 5.3 Chart showing the four heat transfer regimes for natural convection in a two-dimensional enclosure heated from the side.

discussion of the scenarios possible near the adiabatic walls is presented in Ref. [6]. An alternative, more direct approach to predicting the presence of distinct wall layers in the steady state is to regard the circulation loop as a counterflow heat exchanger in which the horizontal branches can communicate thermally over a horizontal distance of order L [6, 7]. The enthalpy flow between the two vertical ends is

$$Q_{\substack{\text{convection} \\ \text{left} \to \text{right} \\ \text{(Fig.5.1)}}} \sim (\rho v \delta_T)_f c_P \Delta T$$

$$\sim k \, \Delta T \, \text{Ra}_H^{1/4} \tag{20}$$

This estimate is the same as the heat transfer rate through each of the vertical walls, $(k/\delta_{T,f})H\Delta T$.

Heat can diffuse vertically from the warm upper branch of the counterflow to the lower branch at a rate

$$Q_{\substack{\text{conduction} \\ \text{top} \rightarrow \text{bottom}}} \sim kL\frac{\Delta T}{H} \qquad (21)$$

$$\text{(Fig.5.1)}$$

The enthalpy carried by the stream $(\rho v \delta_T)_f$ will reach the opposite end intact when the diffusion vertically to the counterflowing stream is negligible.

$$kL\frac{\Delta T}{H} < k\,\Delta T\,\text{Ra}_H^{1/4}$$

In other words, when

$$\frac{H}{L} > \text{Ra}_H^{-1/4} \qquad (22)$$

When condition (22) is met, the horizontal streams along the adiabatic walls retain their temperature identity, as is shown in the sketch accompanying the $H/L = \text{Ra}_H^{-1/4}$ line on Fig. 5.3.

The two criteria for distinct thermal layers [eqs. (18) and (22)] and the convective heat transfer requirement [$\text{Ra}_H > 1$] divide the $H/L - \text{Ra}_H$ field into four sections. Each section corresponds to a certain regime in the steady state:

(I) Conduction limit. The temperature varies linearly across the cavity; hence, the heat transfer rate between the two side walls is of order $kH\Delta T/L$. The horizontal temperature gradient $\Delta T/L$ gives rise to a slow, clockwise circulation, however, the heat transfer contribution of this flow is insignificant.

(II) Tall enclosure limit. For most of the enclosure height, the temperature varies linearly between the two side walls. The heat transfer rate is of order $kH\Delta T/L$, as in the preceding case. The clockwise circulation pattern is characterized by distinct layers in the vicinity of the top and bottom walls.

(III) High-Ra_H limit (the boundary layer regime). Vertical thermal boundary layers form distinctly along the differentially heated side walls. The heat transfer rate across the cavity scales as $(k/\delta_{T,f})H\Delta T$. The adiabatic horizontal walls are lined by distinct thermal layers. Most of the cavity fluid (the core) is relatively stagnant and thermally stratified.

(IV) Shallow enclosure limit. The heat transfer mechanism is dominated by the presence of vertical thermal layers; hence, it scales again as $(k/\delta_{T,f})H\Delta T$. This scale represents an upper bound, because an additional insulation effect is provided by the long horizontal core of the cavity. In this region, the two branches of the horizontal counterflow make good thermal contact, rendering the counterflow an effective insulation in the left–right direction [8].

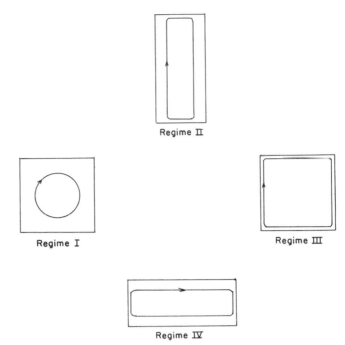

Figure 5.4 Schematic of the circulation patterns possible under regimes (I)–(IV) shown in Fig. 5.3.

The circulation patterns corresponding to the four regimes (I)–(IV) are sketched in Fig. 5.4. These patterns have been confirmed by numerous experimental studies involving Pr > 1 fluids, for example, the numerical simulations reported in Refs. 23 and 30 and the laboratory measurements reported in Refs. 12 and 31.

THE BOUNDARY LAYER REGIME

In the next two sections we focus on two classes of analytical advances in the direction of predicting the heat transfer rate under regimes (III) and (IV). It is worth pointing out that the first major theoretical work on natural convection in enclosures was reported by Batchelor [9], who considered regimes (I), (II), and (III). Although Batchelor's analysis served as a stimulus for the theoretical work that followed, it is not as critical to heat transfer engineering because under regimes (I) and (II) the heat transfer rate is practically equal to the pure conduction estimate. Batchelor's work on the high-Ra_H regime (III) suffers from an inappropriate choice (guess) for the core flow in the steady state, as explained by Ostrach (Ref. 1, p. 176).

When vertical thermal boundary layers are present, the heat transfer rate is controlled by the thermal resistance of order $\delta_{T,f}/k$ which coats each side wall. To calculate this thermal resistance we must determine the *temperature field* in the vertical region of thickness $\delta_{T,f}$. This engineering task justifies the following nondimensionalization of the governing equations (1), (4), and (7):

$$x_* = \frac{x}{\delta_{T,f}}, \qquad y_* = \frac{y}{H}$$

$$T_* = \frac{T}{\Delta T}$$

$$u_* = \frac{u}{(\delta_{T,f}/H)v_f}, \qquad v_* = \frac{v}{v_f} \tag{23}$$

In these definitions $\delta_{T,f}$ is the final thermal boundary layer thickness [eq. (14)] and v_f is the velocity scale (11') evaluated at $t = t_f$. Substituting the new dimensionless variables (23) into the steady state conservation equations yields:

$$\frac{\partial u_*}{\partial x_*} + \frac{\partial v_*}{\partial y_*} = 0 \tag{24}$$

$$u_* \frac{\partial T_*}{\partial x_*} + v_* \frac{\partial T_*}{\partial y_*} = \frac{\partial^2 T_*}{\partial x_*^2} + \mathrm{Ra}_H^{-1/2} \frac{\partial^2 T_*}{\partial y_*^2} \tag{25}$$

$$\frac{1}{\mathrm{Pr}} \left[\frac{\partial}{\partial x_*} \left(u_* \frac{\partial v_*}{\partial x_*} + v_* \frac{\partial v_*}{\partial y_*} \right) - \mathrm{Ra}_H^{-1/2} \frac{\partial}{\partial y_*} \left(u_* \frac{\partial u_*}{\partial x_*} + v_* \frac{\partial u_*}{\partial y_*} \right) \right]$$

$$= \frac{\partial}{\partial x_*} \left(\frac{\partial^2 v_*}{\partial x_*^2} + \mathrm{Ra}_H^{-1/2} \frac{\partial^2 v_*}{\partial y_*^2} \right) - \mathrm{Ra}_H^{-1/2} \frac{\partial}{\partial y_*} \left(\frac{\partial^2 u_*}{\partial x_*^2} + \mathrm{Ra}_H^{-1/2} \frac{\partial^2 u_*}{\partial y_*^2} \right)$$

$$+ \frac{\partial T_*}{\partial x_*} \tag{26}$$

In the high-Ra_H limit and for $\mathrm{Pr} > 1$ fluids, eqs. (24)–(26) reduce to

$$\frac{\partial u_*}{\partial x_*} + \frac{\partial v_*}{\partial y_*} = 0 \tag{24'}$$

$$u_* \frac{\partial T_*}{\partial x_*} + v_* \frac{\partial T_*}{\partial y_*} = \frac{\partial^2 T_*}{\partial x_*^2} \tag{25'}$$

$$0 = \frac{\partial^3 v_*}{\partial x_*^3} + \frac{\partial T_*}{\partial x_*} \tag{26'}$$

Gill [10] solved these equations approximately subject to the side wall conditions

$$u_* = v_* = 0, T_* = \tfrac{1}{2} \text{ at } x_* = 0 \tag{27}$$

and the outer (far from the wall) conditions

$$u_* \to u_{*\infty}(y_*), v_* \to 0, \quad \text{as } x_* \to \infty \tag{28}$$

$$T_* \to T_{*\infty}(y_*), \quad \text{as } x_* \to \infty \tag{29}$$

where $u_{*\infty}$ and $T_{*\infty}$ are the unknown flow and temperature stratification of the core. To circumvent the nonlinearity of the energy equation (25'), Gill [10] relied on the Oseen-linearization technique suggested by Carrier [11]. He replaced the u_* and $\partial T_* / \partial y_*$ factors appearing on the convective side of eq. (25') with two unknown functions of altitude only, $u_A(y_*)$ and $T_A'(y_*)$.

$$(u_A)\frac{\partial T_*}{\partial x_*} + (T_A') v_* = \frac{\partial^2 T_*}{\partial x_*^2} \tag{25''}$$

Eliminating T_* between this equation and the momentum equation (26') yields

$$\frac{\partial^4 v_*}{\partial x_*^4} - (u_A)\frac{\partial^3 v_*}{\partial x_*^3} + (T_A') v_* = 0 \tag{30}$$

This equation can be integrated in x_*. The general solution has the form

$$v_* = \sum_{i=1}^{4} a_i(y_*) e^{-\lambda_i(y_*) x_*} \tag{31}$$

where the λ_i's are the four roots of the characteristic equation

$$\lambda^4 + u_A \lambda^3 + T_A' = 0 \tag{31'}$$

Applying the boundary conditions (27)–(29), the solution takes the form

$$v_* = \frac{\tfrac{1}{2} - T_{*\infty}}{\lambda_2^2 - \lambda_1^2}\left(-e^{-\lambda_2 x_*} + e^{-\lambda_1 x_*} \right) \tag{32}$$

$$T_* = \frac{\tfrac{1}{2} - T_{*\infty}}{\lambda_2^2 - \lambda_1^2}\left(\lambda_2^2 e^{-\lambda_2 x_*} - \lambda_1^2 e^{-\lambda_1 x_*} \right) + T_{*\infty} \cdots \tag{33}$$

where λ_1 and λ_2 are the two roots with positive real parts of eq. (31').

The solution expressed by eqs. (32) and (33) depends on four[†] unknown functions of altitude, λ_1, λ_2, $u_{*\infty}$, and $T_{*\infty}$. Gill determined these functions uniquely by invoking the energy integral condition[‡]

$$\int_0^\infty \left(u_* \frac{\partial T_*}{\partial x_*} + v_* \frac{\partial T_*}{\partial y_*} \right) dx_* = -\left(\frac{\partial T_*}{\partial x_*} \right)_{x_*=0} \tag{34}$$

plus two centrosymmetry conditions, meaning that the cold-side boundary layer solution must approach the *same* core solution (28) and (29). The remaining algebraic details of [10] are not reproduced here; the final results for the auxiliary functions can be expressed as

$$\lambda_{1,2} = \frac{1}{4} p(1 - q)\left[1 \pm i(1 + 2q)^{1/2}\right] \tag{35}$$

$$T_{*\infty} = \frac{q}{1 + q^2} \tag{36}$$

where $p(y_*)$ is an even function and $q(y_*)$ is an odd function resulting from the system:

$$p = \frac{2(1 + 3q^2)^{11/9}}{C(1 + q^2)^{2/3}(1 - q^2)} \tag{37}$$

$$\frac{dq}{dy_*} = \frac{2(1 + 3q^2)^{53/9}}{C^4(7 - q^2)(1 - q^2)^3(1 + q^2)^{2/3}} \tag{38}$$

For this formulation the origin of y_* was taken at the midheight of the enclosure, so that y_* varies from $-1/2$ to $1/2$. Gill integrated eqs. (37) and (38) numerically and determined the constant C from the *arbitrary* condition that the vertical velocity in the boundary layers (v_*) is zero at the two corners ($y_* = \pm 1/2$). Representative boundary layer isotherms and streamlines based on this solution are reproduced in Fig. 5.5 (note the different notation: on Fig. 5.5 z is y_*, T is T_*, and ψ is the streamfunction defined as $u_* = -\partial\psi/\partial y_*$, $v_* = \partial\psi/\partial x_*$). The complete solution was compared with the temperature and velocity profiles reported experimentally by Elder [12]. The agreement between theory and experiment proved adequate, although questions have persisted in connection with the arbitrary choice of impermeable top and bottom walls made in order to determine the constant C [13, 14]. The basis for

---

[†] The fourth unknown, $u_{*\infty}$, appears in the expression for u_*, which is obtained by combining eqs. (32) and (24).
[‡] The integral condition (34) must be an energy condition because the inexact character of solution (32, 33) stems from the Oseen-linearization (approximation) of the energy equation (25′).

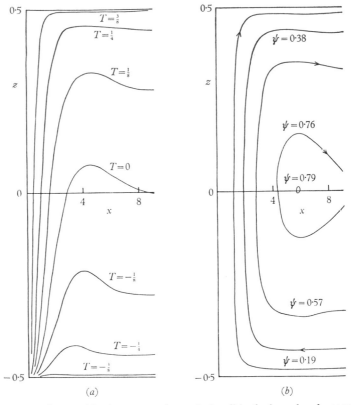

Figure 5.5 Streamlines and isotherms near the vertical wall in the boundary layer regime [10].

this criticism is the fact that the solution for v_* is valid in the *boundary layer* only, therefore, it is improper to use it in the corners, where the boundary layer scaling (23) breaks down.

The important heat transfer engineering result of the preceding analysis, not reported in Ref. 10, is the *overall* heat transfer rate across the enclosure.

$$Q = k \int_{-H/2}^{H/2} \left(-\frac{\partial T}{\partial x} \right)_{x=0} dy$$

$$= 0.364 k \, \Delta T \, \mathrm{Ra}_H^{1/4} \tag{39}$$

This result was reported by Bejan [14]. Noting that the majority of experimental and numerical studies have reported their findings in the form of Nusselt number correlations where Nu is defined as

$$\mathrm{Nu} = \frac{Q}{Q_{\text{pure conduction}}}$$

the present result can be rewritten as

$$\text{Nu} = \frac{Q}{kH\Delta T/L} = 0.364\frac{L}{H}\text{Ra}_H^{1/4} \tag{40}$$

The analytical heat transfer result developed above was improved and extended in more recent studies. As a substitute for Gill's choice of impermeable wall conditions at $y_* = \pm 1/2$, Bejan [14] proposed the condition of zero net energy flow (by convection and conduction) through the top and bottom walls,

$$Q_y = \int_0^L \left(\rho c_p vT - k\frac{\partial T}{\partial y}\right) dx = 0, \quad \text{at } y = \pm H/2 \tag{41}$$

This statement takes into account, in an integral sense, the conditions of both impermeable *and* adiabatic horizontal walls. The heat transfer result based on this approach is shown in Fig. 5.6: the numerical coefficient in the Nu $\sim$ $(L/H)\text{Ra}_H^{1/4}$ relation (40) is actually a function of the new group $\text{Ra}_H^{1/7}(H/L)^{4/7}$. Equation (40) emerges as a limiting result, valid for the boundary layer regime in the high-Ra_H limit. The general result of Fig. 5.6 shows that the $\text{Nu}(H/L, \text{Ra}_H)$ relationship is more complicated than the one suggested by eq. (40). As is shown by the solid curves of Figs. 5.7 and 5.8, this relationship cannot be expressed as a power law of the type Nu $=$ $a(L/H)^b\text{Ra}_H^c$, with constant a, b, and c (note that, in general $c > 1/4$ and $b < 1$). These observations are consistent with the conclusion of numerous attempts to correlate existing experimental and heat transfer results [1, 15].

We end this section with a survey of the overall Nu correlations often quoted in the literature. Figures 5.7 and 5.8 show a selection of Nusselt number correlations *vis-à-vis* the theoretical result of Fig. 5.6. The experimental correlations displayed on Fig. 5.7 were developed by Eckert and Carlson

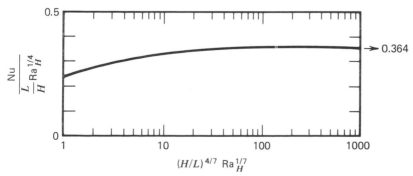

Figure 5.6 Nusselt number for the boundary layer regime in an enclosure heated from the side [14].

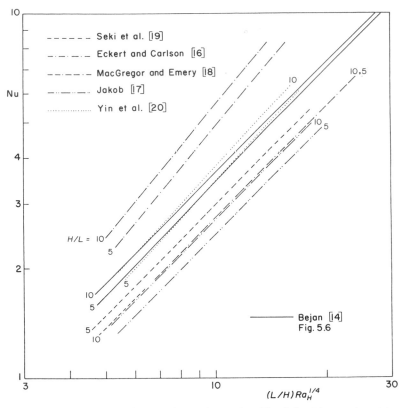

Figure 5.7 Comparison of the theoretical Nusselt number (Fig. 5.6) with experimental correlations.

[16], Jakob [17], MacGregor and Emery [18], Seki, Fukusako and Inaba [19], and Yin, Wung, and Chen [20]. Correlations based on simulations of the natural convection heat transfer problem (Fig. 5.8) were reported by De Vahl Davis (see Landis and Rubel [21]), Newell and Schmidt [22], and Pepper and Harris [23]. As should be expected, the numerical correlations of Fig. 5.8 are in superior mutual agreement compared with the experimental results in Fig. 5.7. In either case, the analytical results for the overall Nusselt number (the solid lines) split the field covered by these correlations right through the middle.

The agreement between theory and correlations based on numerical correlations is excellent, particularly near $(L/H)\mathrm{Ra}^{1/4} \sim 10$, which is the range where the boundary layer model $(L/\delta_{T,f} > 1)$ is an acceptable approximation. Below this range the heat transfer mechanism is slowly replaced by direct conduction in the horizontal direction (see Fig. 5.3). Above this range, the boundary layer picture becomes considerably more complicated due to the transition to turbulent flow (see Fig. 6.7 and the discussion concluding Chapter 6). The theoretical Nusselt number of Fig. 5.6 and eq. (40) can be used with the same,

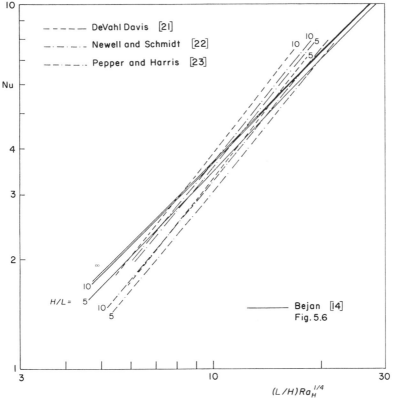

Figure 5.8 Comparison of the theoretical Nusselt number (Fig. 5.6) with numerical correlations.

if not higher degree of confidence than the heat-transfer correlations available today.

In this section we focused on the boundary layer regime in enclosures filled with Pr > 1 fluids such as water, oils, and, as a limiting case, gases (air). The Oseen-linearized solution based on the zero energy flow condition (41) [14] was most recently extended to Pr < 1 fluids by Graebel [24]. This extension was not compared with numerical or experimental results, which were unavailable; however, judging from the success of the Pr > 1 analysis, Graebel's results can be used with confidence. A sample of numerical heat transfer results in the Pr < 1 range was just published by Shiralkar and Tien [25].

Finally, the problem considered originally by Gill [10] (i.e., Pr > 1 fluids, Ra_H > 1 and impermeable tops and bottom walls) was solved by Blythe and Simpkins [26] using an integral method instead of the Oseen-linearization approach. Like Gill, Blythe and Simpkins did not report the overall Nusselt number predicted by their method of solution. However, the *local* heat flux reported by Blythe and Simpkins [26] is very similar to the local heat flux calculated by Gill [10], which inspires confidence in both analyses.

THE SHALLOW ENCLOSURE LIMIT

Another convection-dominated regime that has numerous engineering and geophysical applications is regime (IV): according to Figs. 5.3 and 5.4, if H/L decreases and Ra_H and Pr remain fixed, the cavity becomes dominated by a horizontal counterflow in which the two branches are in very good thermal contact. Of engineering interest is the end-to-end insulation effect produced by the horizontal counterflow sandwiched by the two adiabatic walls of the enclosure. For this reason, we focus on the core region—that is, the region sufficiently far from both vertical walls—where the proper scales for x and y are L and H, respectively (see Fig. 5.9).

In the shallow enclosure limit $H/L \to 0$, the scales of the terms that dominate the steady state mass, energy, and momentum conservation statements [eqs. (1), (4), and (7)] are

$$\frac{u}{L} \sim \frac{v}{H} \tag{42}$$

$$\underbrace{u\frac{\Delta T}{L}}_{\text{Convection}} \quad \sim \quad \underbrace{\alpha\frac{\Delta T}{H^2}}_{\substack{\text{Vertical} \\ \text{conduction}}} \tag{43}$$

$$\underbrace{\frac{u^2}{HL}}_{\text{Inertia}} \text{ or } \underbrace{v\frac{u}{H^3}}_{\text{Friction}} \sim \underbrace{g\beta\frac{\Delta T}{L}}_{\text{Buoyancy}} \tag{44}$$

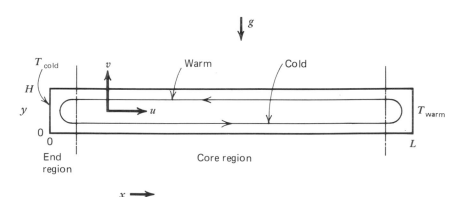

Figure 5.9 The core and end regions of a shallow enclosure heated in the horizontal direction.

This display is instructive, because it unmasks the potential for erroneously evaluating the scales of u and v: we have three apparent balances [eqs. (42)–(44)] for determining only two unknowns, u and v. The correct way to proceed is to recognize that physically the velocity scale is the result of a *force* balance. This means that the u scale follows from eq. (44), not from eq. (43). Now, in the momentum equation (44) we distinguish two possibilities:

1. Friction ~ buoyancy.
2. Inertia ~ buoyancy.

In both cases, the buoyancy effect is present because it is the driving effect (without it, i.e., when $g = 0$, we have no reason to expect a flow). *Assuming the first possibility, friction ~ buoyancy,*[†] we find

$$u \sim \frac{g\beta H^3 \Delta T}{\nu L} \tag{45}$$

and, from eq. (42),

$$v \sim \frac{g\beta H^4 \Delta T}{\nu L^2} \tag{46}$$

Using these scales, the relative order of magnitude of the terms appearing in the energy and momentum equations is

$$\underbrace{\left(\frac{H}{L}\right)^2 \mathrm{Ra}_H}_{\substack{\text{Convection} \\ (\to 0)}} \sim \underbrace{1}_{\substack{\text{Vertical} \\ \text{conduction}}} \tag{43'}$$

$$\underbrace{\left(\frac{H}{L}\right)^2 \frac{\mathrm{Ra}_H}{\mathrm{Pr}}}_{\substack{\text{Inertia} \\ (\to 0)}}, \quad \underbrace{1}_{\text{Friction}} \sim \underbrace{1}_{\text{Buoyancy}} \tag{44'}$$

Examining eq. (44') we conclude that the assumed balance between friction and buoyancy is the correct choice for the flow regime of interest ($H/L \to 0$). Second, expression (43') states that in this regime the heat transfer by thermal diffusion in the vertical direction is far greater than the enthalpy flow in the horizontal (end-to-end) direction. Indeed, the sharp imbalance revealed by expression (43') is the meaning of "good thermal contact in the vertical direction," the distinguishing feature of regime (IV) (Fig. 5.3).

[†] The second choice, inertia ~ buoyancy, is incompatible with the $H/L \to 0$ limit (see Problem 5).

Scales (45) and (46) recommend the following dimensionless variables for the core region

$$u_c = \frac{u}{g\beta H^3 \Delta T/(\nu L)}, \qquad v_c = \frac{v}{g\beta H^4 \Delta T/(\nu L^2)}$$

$$x_c = \frac{x}{L}, \qquad y_c = \frac{y}{H} \tag{47}$$

$$T_c = \frac{T - T_{cold}}{\Delta T}$$

where T_{cold} denotes the cold end temperature and $\Delta T = T_{warm} - T_{cold}$ is the overall, end-to-end temperature difference. Expressions (43') and (44') indicate that in the $H/L \to 0$ limit the governing equations reduce to the following:

Mass: $$\frac{\partial u_c}{\partial x_c} + \frac{\partial v_c}{\partial y_c} = 0 \tag{48}$$

Momentum:

$$\varepsilon \frac{Ra_H}{Pr} \left[\varepsilon \frac{\partial}{\partial x_c} \left(u_c \frac{\partial v_c}{\partial x_c} + v_c \frac{\partial v_c}{\partial y_c} \right) - \frac{\partial}{\partial y_c} \left(u_c \frac{\partial u_c}{\partial x_c} + v_c \frac{\partial u_c}{\partial y_c} \right) \right]$$

$$= \varepsilon \frac{\partial}{\partial x_c} \left(\varepsilon \frac{\partial^2 v_c}{\partial x_c^2} + \frac{\partial^2 v_c}{\partial y_c^2} \right) - \frac{\partial}{\partial y_c} \left(\varepsilon \frac{\partial^2 u_c}{\partial x_c^2} + \frac{\partial^2 u_c}{\partial y_c^2} \right) + \frac{\partial T_c}{\partial x_c} \tag{49}$$

Energy: $$\varepsilon Ra_H \left(u_c \frac{\partial T_c}{\partial x_c} + v_c \frac{\partial T_c}{\partial y_c} \right) = \varepsilon \frac{\partial^2 T_c}{\partial x_c^2} + \frac{\partial^2 T_c}{\partial y_c^2} \tag{50}$$

where $\varepsilon = (H/L)^2$ is a number considerably smaller than unity. The smallness of ε justifies the pursuit of solutions of the type [27]

$$u_c, v_c, T_c = \underbrace{(u_c, v_c, T_c)_0}_{\sim 1} + \underbrace{\varepsilon(u_c, v_c, T_c)_1}_{\sim \varepsilon} + \underbrace{\varepsilon^2 (u_c, v_c, T_c)_2}_{\sim \varepsilon^2} + \cdots \tag{51}$$

The solution for u_c, v_c, and T_c is developed systematically by substituting the series expansions (51) into the three governing equations (48)–(50). Next, the terms multiplied by the same power of ε are grouped together and, as a group, set equal to zero. Corresponding to each power of ε, say, ε^k, we must solve a system of three equations, subject to solid adiabatic wall conditions, yielding as a solution $(u_c, v_c, T_c)_k$. In order for this procedure to work, we must start with the $k = 0$ solution and *sequentially* work our way down the right-hand side of the series expansion (51). This asymptotic expansion procedure is described in some detail in Ref. 27 which shows that, except for $(u_c, v_c, T_c)_0$, all the functions $(u_c, v_c, T_c)_k$ have the same analytical form, regardless of order k. The core solution emerges as

$$\bar{u}_c(y_c) = K_1\left(\frac{y_c^3}{6} - \frac{y_c^2}{4} + \frac{y_c}{12}\right), \qquad v_c = 0 \tag{52}$$

$$T_c(x_c, y_c) = K_1 x_c + K_2 + K_1^2\left(\frac{H}{L}\right)^2 \text{Ra}_H\left(\frac{y_c^5}{120} - \frac{y_c^4}{48} + \frac{y_c^3}{72}\right) \tag{53}$$

Parameters K_1 and K_2 must be determined from end conditions in the x direction, in order to account for the flow and temperature patterns prevailing in the two end regions (Fig. 5.9). The velocity and temperature profiles across the core region are shown plotted in Fig. 5.10: the core flow consists of a thermally stratified counterflow whose velocity profile and degree of thermal stratification are independent of longitudinal position x.

The net heat transfer rate from T_{warm} to T_{cold} (Fig. 5.9) follows from the energy flux integral at any x across the core counterflow

$$Q = \int_0^H \left(k\frac{\partial T}{\partial x} - \rho c_p u T\right) dy \tag{54}$$

Combining this statement with the core solution [Eqs. (52) and (53)] and using the conduction-referenced Nusselt number definition (40), we obtain the general result [28]

$$\text{Nu} = K_1 + \frac{K_1^3}{362880}\left(\frac{H}{L}\text{Ra}_H\right)^2 \tag{55}$$

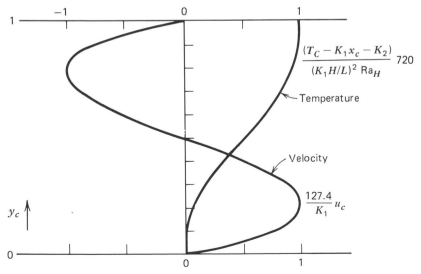

Figure 5.10 Velocity and temperature profiles in the core region of a shallow enclosure [28].

We find that the ability to predict the heat transfer rate is intimately connected to the task of determining the core axial gradient K_1. There are at least two ways of determining K_1. The first involves a formal procedure of matching the core flow to numerically derived solutions for the end regions [27]. A much more direct approach is to reason that as $(H/L)^2 \mathrm{Ra}_H$ decreases the core temperature distribution (53) becomes independent of y_c; hence, the core temperature must decrease linearly between the extreme ends of the cavity [28]. Writing

$$T_c = 0 \text{ at } x_c = 0 \quad \text{and} \quad T_c = 1 \text{ at } x_c = 1 \tag{56}$$

in the limit $(H/L)^2 \mathrm{Ra}_H \to 0$ we find

$$K_1 = 1 \quad \text{and} \quad K_2 = 0 \tag{57}$$

Therefore

$$\mathrm{Nu} = 1 + \frac{1}{362880}\left(\frac{H}{L}\mathrm{Ra}_H\right)^2, \quad \text{as} \left(\frac{H}{L}\right)^2 \mathrm{Ra}_H \to 0 \tag{55'}$$

Comparing this expression with the corresponding result obtained via the first method [27], we find that the coefficient $1/362880 = 2.76 \times 10^{-6}$ corrects the approximate coefficient 2.86×10^{-6} found numerically in Ref. 27.

The asymptotic heat transfer result (55′) is of limited applicability; the real challenge lies in evaluating parameters $K_{1,2}$ for the general Nu expression (55) when $(H/L)^2 \mathrm{Ra}_H$ is finite, that is, when the longitudinal temperature drop across the core region is *less* than the overall end-to-end temperature difference $T_{\mathrm{warm}} - T_{\mathrm{cold}}$. The general situation is shown schematically in Fig. 5.11 where,

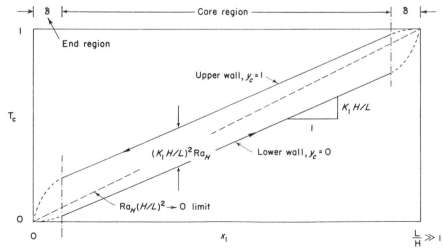

Figure 5.11 Temperature distribution along the top and bottom walls of a shallow enclosure [28].

in order to make visible the end regions, the horizontal coordinate is redefined as

$$x_1 = \frac{x}{H} \tag{58}$$

Figure 5.11 shows that as $(H/L)^2 \mathrm{Ra}_H$ increases, the thermal stratification of the core must be taken into account. One way of treating this general case analytically is to match the core solution [Eqs. (52) and (53)] to integral solutions for flow and temperature in the end regions [28]. This analytical technique was originally developed by Bejan and Tien [29] for calculating the heat transfer by natural convection in a horizontal slot filled with a fluid-saturated porous medium (see Chapter 11).

We define the *end region* as that portion of the horizontal enclosure in which the core solution [Eqs. (52) and (53)] breaks down. Inside the end region, $0 < x_1 < \delta$, the flow is turned around and cooled as it comes in contact with the vertical wall at $x_1 = 0$. We seek two equations for the unknown parameters K_1 and K_2. The first equation follows from integrating the steady state energy equation (4) twice, the first time from $y_c = 0$ to $y_c = 1$ and the second time from $x_1 = 0$ to $x_1 = \delta$. To obtain the second equation we integrate the momentum equation (7) twice, like the energy equation. We obtain

$$\int_0^1 \left.\frac{\partial T_e}{\partial x_1}\right|_{x_1=0} dy_c = \frac{H}{L} K_1 - \int_0^1 |u_e T_e|_{x_1=\delta} dy_c \tag{59}$$

$$\left| \int_0^1 T_e \, dy_c \right|_{x_1=0}^{x_1=\delta} = \int_0^\delta \left.\frac{\partial^2 u_e}{\partial y_c^2}\right|_{y_c=0}^{y_c=1} dx_c - \left.\frac{d^2}{dx_1^2}\int_0^1 v_e \, dy_c\right|_{x_1=0}^{x_1=\delta} \tag{60}$$

The next step is the selection of reasonable profiles for the velocity and temperature distributions inside the end region (these functions are denoted by subscript e). The general rule in performing this operation is to select profiles that satisfy the boundary conditions along the solid walls ($y_c = 0, 1$ and $x_1 = 0$) and match the value and slope of the core profiles at $x_1 = \delta$, as is shown on Fig. 5.12. Thus, using

$$u_e = \frac{H}{L} K_1 \left(\frac{x_1}{\delta}\right)^2 \left[6 - 8\frac{x_1}{\delta} + 3\left(\frac{x_1}{\delta}\right)^2\right] \left(\frac{y_c^3}{6} - \frac{y_c^2}{4} + \frac{y_c}{12}\right)$$

$$\left.\begin{aligned} v_e &= -\frac{H}{L}\frac{K_1}{\delta}\left(\frac{x_1}{\delta}\right)\left(1 - \frac{x_1}{\delta}\right)^2\left(\frac{y_c^4}{2} - y_c^3 + \frac{y_c^2}{2}\right) \end{aligned}\right\} \tag{61}$$

$$T_e = \left(T_c - \delta K_1 \frac{H}{L}\right)\left[2\frac{x_1}{\delta} - \left(\frac{x_1}{\delta}\right)^2\right] + \frac{H}{L} K_1 x_1$$

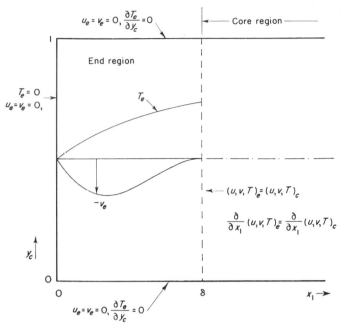

Figure 5.12 Boundary conditions and the matching conditions for the integral analysis of the end region [28].

the energy and momentum integrals (59) and (60) yield

$$\left(\frac{H}{L}K_1\right)^3 \frac{\delta \, \mathrm{Ra}_H^2}{725760} = K_2 + \left(\frac{H}{L}K_1\right)^2 \frac{\mathrm{Ra}_H}{1440} \qquad (59')$$

$$\frac{2}{5}\frac{H}{L}K_1\delta\left(\frac{1}{4\delta^4} - 1\right) = K_2 + \left(\frac{H}{L}K_1\right)^2 \frac{\mathrm{Ra}_H}{1440} \qquad (60')$$

The third equation necessary for uniquely determining K_1, K_2, and δ must come from the equivalent integral analysis of the warm end region, $(L/H - \delta)$ $< x_1 < L/H$. This procedure is equivalent to noticing that the flow in the entire cavity is symmetric about the geometric center of the cavity; the centrosymmetry condition can be expressed as

$$T_c = \tfrac{1}{2} \quad \text{at} \quad x_c = y_c = \tfrac{1}{2} \qquad (62)$$

or, substituting into the core temperature expression (53),

$$\frac{K_1}{2} + K_2 + \left(\frac{H}{L}K_1\right)^2 \frac{\mathrm{Ra}_H}{1440} = \frac{1}{2} \qquad (62')$$

Equations (59'), (60'), and (62') constitute a parametric solution for the result of interest, the function $K_1(H/L, \mathrm{Ra}_H)$. Substituting this result into the general Nu expression (55) leads to the desired relationship $Nu(H/L, Ra_H)$, which is displayed graphically in Fig. 5.13. The presentation of this relationship is made on a $Nu - (H/L)Ra_H$ field, so that the asymptotic result (55') can be plotted as a single line. Figure 5.13 also shows that the Nusselt number predicted by the integral analysis of the end regions [28] is in excellent agreement with numerical and experimental results [30, 31] for a wide variety of shallow enclosures, including the limiting square geometry ($H/L = 1$). A number of subsequent numerical and experimental studies [32–34] produced new heat transfer results that support Fig. 5.13 as an engineering tool for predicting the heat transfer rate in shallow enclosures. The usefulness of Fig. 5.13 is stressed further by the fact that the theoretical curves agree with the empirical data not only in domain (IV) (Fig. 5.3), but also well into domain (III): this would be easy to see if the (III)–(IV) frontier $Ra_H \sim (H/L)^{-4}$ was plotted on Fig. 5.13 (the frontier is not shown, because it would interfere with the reading of the Nu information). The success of the theoretical curves in domain (III) is explained by the fact that the theory takes into account the thermal resistances associated with the vertical end walls: earlier in this chapter, we concluded that regime (III) is one where the vertical boundary layers dominate the overall $T_{\text{warm}} \rightarrow T_{\text{cold}}$ thermal resistance.

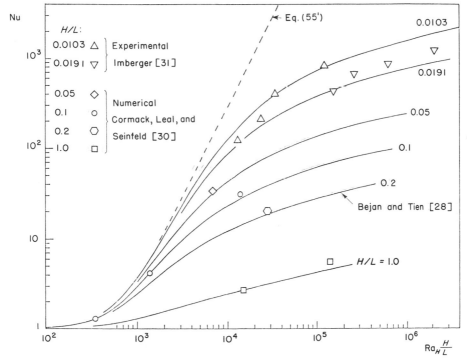

Figure 5.13 Nusselt number chart for natural convection in a shallow enclosure heated in the end-to-end direction [28].

SUMMARY OF HEAT TRANSFER RESULTS

In the preceding two sections we stressed the method behind two theoretical results designed to predict the heat transfer through a two-dimensional rectangular space heated from the side. From an engineering standpoint, perhaps the best way to use these results is to think of three classes of possible applications:

(i) **Tall enclosures ($H/L > 1$).** The Nusselt number formula recommended for this class is eq. (40) or Fig. 5.6. As is shown by the experimental and numerical data cited in Figs. 5.7 and 5.8, this calculation procedure is applicable when $(L/H)\text{Ra}_H^{1/4} \gtrsim 5$, that is, when the convective heat transfer effect is significant (Nu > 1).

(ii) **Shallow enclosures ($H/L < 1$).** For this class of applications, the theoretical curves shown in Fig. 5.13 are adequate. For cases not plotted on Fig. 5.13, the Nusselt number can be calculated using eqs. (55), (59′), (60′), and (62′).

(iii) **Square enclosures ($H/L = 1$).** Figure 5.13 showed that the integral matching analysis of Ref. 28 predicts the Nusselt number correctly, provided $(H/L)\text{Ra}_H$ does not exceed 10^5. A better way is to use the formula recommended for tall cavities [eq. (40)]; Figure 5.14 shows that eq. (40) agrees very well with the numerical and experimental data gathered by Gadgil et al. [35] from eight independent sources [36–42].

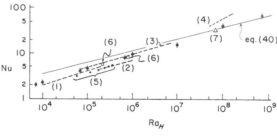

CURVE (1):   de VAHL DAVIS CALCULATIONS (Ref. 36)

CURVE (2):   FROM RESULTS OF EMERY, EXPERIMENTAL (Ref. 37)

CURVE (3):   PORTIER et al., CALCULATIONS, Pr = 0.7 (Ref. 38)

CURVE (4):   BURNAY et al., DATA REDUCED FROM EXPERIMENTS, Pr = 0.7 (Ref. 39)

RESULTS (5) ARE FROM RUBEL AND LANDIS, CALCULATIONS (Ref. 40)

RESULTS (6) ARE FROM QUON, CALCULATIONS (Ref. 41)

RESULT (7) IS FROM FROMM (Ref. 42)

POINTS INDICATED BY φ ARE FROM GADGIL et al. (Ref. 35)

Figure 5.14 Summary of experimental and numerical heat transfer results for natural convection in a square enclosure [35].

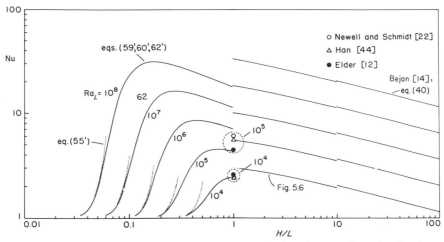

Figure 5.15 The effect of aspect ratio H/L on heat transfer through a two-dimensional enclosure heated from the side [43]; note that Ra_L is equal to $Ra_H(L/H)^3$.

Figure 5.15 presents an alternative summary of the heat transfer methodology discussed in this chapter. Plotting the conduction-referenced Nusselt number Nu versus the geometric ratio H/L, we learn that the convective heat transfer effect reaches a maximum in the vicinity of $H/L \sim 1$, that is, when the enclosure geometry does not suppress the fluid circulation. Conversely, the convective heat transfer contribution vanishes (Nu $\rightarrow$ 1) as the buoyancy-driven loop is snuffed-out (flattened) into a counterflow whose two branches are in excellent thermal contact ($H/L \rightarrow 0$ or $H/L \rightarrow \infty$).

The enclosure flows discussed so far are all laminar. As is shown in the next chapter, the transition to turbulence occurs when the wall jet Reynolds number based on the vertical velocity scale and the local thickness of the jet exceeds $O(10^2)$. The calculation of heat transfer in the turbulent regime is discussed in Chapter 7.

FLUID LAYERS HEATED FROM BELOW

In this section we focus briefly on enclosed fluid layers heated from below. Natural convection in this basic mode was identified as configuration 2 in the introduction to this chapter. The fundamental difference between configuration 1 (enclosures heated from the side) and configuration 2 (enclosures heated from below) is that in enclosures heated from the side convection is present as soon as a very small ΔT is imposed across the enclosure. By contrast, in enclosures heated from below, the imposed ΔT must exceed a critical value before the first signs of fluid motion are detected.

The onset of convection in infinitely wide horizontal fluid layers heated from below was studied for the first time analytically by Rayleigh [45] in the highly idealized case of a layer with free boundaries. Employing the arguments of hydrodynamic stability analysis (Chapter 6), Rayleigh showed that convection may occur if the dimensionless group known since the 1950s as the *Rayleigh number* exceeds a certain critical value. The subsequent study of the onset of convection in a layer sandwiched between rigid boundaries determined that cellular natural convection is possible if [46, 47]

$$\mathrm{Ra}_H > \approx 1108 \tag{63}$$

where H is the thickness of the horizontal layer. (See Fig. 11.15 in Chapter 11.) The analysis responsible for criterion (63) is not shown here, because it is conceptually the same as the stability analysis outlined in Chapter 11 for fluid-saturated fluid layers heated from below. It is worth noting, however, that in both analyses one questions the stability of a thermally stratified *motionless* layer of fluid, in other words, one investigates the stability of the pure conduction solution to the complete Navier–Stokes equations. This sort of stability analysis differs fundamentally from the hydrodynamic stability analysis of postulated laminar-like *flows* of the type discussed in connection with Fig. 6.5 in Chapter 6.

The heat transfer rate through the layer is of interest in heat transfer practice and in theory as well. Below the critical Rayleigh number of approximately 1108 [eq. (63)], the fluid is motionless and the heat flux is the one ruled by pure conduction, $k\,\Delta T/H$. Above the critical Rayleigh number, the cellular convection pattern enhances the heat flux, so that the conduction-referenced Nusselt number is appropriately described by the empirical correlation [48]

$$\mathrm{Nu} = 0.069\,\mathrm{Ra}_H^{1/3}\mathrm{Pr}^{0.074} \tag{64}$$

Note that for eq. (64) to apply, Ra_H and Pr must be such that $\mathrm{Nu} > 1$. A larger selection of high-Ra_H correlations for heat transfer in Bénard convection appears in Ref. 3.

Whether the $\mathrm{Ra}_H^{1/3}$ dependence exhibited by the measured Nusselt number is correct is currently the subject of debate [49]. In the classroom, the questions raised by eq. (64) offer a very good opportunity to achieve a theoretical understanding of the scales of cellular natural convection, an understanding of how cellular convection becomes an increasingly effective heat transfer mechanism as the temperature difference ΔT (or Ra_H) increases. To begin with, it has been observed that the Bénard cells that occupy the H-thick layer *multiply* as the Rayleigh number increases. Assuming that H is held constant, this observation means that each cell becomes taller (more slender) as Ra_H increases.

To understand why the cells must become more slender as Ra_H increases, think of the path followed by the energy carried by one cell from the bottom wall to the top wall. The energy is carried away from the bottom wall by a vertical plume moving faster as ΔT increases. Eventually, the plume hits the fluid adjacent to the top wall, in a way similar to a jet hitting the fluid held in a glass. The post-plume flow breaks up into thin enough cells; hence, into a sufficient number of pieces (new downward plumes) in order to be able to dissipate the kinetic energy gathered by plume fluid. The proliferation of Bénard-like cells is visible in the process of filling a glass with dark Bavarian beer, while pouring as closely as possible to the center of the glass (Fig. 5.16). The bubbles present in the beer are the flow visualization device: they collect along the surface only in those regions that correspond to downward flow. The photographs of Fig. 5.16 show clearly the formation of surprisingly regular cells, as each downflow zone is sandwiched by upflow zones (dark areas). After some practice, it becomes possible to vary the pouring rate and to discover that the cell number decreases as the pouring is done more gently. It seems that the flow has the natural property to select its own length scales.

With the experiment of Fig. 5.16 in mind, consider the scale analysis of a single cell of height H (fixed) and horizontal dimension L (unknown). As is shown in Fig. 11.18, which is also applicable to the present discussion, the cell is made up of two types of flows, vertical plumes of height H and thickness L, and horizontal boundary layers of length L and thickness δ_H (unknown). From the outset, we should keep in mind that the structure shown in Fig. 11.18 *is correct only in the case of* $\text{Pr} = O(1)$ *fluids*, because only for such fluids do the two transversal length scales of forced and natural boundary layers collapse into a single length scale (see Chapters 2 and 4). Assuming that each plume is vertically slender and that each wall layer is horizontally slender, the

Figure 5.16 The formation of regularly spaced cells while filling a glass with dark beer.

governing equations recommend the following scaling laws:

$$\nu\frac{v}{L^2} \sim g\beta\Delta T,$$ Plume momentum balance; friction $\sim$ (65)
buoyancy

$$\Delta P \sim \rho v^2,$$ Pressure rise on the wall region in   (66)
which the plume stagnates

$$\frac{\Delta P}{\rho L} \sim \nu\frac{u}{\delta_H^2},$$ Boundary layer momentum balance;   (67)
pressure gradient $\sim$ friction

$$u\frac{\Delta T}{L} \sim \alpha\frac{\Delta T}{\delta_H^2},$$ Boundary layer energy balance; longi-   (68)
tudinal convection $\sim$ transversal con-
duction

$$vL \sim u\delta_H,$$ Mass continuity   (69)

Equations (65)–(69) are sufficient for determining the five unknown scales $(u, v, L, \delta_H, \Delta P)$; of particular interest are the resulting scaling laws for the slenderness ratio

$$\frac{H}{\delta_H} \sim \mathrm{Gr}_H^{1/3}, \quad (\mathrm{Pr} \sim 1) \tag{70}$$

and the conduction-referenced Nusselt number

$$\mathrm{Nu} \sim \frac{k\,\Delta T/\delta_H}{k\,\Delta T/H} \sim \mathrm{Gr}_H^{1/3}, \quad (\mathrm{Pr} \sim 1) \tag{71}$$

These laws support the heat transfer measurements correlated as eq. (64): note that eqs. (64) and (71) represent the same scaling law in the case considered here (Pr $\sim$ 1).

OTHER CONFIGURATIONS

Throughout this chapter we made use of a very simple enclosure geometry—the two-dimensional rectangular box of Fig. 5.1—in order to illustrate the main properties of the phenomenon of natural convection in a confined space. We consistently modeled the differentially heated walls as isothermal, and we assumed that the walls of the enclosure are either vertical or horizontal. Furthermore, we regarded the confined space as empty, that is, free of any possible flow obstructions. From a pedagogical standpoint, all these assumptions were necessary in order to isolate the natural convection phenomenon and to illustrate the basic difference between (1) heating from the side and (2) heating from below (see the introduction to this chapter). From an engineering standpoint, however, the study of the simple model of Fig. 5.1 is only the beginning: in many engineering applications, the confined fluid departs substantially from the vertical box model of Fig. 5.1. In this last section we review four directions in which the enclosure model may be improved in order to answer specific engineering applications of the natural convection phenomenon.

Tilted Rectangular Enclosures

If we think of the design of solar collectors as a possible application of the things learned in this chapter, we realize that—as a rule—the collector enclosure is not oriented vertically. In general, the fluid space makes an angle τ with the horizontal plane (Fig. 5.17), where τ is defined such that in the range $0° < \tau < 90°$ the heated surface is positioned *below* the cooled surface. The tilted enclosure geometry has received considerable attention in the heat transfer literature, attention undoubtedly stimulated by the growing interest in solar collector technology. The basic research in this area has been reviewed by Catton [2] and, in a highly comprehensive handbook chapter by Raithby and Hollands [50].

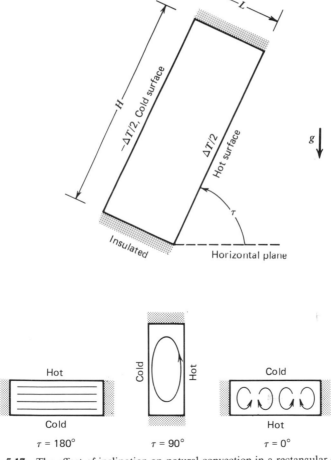

Figure 5.17   The effect of inclination on natural convection in a rectangular enclosure.

The angle of tilt τ has a dramatic impact on the flow housed by the enclosure. Figure 5.17 suggests that as τ decreases from 180° to 0°, the heat transfer mechanism switches from conduction at $\tau = 180°$ to single-cell convection at $\tau = 90°$ and to Bénard convection at $\tau = 0°$. The competition between various modes of heat transfer and the coexistence of competing modes at intermediate values of τ makes the tilted geometry both difficult analytically and challenging. For this reason, most of the heat transfer information in this area has been obtained by means of experimental measurements and numerical simulations. On the analytical side, scaling arguments have been proposed to correlate the available empirical data in order to provide engineers with easy-to-use formulas. A table (Table 5.1) of such formulas can be constructed by following Catton's recommendations [2].

In the formulas of Table 5.1, $Nu(\tau)$ is the conduction-referenced Nusselt number, in other words, the actual wall-to-wall heat transfer rate divided by the pure conduction estimate, $kH\Delta T/L$. With reference to the notation shown in Fig. 5.17, the Rayleigh number Ra_L is based on the wall-to-wall spacing L. Note also that $Nu(90°)$ and $Nu(0°)$ represent the two limiting cases highlighted in this chapter, enclosures heated from the side and enclosures heated from below.

The critical angle τ^* denotes the special position for which the convection heat transfer rate across the enclosure is minimum. It was found that if a box of fixed size (H, L) and fixed ΔT or Ra_L is rotated clockwise from $\tau = 180°$ all the way to $\tau = 0°$, the conduction-referenced Nusselt number rises from

Table 5.1 Correlations for Laminar Natural Convection Heat Transfer in Tilted Rectangular Enclosures

Tilt Angle	Nusselt Number	References
$180° > \tau > 90°$	$Nu(\tau) = 1 + [Nu(90°) - 1]\sin\tau$	51
$90° > \tau > \tau^*$	$Nu(\tau) = Nu(90°)(\sin\tau)^{1/4}$	52
$\tau^* > \tau > 0°$	$Nu(\tau) = \left[\dfrac{Nu(90°)}{Nu(0°)}(\sin\tau^*)^{1/4}\right]^{\tau/\tau_*}$	53
	Valid for $H/L < 10$	
	$Nu(\tau) = 1 + \left(1 - \dfrac{1708}{Ra_L\cos\tau}\right)^*\left[1 - \dfrac{(\sin 1.8\tau)^{1.6}1708}{Ra_L\cos\tau}\right]$	
	$+\left[\left(\dfrac{Ra_L\cos\tau}{5830}\right)^{1/3} - 1\right]^*$	53
	Valid for $H/L > 10$; the quantities denoted by [ ]* should be set to zero if they become negative.	

Table 5.2 The Critical Angle τ^* Associated with Minimum Convection Heat Transfer in a Tilted Rectangular Enclosure [54]

H/L	1	3	6	12	> 12
τ^*	25°	53°	60°	67°	70°

Nu(180°) = 1 to a maximum at $\tau \cong 90°$. As τ decreases below 90° (Fig. 5.17), the heat transfer rate first decreases and passes through a local minimum at $\tau = \tau^*$; as τ decreases below τ^*, the heat transfer rate rises toward another peak value associated with Bénard convection, Nu(0°). The critical angle τ^* was determined by Arnold et al. [54] (see Table 5.2).

Enclosures with Uniform Heat Flux from the Side

If the study of natural convection in enclosures is to enhance our understanding of how the air and energy carried by air circulate through buildings, then the isothermal walls model of Fig. 5.1 is not the best model. The temperature of the great majority of walls encountered in architectural and solar energy applications is not uniformly maintained; rather, it is the consequence of the heat flux administered to the wall. The temperature of a wall separating fluid chambers at different temperature "floats" such that the wall becomes increasingly warmer with altitude and the heat flux through the wall is essentially uniform [55–57] (see also Fig. 4.9).

It seems that a more appropriate model for the study of natural convection in buildings is the rectangular cavity with uniform heat flux along the two vertical sides (Fig. 5.18). This problem was formulated and solved just recently [58]: in the form of a relatively compact analytical solution, it was shown that the natural circulation inside a vertical enclosure with constant heat flux q'' is characterized by vertical boundary layers whose thickness does not vary with altitude (Fig. 5.18) and a motionless and thermally stratified core region. It was also shown that the vertical temperature gradient through the core is constant and equal to the temperature gradient along both vertical walls,

$$\frac{\partial T}{\partial y} = \frac{(8192)^{1/9}}{64} \frac{\alpha \nu}{g\beta H^4} \left(\frac{H}{L} \right)^{4/9} \mathrm{Ra}_{*H}^{8/9}, \quad \text{constant} \tag{72}$$

where Ra_{*H} is the Rayleigh number based on the imposed heat flux,

$$\mathrm{Ra}_{*H} = \frac{g\beta q'' H^4}{\alpha \nu k} \tag{73}$$

The overall Nusselt number or the resulting wall-to-wall temperature difference

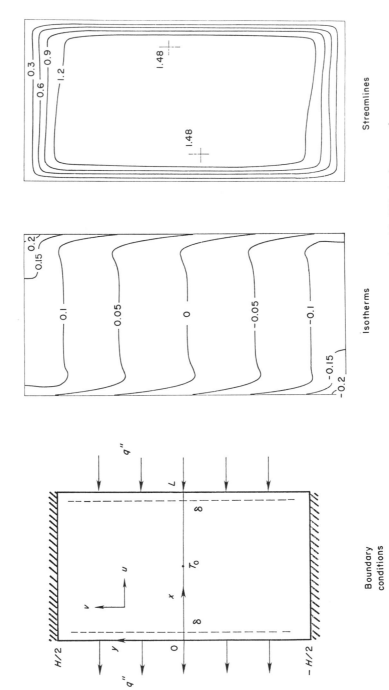

Boundary
conditions

Isotherms

Streamlines

Figure 5.18 Rectangular enclosure with uniform heat flux from the side [58]: isotherms and streamlines for $Ra_{*H} = 3.5 \times 10^6$, $Pr = 7$, $H/L = 2$.

ΔT was found to obey the relationship

$$\text{Nu} = \frac{q''}{\Delta T}\frac{H}{k} = \frac{(8192)^{1/9}}{8}\left(\frac{H}{L}\right)^{1/9}\text{Ra}_{*H}^{2/9} \tag{74}$$

Note that in this problem the wall-to-wall temperature difference ΔT is independent of y, and that along both walls the temperature increases linearly with y [see eq. (72)]. All these theoretical features were confirmed strongly by numerical simulations conducted in the range $\text{Ra}_{*H} = 3 \times 10^5 - 3 \times 10^8$, $H/L = 1-3$, $\text{Pr} = 7$ [58]. A representative set of streamlines and isotherms is reproduced in Fig. 5.18.

Partially Divided Enclosures

Real-life systems such as buildings, lakes, and solar collectors rarely conform to the single-enclosure model used throughout this chapter and in much of the natural convection literature. A very basic model for the study of natural convection in such systems is the association of two enclosures communicating laterally through a doorway, window, corridor, or over an incomplete dividing wall. The partially divided enclosure model has only recently come under scrutiny [59–62], in the form of basic experimental studies involving the two-dimensional geometries sketched in Fig. 5.19.

The new flow feature caused by the presence of a vertical obstacle inside the cavity is the *trapping* of the fluid on one side of the obstacle [59, 60]. For example, if the partial wall is mounted on the floor of the cavity, the fluid on the cold side of the obstacle becomes trapped and inactive with respect to convection heat transport. Relative to convection in a box without internal flow obstructions, where the flow fills the entire cavity, the presence of sizeable pools of inactive fluid in Fig. 5.19 has a significant effect on the overall heat transfer rate between the far ends of the cavity.

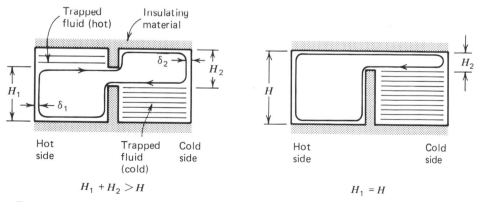

Figure 5.19 Laminar natural convection patterns and the "fluid trap" effect in enclosures communicating through an opening.

The reduction in the end-to-end heat transfer rate is documented by the experimental measurements reported in Refs. 59, 60, and 62. The same effect can also be predicted based on scale analysis, as is shown in Refs. 59 and 62. Assuming that no heat transfer takes place through the incomplete partition (obstacle), the end-to-end heat transfer is impeded by the two thermal resistances associated with the two vertical boundary layers driven by the differentially heated end walls. Let (δ_1, H_1) and (δ_2, H_2) be the length scales of the two boundary layers, as is shown in Fig. 5.19. The two heights, H_1 and H_2, are known from the geometry of the internal partition. The thermal boundary layer thickness scales are known from arguments presented earlier in this chapter and in Chapter 4 for $Pr > 1$,

$$\delta_1 \sim H_1 \mathrm{Ra}_{H_1}^{-1/4}, \qquad \delta_2 \sim H_2 \mathrm{Ra}_{H_2}^{-1/4} \tag{75}$$

where the subscripts H_1, H_2 indicate that in each case the Rayleigh number is based on the actual height of the boundary layer, not on the height of the enclosure, H. Noting that the thermal resistances of the two boundary layers scale as

$$\frac{\delta_1}{kH_1} \quad \text{and} \quad \frac{\delta_2}{kH_2} \tag{76}$$

the end-to-end heat transfer rate Q may be calculated as

$$Q = \frac{\Delta T}{C_1 \delta_1 / k H_1 + C_2 \delta_2 / k H_2} \tag{77}$$

where C_1 and C_2 are by definition numerical coefficients of order $O(1)$. This heat transfer result may be cast in dimensionless form to show specifically the effect of obstacle geometry.

$$\frac{Q}{k\,\Delta T} = \frac{\mathrm{Ra}_H^{1/4}}{C_1 (H/H_1)^{3/4} + C_2 (H/H_2)^{3/4}} \tag{78}$$

Looking at the right side of Fig. 5.19 and at eq. (78), we see that as the opening left above the partition decreases, the ratio H/H_2 becomes very large and the heat transfer rate drops sharply. Setting $C_1 = 1.5$ and $C_2 = 3$, eq. (78) correlates the heat transfer rates measured in a box with a single internal partition [62] (as on the right side of Fig. 5.19, where $H = H_1$), in the Ra_H range 10^9–10^{10}.

The *fluid trap* phenomenon created by the partitions considered above is capable of totally shutting off the natural circulation in the enclosure. Figure 5.20 shows the two ways in which one could install two incomplete internal walls whose heights add to more than H. If the bottom-floor obstacle is on the

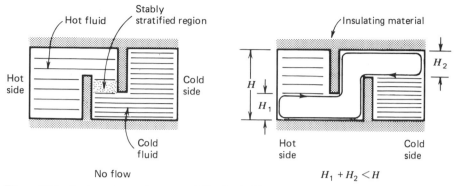

Figure 5.20 Rectangular enclosure partially divided by two incomplete vertical walls: the left drawing shows the arrangement that suppresses natural convection.

hot side of the enclosure, then the fluid sandwiched between the two obstacles is stably stratified and natural convection is prohibited. If the bottom-floor obstacle is on the cold side, Fig. 5.20 shows that natural convection is possible as a single cell strangled by the two obstacles. The enclosure geometries of Fig. 5.20 remain to be investigated experimentally and numerically; note that the heat transfer scaling law (78) does apply to the flow sketched on the right side of Fig. 5.20.

Triangular Enclosures

Another basic enclosure geometry that has important applications is the triangular enclosure with different temperatures maintained along the horizontal wall and the sloped wall (see Fig. 12.6). The interest in this geometry stems from problems involving the natural circulation in shallow waters with sloping bottoms heated by solar radiation [63], and energy conservation in solar collectors and attics [64]. Heat transfer experiments in triangular enclosures were first reported by Flack, Konopnicki, and Rooke [65] and Flack [66] who, using an enclosure filled with air, achieved Rayleigh numbers based on height in the range 7.5×10^4–10^6. This range of Rayleigh numbers characterizes solar collectors of triangular shape, however, it is too low to be relevant to building-size spaces (attics). Experiments in the Rayleigh number range 10^8–10^9 were carried out in a triangular enclosure filled with water [67]; numerical simulations for $1 < \mathrm{Ra}_H < 10^6$ and an asymptotic analytical solution valid for very shallow attic spaces are also available [68].

The natural convection picture painted by the study of triangular enclosures constitutes an interesting interaction between the two basic configurations identified in the introduction to this chapter, namely, enclosures heated from the side and enclosures heated from below. If the sloped wall is situated above the horizontal wall, and if the sloped wall is heated and the bottom wall is cooled, the downward heat transfer through the enclosure is ruled by pure conduction. If, on the other hand, the sloped wall is cooled and the bottom wall is heated, the enclosure is ruled by a single-cell flow driven by the sloped wall: the same type of flow occurs in a triangular space filled with porous material, as illustrated in Fig. 12.6. The single-cell circulation persists (in a time-averaged sense) even at high Rayleigh numbers, where the flow is turbulent [67]. The interesting aspect of this conclusion is that, while the enclosure is cooled from above *and* from the side (along the sloped wall), the observed natural circulation is of the type associated with enclosures heated from the side (i.e., single-cell). However, this conclusion is based on a limited set of observations [66–68]. It may very well be that at high enough Rayleigh numbers and in shallow enough triangular spaces cooled from above, the flow will opt for the multicellular Bénard pattern characteristic of layers heated from below. This basic aspect of triangular enclosure convection remains to be clarified in future studies.

SYMBOLS

c_P	specific heat at constant pressure
g	gravitational acceleration
H	the vertical dimension of the enclosure
k	fluid thermal conductivity
$K_{1,2}$	constants [eqs. (52, 53)]
L	the horizontal dimension of the enclosure
q	odd function [eq. (37)]
Q	heat transfer rate
Q_y	heat transfer rate through the horizontal walls [eq. (41)]
p	even function [eq. (37)]
P	pressure
Pr	Prandtl number
Ra_H	Rayleigh number based on enclosure height [eq. (15)]
Ra_{*H}	Rayleigh number based on heat flux [eq. (73)]
t	time
t_f	time of convective boundary layer development [eq. (13')]
T	temperature
T_{cold}	temperature of cold vertical wall
T_{warm}	temperature of warm vertical wall
ΔT	overall temperature difference ($T_{\text{warm}} - T_{\text{cold}}$)
T_0	reference temperature
T_A'	Oseen-linearization function [eq. (25″)]
$T_{*\infty}$	core temperature in the high Ra_H regime
u	horizontal velocity
u_A	Oseen-linearization function
$u_{*\infty}$	core velocity in the high Ra_H regime

v	vertical velocity
x	horizontal coordinate
x_1	dimensionless horizontal coordinate in the end-turn region [eq. (58)]
y	vertical coordinate
α	fluid thermal diffusivity
β	coefficient of thermal expansion
δ	dimensionless thickness of the end region in a shallow enclosure
δ_T	thermal boundary layer thickness
$\delta_{T,f}$	thermal boundary layer thickness at the end of its development [eq. (14)]
δ_v	velocity boundary layer thickness [eq. (17)]
ε	small parameter $(H/L)^2$ for shallow enclosures
$\lambda_{1,2}$	functions of altitude [eq. (31)]
ρ	fluid density
τ	angle of tilt (Fig. 5.17)
$(\ )_c$	dimensionless variables for the shallow core solution [eq. (47)]
$(\ )_e$	expressions for the integral analysis of the end region [eq. (61)]
$(\ )_*$	dimensionless variables for the high-Ra_H solution [eq. (23)]

REFERENCES

1. S. Ostrach, Natural convection in enclosures, *Adv. Heat Transfer*, Vol. 8, 1972, pp. 161–227.

2. I. Catton, Natural convection in enclosures, *6th International Heat Transfer Conference, Toronto, 1978*, Vol. 6, 1979, pp. 13–43.

3. Y. Jaluria, *Natural Convection Heat and Mass Transfer*, Pergamon, Oxford, 1980.

4. S. Chandrasekhar, *Hydrodynamic and Hydromagnetic Stability*, Clarendon Press, Oxford, 1961.

5. J. S. Turner, *Buoyancy Effects in Fluids*, Cambridge University Press, Cambridge, 1973.

6. J. Patterson and J. Imberger, Unsteady natural convection in a rectangular cavity, *J. Fluid Mech.*, Vol. 100, 1980, pp. 65–86.

7. A. Bejan, A. A. Al-Homoud, and J. Imberger, Experimental study of high Rayleigh number convection in a horizontal cavity with different end temperatures, *J. Fluid Mech.*, Vol. 109, 1981, pp. 283–299.

8. A. Bejan, A general variational principle for thermal insulation system design, *Int. J. Heat Mass Transfer*, Vol. 22, 1979, pp. 219–228.

9. G. K. Batchelor, Heat transfer by free convection across a closed cavity between vertical boundaries at different temperature, *Q. Appl. Math.*, Vol. 12, 1954, pp. 209–233.

10. A. E. Gill, The boundary layer regime for convection in a rectangular cavity, *J. Fluid Mech.*, Vol. 26, 1966, pp. 515–536.

11. G. F. Carrier, *Proceedings of the 10th Conference on Applied Mechanics*, Elsevier, Amsterdam, 1962.

12. J. W. Elder, Laminar convection in a vertical slot, *J. Fluid Mech.*, Vol. 23, 1965, pp. 77–98.

13. C. Quon, Free convection in an enclosure revisited, *J. Heat Transfer*, Vol. 99, 1977, pp. 340–342.

14. A. Bejan, Note on Gill's solution for free convection in a vertical enclosure, *J. Fluid Mech.*, Vol. 90, 1979, pp. 561–568.

15. I. Catton, Oral Discussion during the 19th National Heat Transfer Conference, Orlando, FL, July 27–30, 1980.

16. E. R. G. Eckert and W. O. Carlson, Natural convection in an air layer enclosed between two vertical plates with different temperatures, *Int. J. Heat Mass Transfer*, Vol. 2, 1961, pp. 106–120.

17. M. Jakob, *Heat Transfer*, Wiley, New York, 1957.

18. R. K. MacGregor and A. F. Emery, ASME Paper No. 68-WA HT-4, 1968.

19. N. Seki, S. Fukusako, and H. Inaba, Visual observation of natural convective flow in a narrow vertical cavity, *J. Fluid Mech.*, Vol. 84, 1978, pp. 695–704.

20. S. H. Yin, T. Y. Wung, and K. Chen, Natural convection in an air layer enclosed within rectangular cavities, *Int. J. Heat Mass Transfer*, Vol. 21, 1978, pp. 307–315.

21. F. Landis and A. Rubel, Discussion of Ref. 22, *J. Heat Transfer*, Vol. 92, 1970, pp. 167–168.

22. M. E. Newell and F. W. Schmidt, Heat transfer by natural convection within rectangular enclosures, *J. Heat Transfer*, Vol. 92, 1970, pp. 159–167.

23. D. W. Pepper and S. D. Harris, Numerical simulation of natural convection in closed containers by a fully implicit method, *J. Fluids Eng.*, Vol. 99, 1977, pp. 649–656.

24. W. P. Graebel, The influence of Prandtl number on free convection in a rectangular cavity, *Int. J. Heat Mass Transfer*, Vol. 24, 1981, pp. 125–131.

25. G. S. Shiralkar, and C. L. Tien, A numerical study of laminar natural convection in shallow cavities, *J. Heat Transfer*, Vol. 103, 1981, pp. 226–231.

26. P. A. Blythe and P. G. Simpkins, Thermal convection in a rectangular cavity, *Physicochemical Hydrodynamics*, Vol. 2, 1977, pp. 511–524.

27. D. E. Cormack, L. G. Leal, and J. Imberger, Natural convection in a shallow cavity with differentially heated end walls. Part 1: Asymptotic theory, *J. Fluid Mech.*, Vol. 65, 1974, pp. 209–229.

28. A. Bejan and C. L. Tien, Laminar natural convection heat transfer in a horizontal cavity with different end temperatures, *J. Heat Transfer*, Vol. 100, 1978, pp. 641–647.

29. A. Bejan and C. L. Tien, Natural convection in a horizontal porous medium subjected to an end-to-end temperature difference, *J. Heat Transfer*, Vol. 100, 1978, pp. 191–198.

30. D. E. Cormack, L. G. Leal, and J. H. Seinfeld, Natural convection in a shallow cavity with differentially heated end walls. Part 2: Numerical solutions, *J. Fluid Mech.*, Vol. 65, 1974, pp. 231–246.

31. J. Imberger, Natural convection in a shallow cavity with differentially heated end walls. Part 3: Experimental results, *J. Fluid Mech.*, Vol. 65, 1974, pp. 247–260.

32. R. A. Wirtz and W. F. Tseng, Natural convection across tilted, rectangular enclosures of small aspect ratio, *Natural Convection in Enclosures*, ASME publication HTD-Vol. 8, 1980, pp. 47–54.

33. E. I. Lee and V. Sernas, Numerical study of heat transfer in rectangular air enclosures of aspect ratios less than one, ASME Paper No. 80-WA/HT-43, 1980.

34. Y. Kamotani, L. W. Wang, and S. Ostrach, Experiments on Natural Convection Heat Transfer in Low Aspect Ratio Enclosures, Paper no. AIAA-81-1066, AIAA 16th Thermophysics Conference, Palo Alto, California, June 23–25, 1981.

35. A. Gadgil, F. Bauman, and R. Kammerud, Natural convection in passive solar buildings: Experiments, analysis, and results, *Passive Solar J.*, Vol. 1, 1982, pp. 28–40.

36. G. de Vahl Davis, Laminar natural convection in an enclosed rectangular cavity, *Int. J. Heat Mass Transfer*, Vol. 11, 1968, pp. 1675–1693.

37. A. F. Emery, The effect of a magnetic field upon the free convection in a conducting fluid, *Trans. ASME, J. Heat Transfer*, Vol. 85, No. 2, 1963, pp. 119–124.

38. J. J. Portier and O. A. Arnas, *Heat Transfer and Turbulent Buoyant Convection*, Vol. II, Hemisphere, Washington, D.C., 1977.

39. G. Burnay, J. Hannay, and J. Portier, *Heat Transfer and Turbulent Buoyant Convection*, Vol. II, Hemisphere, Washington, D.C., 1977.

40. A. Rubel and R. Landis, Numerical study of natural convection in a vertical rectangular enclosure, *Phys. Fluids*, Suppl. II, Vol. 12-II, 1969, pp. 208–213.

41. C. Quon, High Rayleigh number convection in an enclosure: A numerical study, *Phys. Fluids*, Vol. 15-I, 1972, pp. 12–19.

42. J. E. Fromm, A Numerical Method for Computing the Nonlinear, Time Dependent, Buoyant Circulation of Air in Rooms, *Building Science Series No. 39*, National Bureau of Standards, Washington, D.C., 1971.

43. A. Bejan, A synthesis of analytical results for natural convection heat transfer across rectangular enclosures, *Int. J. Heat Mass Transfer*, Vol. 23, 1980, pp. 723–726.

44. J. T. Han, M.A. Sc. Thesis, Department of Mechanical Engineering, University of Toronto, 1967.

45. Lord Rayleigh, On convection currents in a horizontal layer of fluid when the higher temperature is on the underside, *Philos. Mag.*, Vol. 6, No. 32, 1916, pp. 529–546.

46. H. Jeffreys, The stability of a layer of fluid heated below, *Philos. Mag.*, Vol. 2, 1926, pp. 833–844.

47. H. Jeffreys, Some cases of instability in fluid motion, *Proc. R. Soc. London Ser. A*, Vol. 118, 1928, pp. 195–208.

48. S. Globe and D. Dropkin, Natural convection heat transfer in liquids confined by two horizontal plates and heated from below, *J. Heat Transfer*, Vol. 81, 1959, pp. 24–28.

49. I. Catton and K. G. T. Hollands, Discussion during the NSF Natural Convection Workshop, Breckenridge, Colorado, July 18–21, 1982.

50. G. D. Raithby and K. G. T. Hollands, Natural convection, In W. M. Rohsenow, J. P. Hartnett, and E. Ganic, eds. *Handbook of Heat Transfer*, 2nd ed., McGraw-Hill, New York, 1984.

51. J. N. Arnold, P. N. Bonaparte, I. Catton, and D. K. Edwards, Proceedings of the 1974 Heat Transfer and Fluid Mechanics Institute, Stanford University Press, Stanford, CA, 1974.

52. P. S. Ayyaswamy and I. Catton, The boundary layer regime for natural convection in a differentially heated tilted rectangular cavity, *J. Heat Transfer*, Vol. 95, 1973, pp. 543–545.

53. K. G. T. Hollands, T. E. Unny, G. D. Raithby, and L. J. Konicek, Free convection heat transfer across inclined air layers, *J. Heat Transfer*, Vol. 98, 1976, pp. 189–193.

54. J. N. Arnold, I. Catton, and D. K. Edwards, Experimental investigation of natural convection in inclined rectangular regions of differing aspect ratios, *J. Heat Transfer*, Vol. 98, 1976, pp. 67–71.

55. E. M. Sparrow and C. Prakash, Interaction between internal natural convection in an enclosure and an external natural convection boundary layer flow, *Int. J. Heat Mass Transfer*, Vol. 24, 1981, pp. 895–907.

56. R. Viskanta and D. W. Lankford, Coupling of heat transfer between two natural convection systems separated by a wall, *Int. J. Heat Mass Transfer*, Vol. 24, 1981, pp. 1171–1177.

57. R. Anderson and A. Bejan, Natural convection on both sides of a vertical wall separating fluids at different temperatures, *J. Heat Transfer*, Vol. 102, 1980, pp. 630–635.

58. S. Kimura and A. Bejan, The boundary layer natural convection regime in a rectangular cavity with uniform heat flux from the side, *J. Heat Transfer*, Vol. 106, 1984, pp. 98–103.

59. A. Bejan and A. N. Rossie, Natural convection in horizontal duct connecting two fluid reservoirs, *J. Heat Transfer*, Vol. 103, 1981, pp. 108–113.

60. M. W. Nansteel and R. Greif, Natural convection in undivided and partially divided rectangular enclosures, *J. Heat Transfer*, Vol. 103, 1981, pp. 623–629.

61. S. M. Bajorek and J. R. Lloyd, Experimental investigation of natural convection in partitioned enclosures, *J. Heat Transfer*, Vol. 104, 1982, pp. 527–532.

62. N. N. Lin and A. Bejan, Natural convection in a partially divided enclosure, *Int. J. Heat Mass Transfer*, Vol. 26, 1983, pp. 1867–1878.

63. H. B. Fischer, E. J. List, R. C. Y. Koh, J. Imberger, and N. H. Brooks, *Mixing in Inland and Coastal Waters*, Academic Press, New York, 1979.

64. D. Poulikakos, Natural convection in a triangular enclosure filled with newtonian fluid or fluid-saturated porous medium, Ph.D. thesis, Mechanical Engineering Department, University of Colorado, Boulder, May 1983.

65. R. D. Flack, T. T. Konopnicki, and J. H. Rooke, The measurement of natural convective heat transfer in triangular enclosures, *J. Heat Transfer*, Vol. 101, 1979, pp. 648–654.

66. R. D. Flack, The experimental measurement of natural convection heat transfer in triangular enclosures heated or cooled from below, *J. Heat Transfer*, Vol. 102, 1980, pp. 770–772.

67. D. Poulikakos and A. Bejan, Natural convection experiments in a triangular enclosure, *J. Heat Transfer*, Vol. 105, 1983, pp. 652–655.

68. D. Poulikakos and A. Bejan, The fluid dynamics of an attic space, *J. Fluid Mech.*, Vol. 131, 1983, pp. 251–269.

69. A. Bejan and S. Kimura, Penetration of free convection into a lateral cavity, *J. Fluid Mech.*, Vol. 103, 1981, pp. 465–478.

70. M. J. Lighthill, Theoretical considerations on free convection in tubes, *Q. J. Appl. Math.*, Vol. 6, 1953, pp. 398–439.

PROBLEMS

1. Employing the proper scales for the development of a thermal boundary layer along the heated vertical walls of a rectangular enclosure, show that the momentum equation (7) is dominated by the three terms whose scales are listed as expression (9).

2. Derive the criterion for distinct vertical thermal boundary layers [expression (18)] by comparing the time of convective layer development (t_f) with the time of penetration by pure conduction over the entire length of the enclosure L.

3. Rely on pure scaling arguments to prove that $Ra_H < 1$ denotes the domain in which the overall heat transfer rate across a square enclosure is dominated by pure conduction.

4. Derive an expression for the horizontal velocity u_* in the vertical boundary layer in the high-Ra_H regime. Recognizing $u_{*\infty} = \lim_{x_* \to \infty} u_*$ as the horizontal velocity through the core region, show that $u_{*\infty}$ is an odd function of altitude. What is the scale of the horizontal velocity in the core region?

5. Prove that (inertia) ~ (buoyancy) in eq. (44) is inadmissible as a momentum balance in the shallow enclosure limit $(H/L \to 0)$. In view of this finding, is the stratified counterflow core solution (52) and (53) valid for all values of Pr? Explain.

6. Machined into a solid wall of temperature T is a slender two-dimensional cavity of height H and length L ($H \ll L$). The cavity communicates laterally with an infinitely large fluid reservoir of temperature $T + \Delta T$. The situation is shown schematically on the left side of Fig. 11.12. Show that if the cavity is slender enough, the buoyancy-driven flow will penetrate into the cavity only to a certain depth whose order of magnitude is

$$L_x \sim H \, \mathrm{Ra}_H^{1/2}$$

Determine also the order of magnitude of the total heat transfer rate between the fluid reservoir and the walls of the cavity. Compare the results of your scale analysis with the similarity solution to the same problem given in Ref. 69.

7. Consider the penetration of natural convection into a vertical slender cavity of height H and gap-thickness L ($H \gg L$). As is shown on the right side of Fig. 11.12, one end of the cavity is closed and the other communicates with a very large fluid reservoir. Rely on scaling arguments to show that when a temperature difference ΔT is established between the fluid reservoir and the walls of the cavity, the flow penetrates vertically to a depth that scales as

$$L_y \sim L \, \mathrm{Ra}_L$$

Estimate also the order of magnitude of the heat transfer rate between the fluid reservoir and the walls of the cavity, assuming that $L_y < H$. Compare your scaling conclusions with Lighthill's integral solution to the same problem [70].

6

Transition to Turbulence

The flows treated in the preceding four chapters are all laminar and, as such, are destined to exist only under special circumstances. It is common knowledge —a fact reinforced daily by direct observations—that laminar flows can come undone and break down to a seemingly more complicated state of affairs called *turbulence*. Yet, there is nothing in the laminar flow solutions of Chapters 2–5 to suggest that these laminar flows do not exist for all Reynolds and Rayleigh numbers imaginable. The special circumstances necessary for the transition from analytically predictable laminar flows to, as is shown in Chapter 7, analytically unpredictable turbulent flows form one of the most active subfields in fluid mechanics research today. From the point of view of convective heat transfer research, however, the study of *transition* is important because the heat transfer potential of a turbulent flow differs vastly from its laminar counterpart.

The practical implications of transition to heat transfer engineering, particularly in the area of design, are to blame for the manner in which this important topic has been treated in the heat transfer literature. If the bulk of presently available heat transfer textbooks is an indication of this treatment, then it is fair to say that the subject of transition in heat transfer is nothing but a disjointed collection and repetition of hearsay on the subject, a handbook-style memorization of what it takes for a laminar flow to break down. Little time is being spent on explaining and, possibly, predicting transition.

In the present chapter we review the available *bagage* of empirical information in the very beginning, via Table 6.1. There are important reasons behind such impatience. First, there is little interesting in repeating after so many others that particular laminar flows break down at certain (critical) Reynolds numbers and Grashof numbers when, as demonstrated in the earlier chapters, the conceptual basis for even speaking about numbers such as Re and Gr is not at all clear. That laminar flows break down according to Table 6.1 is of engineering interest, of course (this is why the table is shown). But, an important story lies beneath the kind of physical observations we usually

Table 6.1 Summary of Physical Observations Concerning the Beginning of Transition from Laminar to Turbulent Flow

Flow Configuration	Condition[a] Necessary for the Existence of Laminar Flow	Source	Observations
Boundary layer flow (without longitudinal pressure gradient)	$Re < 3.5 \times 10^5$ $Re < 2 \times 10^4 - 10^6$	[1] [2,3][c]	Re is the Reynolds number based on wall length and free stream velocity
Duct flow	$Re < 2000$	[1]	Re is based on hydraulic diameter and duct-averaged velocity
Free-jet flow (axisymmetric)	$Re < 10-30$	[4]	Re is based on nozzle diameter and mean velocity through the nozzle
Wake flow (two-dimensional)	$Re < 32$	[5]	Re is based on the cylinder diameter and free-stream velocity
Natural convection boundary layer flow Isothermal wall	$Gr < 1.5 \times 10^9 \ (Pr = 0.71)$ $Gr < 1.3 \times 10^9 \ (Pr = 6.7)$	[6][b] [6,7]	The Grashof number Gr is based on wall height and wall-ambient temperature difference
Constant heat-flux wall	$Gr_* < 1.6 \times 10^{10} \ (Pr = 0.71)$ $Gr_* < 6.6 \times 10^{10} \ (Pr = 6.7)$	[6] [8]	Parameter Gr_* is the Grashof number based on heat flux and wall height. (See eq. (70) in Chapter 4, where $Gr_* = Ra_*/Pr$.)
Plume flow (axisymmetric)	$Ra_q < 10^{10} \ (Pr = 0.71)$	[9]	Ra_q is the Rayleigh number based on heat source strength and plume height [see eq. (6)].
Film condensation on a vertical plate	$\dfrac{4\Gamma}{\mu} < 1800$	[10]	Γ is the condensate mass flowrate per unit film width

[a]All numerical values are order-of-magnitude approximate and vary from one experimental report to another.

[b]Averaged from the data compiled in Ref. 6.

[c]The transition is triggered by disturbances in excess of 18 percent of the free-stream velocity.

associate with transition. The goal of this chapter is to stimulate the reader to search for the meaning behind the numbers of Table 6.1. I believe that only *after* such thinking one is equipped to tackle the more ambitious topics dealing with turbulent heat transfer.

THE SCALING LAWS OF TRANSITION

Nature offers us clear signs that the phenomenon of transition is associated with a fundamental property of fluid flow that is only now beginning to attract attention in the literature. Table 6.1 shows that any special class of laminar flows is characterized by a *critical number* that serves as a landmark for the laminar–turbulent transition. One century of research on transition has shown that these critical numbers are *universal*, in spite of the sometimes sizeable numerical variations from one experimental reporting of transition observations to another. The concept of critical number of transition is empirical in origin: indeed, the real challenge for the researcher is to predict these numbers and in this way to show the universality of the transition phenomenon, that is, the common origin of all the numbers collected in Table 6.1.

Another important clue emphasized in some textbooks (e.g., [11]) is that although the critical transition numbers differ in orders of magnitude from one flow class to another, they all seem to suggest that an appropriate Reynolds number based on the relevant velocity and *transversal* dimension of the flow has in all cases the same order of magnitude, $O(10^2)$. This observation is particularly true not only for forced boundary layer flow, as is shown in Ref. 11, but for wall jet flows and free-jet flows encountered in natural convection phenomena, as well as for jet and wake flows (Table 6.1). It is important to keep in mind that this seemingly universal transition Reynolds number is a number *considerably greater than O(1)*.

To study a phenomenon as common and well-traveled as transition, it is very tempting to automatically turn to the large body of information already published on the subject. A few very good overview articles exist and are highly recommended, particularly Granger [3] and Reshotko [12] for boundary layer flows and Gebhart and Mahajan [13] and Mahajan and Gebhart [6] for vertical natural convection flow. In parallel with such reading, however, it is essential to keep an inquisitive eye on the natural transition phenomena that surround us and to take another look at what our teachers would want us to forever regard as well-documented phenomena. In the case of transition, the danger associated with not looking first at Nature is that the century-old body of experimental observations accumulated in the literature is in fact heavily biased towards the theory that defined and triggered these experiments. That theory—the theory of hydrodynamic stability—is responsible for even the words used by our elders to report their observations. Now, inasmuch as the classical hydrodynamic stability arguments alone have not been entirely successful in accounting for transition, it is essential to recognize that the transition phenomenon is in fact begging for attention.

We can identify the most important features of the laminar–turbulent transition by taking a close look at one of the most common occurrences of the phenomenon, namely, the cigarette-smoke plume shown in Fig. 6.1. This photograph was taken by S. Kimura during a controlled reproduction of the phenomenon, one where the air plume was generated by a concentrated heat source of known strength $q[W]$ and where the smoke was introduced separately only in order to visualize the flow [9]. The laminar plume flow prevails below a certain characteristic plume height y_{tr}. The transition is also marked by the meandering or *buckling* of the plume stream into a sinusoidal shape of characteristic wavelength λ_B. One can take many photographs of the type

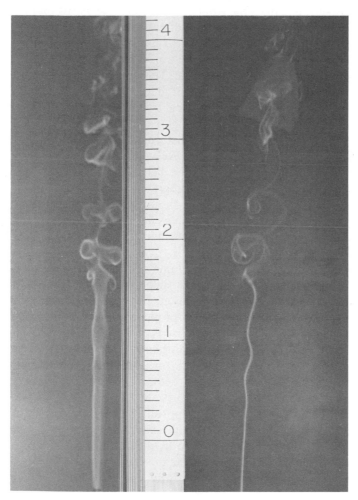

Figure 6.1 Smoke visualization of transition in air plume flow above a concentrated heat source [9]; right side = direct view; left side = side view ($q = 5.1$ W; one division on the vertical scale equals 1 cm).

shown in Fig. 6.1; one can cough or not cough while taking each picture, but the buckling or meandering wavelength λ_B always turns out to be proportional to the transition height y_{tr} [9]. By varying the heat source strength q one has the opportunity to observe more than just one cigarette-smoke plume; as summarized in Fig. 6.2a, although both λ_B and y_{tr} decrease as q increases, the $\lambda_B \sim y_{tr}$ proportionality is preserved. According to Fig. 6.2a, the line

$$y_{tr} \sim 10\lambda_B \tag{1}$$

is an order-of-magnitude curvefit for the observations reported in Ref. 9.

Another interesting aspect of these observations is that the buckled shape of the transition section of the plume is in one plane. By photographing the plume simultaneously from the front and the side, one learns that the meander is most visible from the special viewing direction that happens to be perpendicular to the plane of the meander. This observation is important because it contradicts some people's belief that the transitional shape of the buoyant jet is spiral (helical); hence, three-dimensional. This belief is a good example of how an existing theory influences the end-product (the written record) of experimental observations: in the first analytical treatment of the dynamic instability of an inviscid axisymmetric jet, Batchelor and Gill [14] *postulated* the existence of helical, not plane-sinusoidal disturbances. This postulate was adopted by subsequent theoretical studies (e.g., Refs. 15 and 16) and, soon enough, it became possible to talk about observed helical and corkscrew deformations based on a purely two-dimensional photographic record (e.g., Refs. 17 and 18). Of course, helical disturbances are as possible as plane–sinusoidal disturbances; however, a special effort such as double-angle photography must be made before one can ascertain the three-dimensionality of the flow deformation during *natural* transition.

In the case of the cigarette-smoke plume of Fig. 6.1, the disturbances were unknown (random), yet, the observed shape was plane–sinusoidal and the wavelength was basically the same from one photographed instance to another. It seems that the flow has the natural property to, as Gebhart and Mahajan [13] put it, "sharply filter disturbances for essentially a single frequency" out of a whole spectrum of unspecified disturbances. Put another way, the flow appears to have the natural property to meander with a characteristic wavelength during transition, *regardless* of the nature of the disturbing agent. This observation is important because it illustrates the conflict between hydrodynamic stability thinking, to which the postulate of *disturbances* is a necessity, and the natural meandering[†] tendency of real-life flows during transition.

The characteristic wavelength chosen by the flow during transition is in fact proportional to the local thickness of the flow (the stream). For the air plumes documented in Fig. 6.2a, we know from the scale analysis presented in

---

[†] In the present context, *meandering* means a naturally sinusoidal flow with a universal proportionality (scaling) between longitudinal wavelength and stream thickness.

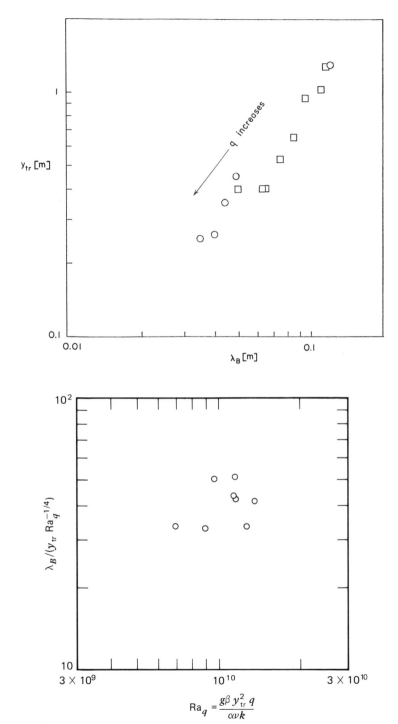

Figure 6.2 Observations on transition in air plume flow; (*a*) transition height versus buckling wavelength [9]; (*b*) the constancy of the transition Rayleigh number and the local proportionality between buckling wavelength and stream thickness scale.

Chapter 4 that (see Table 4.1, and set Pr ~ 1)

$$D \sim y_{tr} \left(\frac{g\beta\Delta T y_{tr}^3}{\alpha\nu} \right)^{-1/4} \tag{2}$$

where D is the transversal length scale of the plume and ΔT is the plume-ambient temperature difference scale. We also know the vertical velocity scale

$$v \sim \frac{\alpha}{y_{tr}} \left(\frac{g\beta\Delta T y_{tr}^3}{\alpha\nu} \right)^{1/2} \tag{3}$$

and, from the argument that the plume carries all the energy released by the heat source

$$q \sim \rho c_p D^2 v \Delta T \tag{4}$$

Combining eqs. (2)–(4) to eliminate v and ΔT, we find

$$D \sim y_{tr} \text{Ra}_q^{-1/4} \tag{5}$$

where Ra_q is the Rayleigh number based on heat source strength,

$$\text{Ra}_q = \frac{g\beta q y_{tr}^2}{\alpha\nu k} \tag{6}$$

Figure 6.2b shows the replotting of the transition data of Fig. 6.2 and Ref. 9 as λ_B/D versus Ra_q. It is clear that regardless of the source strength q, the meander wavelength λ_B always scales with the local plume thickness scale D. In addition, the value of Ra_q at transition oscillates about 10^{10} as q varies: it is easy to show that the local Reynolds number based on scales (3) and (5) is of order $O(10^2)$ at transition (see Problem 3).

To summarize, the all-familiar cigarette-smoke observations suggest that the transition from laminar flow to turbulent flow is characterized by

(i)   A universal proportionality between longitudinal wavelength and stream thickness, that is, by a meander phenomenon.

(ii) A local Reynolds number of order $O(10^2)$, where the Reynolds number is based on the stream velocity scale and the stream thickness scale.

These features can be detected by observing many other flows that undergo transition in Nature: an extensive compilation of such observations is provided in Ref. 19. Perhaps, the most striking transition phenomenon that confirms conclusions (i) and (ii) is the highly regular (buckled) vortex street formed in the wake of a solid obstacle. There, the universal proportionality between wake

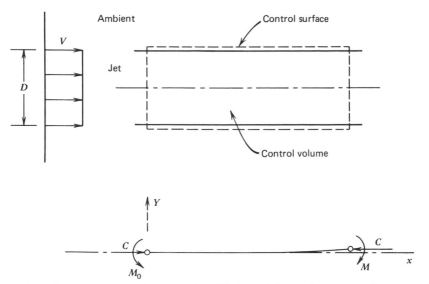

Figure 6.3 The translational and rotational equilibrium of the envelope surrounding a straight stream [19].

wavelength and wake thickness is very obvious [20], and the local Reynolds number is certainly of order $O(10^2)$ (see Table 6.1). Deep down, the rule for being able to see these common characteristics in other naturally behaving flows appears to be Leonardo da Vinci's forgotten advice, "*O miseri mortali aprite li occhi*" [21].[†]

THE BUCKLING OF INVISCID STREAMS

If the transition phenomenon is characterized by the *scaling laws* (i) and (ii), then the challenge to predict the transition reduces to the problem of accounting for these scaling laws *theoretically*. The remainder of this chapter is devoted to the presentation of two alternative theoretical arguments both capable of predicting the transition laws (i) and (ii). The first argument is the most direct and is based on the recently formulated buckling property of inviscid flow [22]. The second approach is based on reviewing the scaling implications of classical results known from the hydrodynamic stability analysis of inviscid flows [9].

   An interesting analogy between the buckling of elastic solid columns and the meandering of inviscid streams results from considering the static equilibrium of a *finite-size* control volume drawn around the stream. If, as is shown in Fig. 6.3, the stream and control volume thickness is of order D, and if the stream cross-section is A, then the control volume (or the thin-walled hose

---

[†]O, wretched mortals, open your eyes! [21].

surrounding the stream) satisfies the two conditions necessary for sinusoidal infinitesimal buckling in elastic systems [22]:

(i) The control volume is in a state of axial compression subject to the impulse and reaction forces (see the force balance discussed also in connection with the bottom of Fig. 2.3)

$$C = \rho V^2 A \tag{7}$$

(ii) If subjected to a *separate* bending test, the control volume develops in its cross-section a resistive bending moment that is directly proportional to the induced curvature (see Problem 4 at the end of this chapter, as well as [19, 22])

$$M = -\rho V^2 I \frac{d^2 Y}{dx^2} \tag{8}$$

In this last expression I is the area moment of inertia of the stream cross-section, $I = \int\int_A z^2 \, dA$, while $(-Y'')$ is the local curvature of the infinitesimally deformed control volume. Note also that eq. (8) is analogous to the expression $M = -EIY''$ derived from applying a separate bending test of prescribed curvature to a slender elastic beam [23]. This means that in inviscid streams, the product ρV^2 plays the role of *modulus of elasticity*, a fact confirmed easily by trying to manually bend a thin-walled hose containing a high Reynolds number stream. The stream control volume possesses *elasticity*, that is, conservative mechanical properties, because in the inviscid flow limit the material that fills the control volume is incapable of generating entropy [19].

Conditions (1) and (2) are essential to the *static equilibrium* of the control volume. The translational equilibrium is evident, as the two forces C balance each other. However, as in Euler's buckling theory of solid columns, the rotational equilibrium must be preserved even when the two forces C are not perfectly collinear; hence,

$$-M(x) + CY + M_0 = 0 \tag{9}$$

or, substituting expressions (7) and (8),

$$(\rho V^2 I) Y'' + (\rho V^2 A) Y + M_0 = 0 \tag{10}$$

This static rotational equilibrium condition indicates that the static equilibrium shape of the nearly straight stream column is a *sinusoid* of vanishingly small amplitude and characteristic (unique) wavelength.

$$\lambda_B = 2\pi \sqrt{\frac{I}{A}} = \begin{cases} \dfrac{\pi}{2} D, & \text{circular cross-section} \\[2mm] \dfrac{\pi}{\sqrt{3}} D, & \text{rectangular cross-section} \end{cases} \tag{11}$$

The *buckling* wavelength λ_B is a geometric property of the finite-size control volume, a length about twice the transversal dimension D. The $\lambda_B \sim D$ scaling predicted by the buckling theory of inviscid streams accounts for the empirical scaling law (i) detected during transition.

Before showing how the buckling property accounts also for the transition scaling law (ii), it is worth making the following observations:

1. The buckling wavelength of an inviscid stream is unique (and of order D) because the compressive load $\rho V^2 A$ is always proportional to the elasticity modulus ρV^2. This feature distinguishes sharply the buckling of inviscid streams from that of elastic solid columns where C and E are *independent*. This is why in solid columns we encounter an infinity of λ_B's (an additional degree of freedom) and, out of these, we must determine a discrete sequence of special λ_B's that satisfy end-clamping conditions. In the case of inviscid streams, the buckling wavelength is unique, and end-boundary-conditions are not an issue (*where* along the jet the first meander appears, depends on the transition scaling law (ii), as is shown later in this section and in Fig. 6.6).

2. The buckling theory of inviscid streams invokes the static equilibrium of a *finite-size* region of the flow field and, as such, represents a dramatic departure from the methodology that prevails in contemporary fluid mechanics. Routine fluid mechanics analysis has as its starting point the Navier–Stokes equations, which account for mechanical equilibrium among infinitesimally small fluid packets $dx\,dy\,dz$ (see Chapter 1). However, the invocation of mechanical equilibrium in finite-size systems is no mystery to those familiar with engineering thermodynamics, especially with the thermodynamics of flow systems such as rotating machines and rockets (see, e.g., Ref. 24). Finite-size systems of the type defined in Fig. 6.3 appear also in the fluid mechanics literature, for example, in Prandtl's famous textbook [25] and in a paper by Kotsovinos [26].

3. Although the proportionality $\lambda_B \sim D$ is universal, the control volume of transversal dimension D has been selected arbitrarily. Any fluid fiber, that is, any control volume of thickness $D' \neq D$ satisfies conditions (1) and (2) for infinitesimal buckling. Out of this infinity of fibers, however, only a special class is in a state of *unstable* equilibrium. The instability of inviscid flow, the discovery that certain fluid fibers are unstable, is an entirely different flow property and the contribution of an entirely different theory (hydrodynamic stability). As is shown in the next section, it is only the fibers thicker than the stream that are unstable, that *resonate* when shaken with a prescribed frequency (wavelength) by the mathematician or the loudspeaker in a laboratory flow experiment.

4. The buckling property or the scaling law $\lambda_B \sim D$ is widely observed in natural flows and can also be visualized in the laboratory. An extensive photographic record of such observations is presented in Refs. 19 and 27: among these, we note the river meandering phenomenon [28, 29], the waving of flags and the meandering fall of paper ribbons [30], the buckling of fast liquid jets shot through the air [31], and the sinuous structure of all turbulent plumes [32]. Practically anyone can visualize the buckling scaling (11) by placing an obstacle under the capillary water column falling from a faucet. Figure 6.4 shows the front and side views of the buckled stream: the sinuous deformation is mainly in one plane, like the cigarette-smoke plume of Fig. 6.1, and the locally measured λ_B/D ratio is consistently of the order of $\pi/2$, as in eq. (11) [33].

We now return to the transition scaling law (ii) armed with the idea that a stream has the $\lambda_B \sim D$ property *if it is inviscid*. The inviscidity (or viscidity) of the stream is a *flow property*, not a fluid property. It is inappropriate to refer to fluids such as honey and lava as viscous when, if the respective streams are wide and fast enough, they buckle (meander) in a way similar to rivers and water columns, as in Fig. 6.4. Experiment with a can of shaving cream in the morning and you will find that by varying the velocity of the foam jet you can control the jet's propensity to buckle, in other words, you can make a seemingly viscous fluid behave *inviscidly as a stream*.

Since buckling is a property of inviscid streams, it can be argued that the laminar–turbulent transition illustrated in Fig. 6.1 is related to the plume's

Figure 6.4 The plane buckled shape of a water column impinging on the flat end of a screwdriver; left side = the direct view; right side = the view through the side mirror [33].

transition from the state of *viscid stream* to that of *inviscid stream* [19, 22]. Imagine the stream sketched in Fig. 6.3: in time, viscous diffusion penetrates in the direction normal to the stream-ambient interface so that in a time of order

$$t_v \sim \frac{D^2}{16\nu} \tag{12}$$

the stream is fully viscous. The above time-scale follows from the error-function solution to the problem of viscous diffusion normal to an impulsively started wall [34] (see also Problem 5); according to this solution, the time of viscous penetration to the stream centerline (to a depth $D/2$) obeys the scaling law

$$\frac{D/2}{2(\nu t_v)^{1/2}} = O(1) \tag{12'}$$

Whether or not the stream becomes viscous depends on how fast it can buckle as an inviscid stream. The end-result of the incipient buckling analyzed early in this section is the birth of eddies, as the crests of the λ_B waves roll at the stream-ambient interface. From symmetry, the λ_B wave moves along the stream with a velocity of order $V/2$; hence, the *buckling time* or the time of eddy formation is

$$t_B \sim \frac{\lambda_B}{V/2} \tag{13}$$

The stream can buckle only if $O(t_B) < O(t_v)$, in other words, if the buckling frequency number $N_B = t_v/t_B$ is greater than $O(1)$.

In conclusion, the time-scale argument presented above recommends the following criterion for transition:

$$N_B = \frac{t_v}{t_B} \begin{cases} < O(1), & \text{laminar flow} \\ = O(1), & \text{transition} \\ > O(1), & \text{buckling or turbulent flow} \end{cases} \tag{14}$$

The fascinating aspect of this criterion is that after replacing t_v and t_B by $D^2/(16\nu)$ and $2\lambda_B/V$, respectively, it reads

$$\frac{VD}{\nu} \begin{cases} < O(10^2), & \text{laminar flow} \\ = O(10^2), & \text{transition} \\ > O(10^2), & \text{buckling or turbulent flow} \end{cases} \tag{15}$$

The group VD/ν is the *local Reynolds number* based on the axial velocity scale

(V) and the transversal dimension of the stream (D). Therefore, the buckling frequency number criterion [(14) and (15)] predicts correctly the transition scaling law (ii) identified empirically in the preceding section. The notion of *critical Reynolds number* for transition emerges as the laboratory (measurable) reflection of the competition between two time scales during transition, $N_B = O(1)$.

To summarize, the buckling property of inviscid flow provides a theoretical basis for the transition scaling laws (i) and (ii), or the proportionality between wavelength and stream thickness, and the local Reynolds number of order $O(10^2)$ during transition. In the next section, we learn that the scales predicted by the buckling theory of inviscid flow are *consistent* with scales also recommended by the hydrodynamic stability theory of inviscid flow. This does not mean that the two theories, buckling and hydrodynamic stability, are equivalent. The purpose of a theory is to explain known physical observations and to forecast future observations. Since no theory is perfect (capable of explaining everything), it is possible that the domains covered by two theories overlap [35]. The next section is about such an overlap, namely the explanation of scaling laws (i) and (ii). However, the discovery of this overlap and, even the stating in English of the scaling laws (i) and (ii), is the contribution of the newer theory.

THE INSTABILITY OF INVISCID FLOW

The issue of whether a parallel inviscid flow is stable or unstable is a century-old problem in modern fluid mechanics, a problem so traveled that it forms the core of one of the most voluminous chapters in the field. The analytical treatment of this problem originated with seminal papers by Helmholtz [36], Kelvin [37], and Rayleigh [38], who focused on the inertial instability of a homogeneous incompressible fluid. This work and its modern extensions to viscid flows and density-stratified flows have been reviewed by a number of authors, notably by Yih [39], Drazin and Howard [15], and Maslowe [40]. In this section we take another look at Rayleigh's analysis of an inviscid jet [38] in order to identify the proper length and time-scales that govern the transition phenomenon.

Consider the parallel flow of homogeneous incompressible fluid shown in Fig. 6.5. Modeling the flow as inviscid and two-dimensional, the continuity and momentum equations reduce to

$$\frac{\partial u}{\partial x} + \frac{\partial v}{\partial y} = 0 \tag{16}$$

$$\frac{\partial \zeta}{\partial t} + u\frac{\partial \zeta}{\partial x} + v\frac{\partial \zeta}{\partial y} = 0 \tag{17}$$

$$\zeta = \frac{\partial v}{\partial x} - \frac{\partial u}{\partial y} \tag{18}$$

where ζ is the vorticity function [41]. Equations (17) and (18) are obtained by eliminating the pressure gradient terms between eqs. (19a) and (19b) of Chapter 1. Next, we imagine that the parallel flow $U(y)$ is disturbed slightly, such that

$$u = U(y) + u'$$

$$v = 0 + v' \tag{19}$$

where the disturbance components (u', v') are regarded as infinitesimally small compared with the base flow $U(y)$. Finally, we seek to find out whether the disturbance amplitude grows with time, that is, whether the flow is unstable *relative* to the postulated disturbance.

Substituting the (base flow) + (disturbance) decomposition (19) into the governing equations and linearizing the result (i.e., neglecting the terms of

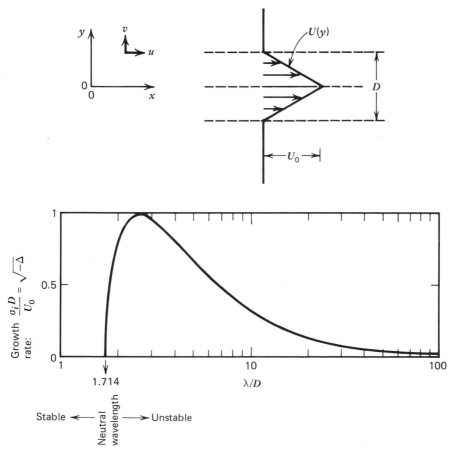

Figure 6.5 The stability characteristics of an inviscid jet of triangular profile.

second order in u', v') yields

$$\frac{\partial u'}{\partial x} + \frac{\partial v'}{\partial y} = 0$$

$$\left(\frac{\partial}{\partial t} + U\frac{\partial}{\partial x}\right)\left(\frac{\partial v'}{\partial x} - \frac{\partial u'}{\partial y}\right) - \frac{d^2 U}{dy^2}v' = 0 \tag{20}$$

Thinking of *sinusoidal* disturbances that may grow or decay in time, Rayleigh replaced u' and v' by the real parts of

$$\hat{u}e^{i(kx+\sigma t)} \quad \text{and} \quad \hat{v}e^{i(kx+\sigma t)} \tag{21}$$

where k is the wave number $2\pi/\lambda$ describing the disturbance periodicity in the x direction. Substituting these expressions into eqs. (20) and solving for $\hat{v}(y)$ yields what is recognized in the literature as the *Rayleigh equation*

$$\left(\frac{\sigma}{k} + U\right)(\hat{v}'' - k^2\hat{v}) - U''\hat{v} = 0 \tag{22}$$

Before proceeding further with the solution for $\hat{v}(y)$, it is worth commenting on the step between eqs. (20) and (21) and the Rayleigh equation (22). The postulate of *sinusoidal* disturbances is nowadays made routinely and without explanation in stability analyses of all kinds. Rayleigh, however, had a very good reason to be curious about the growth of *sinusoidal* disturbances. We have to remember that he lived and created in a period when rooms were lit by gaslight and candlelight and when the University Club did not display a "No Smoking" sign; his mind was stimulated by images very similar to Fig. 6.1, that is, by the flickering of cigarette smoke and candle flames. Rayleigh referred to such flows as *sensitive jets*, because they all appeared to resonate and meander most visibly when exposed to a *particular* sound frequency (musical tone). From the preceding discussion and Fig. 6.2, it is already clear that this natural *sensitivity* is the same as the buckling property of the jet manifested for the first time during transition. For Rayleigh and his epigones the postulate of *sinusoidal* disturbances is pure empiricism, the result of unexplained physical observations. In sharp contrast with stability analysis, the buckling theory predicts the *mathematically* sinusoidal shape of the deformed stream in the earliest stages of transition. Herein lies the aggregate contribution of the two theories: in time, the hydrodynamic stability analysis begins where the buckling theory leaves off.

To solve eq. (22), most insight per unit effort is achieved by modeling the base flow as a broken-line profile (Fig. 6.5). Thus, U'' vanishes in any region of the base flow; hence

$$\hat{v}'' - k^2\hat{v} = 0 \tag{23}$$

with the general solution

$$\hat{v} = C_1 e^{ky} + C_2 e^{-ky} \tag{24}$$

For the four-line jet profile of Fig. 6.5 we write

$$\hat{v} = C_1 e^{ky} + C_2 e^{-ky}, \quad y > D/2$$

$$\hat{v} = C_3 e^{ky} + C_4 e^{-ky}, \quad D/2 > y > 0$$

$$\hat{v} = C_5 e^{ky} + C_6 e^{-ky}, \quad 0 > y > -D/2$$

$$\hat{v} = C_7 e^{ky} + C_8 e^{-ky}, \quad -D/2 > y \tag{25}$$

where, from the condition that $\hat{v}$ does not blow up as $y \to \pm\infty$ we have $C_1 = C_8 = 0$. The remaining six unknowns, C_2–C_7, are determined from the six conditions that account for the continuity of $\hat{v}$ and pressure across the three interfaces, $y = -D/2, 0, D/2$. For example, the condition that $\hat{v}$ is continuous across $y = 0$ yields $C_3 + C_4 = C_5 + C_6$. The three pressure continuity conditions are obtained by integrating the Rayleigh equation (22) across each interface,

$$\left(\frac{\sigma}{k} + U \right) (\hat{v}'_+ - \hat{v}'_-) - \hat{v}(U'_+ - U'_-) = 0 \tag{26}$$

where the $(+)$ and $(-)$ subscripts indicate values calculated at the interface, while approaching the interface from above and below. It is easy to see that these six continuity conditions form a system of homogeneous equations; setting the determinant equal to zero yields the condition necessary for nontrivial solutions [38]

$$(m - \gamma^2)\left[m^2 + (kD - 3 + \gamma^2)m + \gamma^2(1 + kD) \right] = 0$$

where $\gamma = e^{-kD/2}$ and $m = 1 + \dfrac{\sigma D}{U_0}$ $\tag{27}$

Equation (27) contains the information sought by the stability analysis, namely, the growth rate σ compatible with the postulated wave number k. From the exponential forms chosen for u' and v' [eqs. (21)], it is clear that only if σ is complex will the disturbance grow exponentially in time, indicating *instability*. Complex σ's are possible if the discriminant is negative in the quadratic formed by setting the square brackets of eq. (27) equal to zero,

$$\Delta = (kD - 3 + \gamma^2)^2 - 4\gamma^2(1 + kD) < 0 \tag{28}$$

Solving this numerically for kD, we find that the instability condition means

$$kD < 3.666 \quad \text{or} \quad \lambda > 1.714\,D \qquad (29)$$

In conclusion, the disturbance wavelength must exceed a certain multiple of the jet diameter D for the flow to be unstable relative to the postulated disturbance. Consulting eqs. (21) we discover that, during instability, the time rate of growth of the disturbance (the imaginary part of σ) is proportional to $(-\Delta)^{1/2}$.

$$\frac{\sigma_i D}{U_0} = (-\Delta)^{1/2} \qquad (30)$$

The growth rate has been plotted in the lower-half of the Fig. 6.5, showing once again that the neutral wavelength scales with the jet transversal scale D. Beginning with Rayleigh's paper [38], much has been made in the literature of the maximum exhibited by the growth rate curve $\sigma_i D/U_0$. More interesting, however, is the "coincidence" that the neutral wavelength $1.714D$ is only 5 percent smaller than the buckling wavelength scale of a two-dimensional stream $[(\pi/\sqrt{3})D = 1.81\,D$, eq. (11)]. This coincidence seems to be insensitive to the actual shape of the $U(y)$ profile chosen for analysis. For example, in a stack of D-thick counterflow jets of sinusoidal profile ($u = U_0 \sin \pi y/D$) the neutral wavelength is $2D$, which is only 10 percent greater than the buckling length scale $(\pi/\sqrt{3})D$. The same scaling between flow thickness and neutral wavelength is revealed by the stability analysis of other finite-thickness flows [15].

What does this scaling tell us about the laminar–turbulent transition? It tells us that during transition the stream can fluctuate relative to its ambient with a period of the order of $\lambda/(U_0/2)$, where λ is the assumed disturbance wavelength. Since λ is greater than a length nearly identical to λ_B, the fluctuation time scale can only be greater than the buckling time scale t_B [eq. (13)]. Therefore, the fluctuation period exceeds a minimum time-scale that is proportional to the transversal length scale D. The domain of inviscid instability appears to the right of the $t_B \sim D$ line sketch in Fig. 6.6. However, as was argued in the preceding section and in Refs. 19 and 22, any stream of finite thickness becomes viscid at times greater than the viscous penetration scale t_v given by eq. (12); the viscid flow domain appears to the right of the $t_v \sim D^2$ parabola sketched in Fig. 6.6.

To read Fig. 6.6, imagine the upward development of a plume like the cigarette smoke shown in Fig. 6.1. The stream thickness increases monotonically with altitude (this is why D is plotted on the ordinate in Fig. 6.6) and remains laminar as long as $t_v < t_B$. Transition becomes possible beyond a critical plume diameter (or plume height) marked by $t_v \sim t_B$ or by criterion (14, 15) established earlier. Due to the sharply pointed shape of the inviscid instability domain on Fig. 6.6, the first wave during transition has a length about twice the local stream thickness.

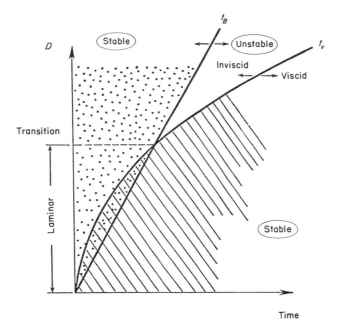

Figure 6.6 The transition as the internal competition between the minimum period for inviscid instability t_B and the viscous communication time t_V.

In conclusion, the scaling revealed by the stability analysis of an inviscid jet (Fig. 6.5) leads back to the buckling frequency number criterion for transition [eqs. (14) and (15)]. In this way, the inviscid stability scaling accounts for the transition laws (i) and (ii) stated in the first section of this chapter.

One final observation: the existence of a semiinfinity of wavelengths for which the D-thick jet is unstable (Fig. 6.5) would seem to contradict the uniqueness of wavelength scale predicted by the buckling theory. In fact, there is no contradiction. The scale analysis of the Rayleigh equation (22) or (23) indicates that the disturbance amplitude is always felt to a y thickness of order $k^{-1} \sim \lambda$. This means that the thickness of the fluid layer that resonates to the imposed disturbance wavelength λ always scales with λ, regardless of the actual thickness of the base flow region [42].

GENERAL CRITERION FOR TRANSITION TO TURBULENCE

We began this chapter with a discussion of transition in a phenomenon so common that it is found, literally, at everyone's fingertips (Fig. 6.1). We then use a straightforward comparison of two time-scales, t_B and t_v, to predict conditions under what the laminar flow might possibly break down. In order to estimate the buckling or fluctuation cycle time-scale t_B, we used two alternative approaches—the buckling property and the instability of inviscid flow. We saw that the end-result of the $t_B \sim t_v$ scaling during transition [eq. (14)] is the prediction of the observed $O(10^2)$ for the local Reynolds number based on the transversal dimension of the flow.

As we conclude this chapter, it is important to understand that it is the simple scaling argument $t_B \sim t_v$, the simple comparison of two time-scales, that is responsible for the $N_B = O(1)$ transition criterion. This scaling argument was first proposed as part of the buckling theory of inviscid flow [19, 22] and only recently as a powerful addendum to linearized inviscid stability theory [9]. Note that *without* the $t_B \sim t_v$ scaling argument, any inviscid stability analysis stops at a place like the bottom of Fig. 6.5, that is, *without* predicting the transition (the instability of inviscid flow is one thing, and the transition exhibited by a real flow, another). The young researcher may benefit from looking at the century-long history of the stability theory and asking the question: Why has such a basic scaling argument been overlooked for such a long time? One possible answer is that, from the beginning, the method of stability analysis impressed Rayleigh's epigones with mathematical difficulty and sophistication, hence, with an air of unmatched accuracy and goodness (very young, we are taught to expect good things at the end of very hard work). One-hundred years later, this impression appears to explain the "poor cousin" status of pure scaling reasoning[†] in fluid mechanics and heat transfer. Why do something approximately on the back of an envelope when you can do it exactly? Very few people seem to factor into their decision-making the *cost* of engineering research; indeed, we hear very few people discussing the unacceptably low return on investment associated with a generally *accepted* method. From journal referees to research proposal referees, the message to the young researcher is that goodness is associated with sophisticated analysis, computational speed and accuracy, and the sensitivity and on-site processing capability of experimental apparatus. No wonder we hear complaints about insufficient research funds, insufficient time to act as referees, and insufficient library funds and space to store all the new journals!

In order to use the transition criterion $N_B = O(1)$, we must be able to estimate the thickness of velocity scales of the given flow. This task stresses the importance of the scaling analyses developed in the preceding four chapters for laminar flow. *Transition is not a mathematically precise event*: the most we can do is predict the order of magnitude range of experimental conditions under which transition can take place. The beginning of transition, that is, the first buckling event depends on the particular state of agitation of the flow. Guided by some of the problems at the end of this chapter, the student is advised to reconstruct the transition observations of Table 6.1 using criterion (14, 15).

It is worth noting that the $N_B = O(1)$ criterion has been tested successfully in Ref. 19, in the case of free-jet flow, wake flow, and free shear flow. More recently, the same criterion was found to work very well towards predicting transition in buoyant plume flow [9, 32] and vertical natural convection boundary layers of both high-Pr and low-Pr fluids [43].

[†]Improving the status of this cost-effective method is one of the objectives of this textbook.

An interesting example of how the $N_B = O(1)$ transition criterion works is presented in Fig. 6.7. As is shown in Chapter 5, the laminar natural convection pattern in a tall rectangular enclosure heated from the side consists of two vertical wall jets in counterflow (Fig. 5.4, case II). During transition, each wall jet will buckle with a wavelength λ_B that scales with the outer (velocity) thickness of the jet, $H \, \mathrm{Ra}_H^{-1/4} \mathrm{Pr}^{1/2}$ (see the Pr > 1 scales listed in Table 4.1). Using the data furnished by Seki et al. [44] for the flow photographed in Fig. 6.6, the buckling wavelength scale can be calculated and plotted alongside the flow cavity. In this manner we discover that the well-documented *secondary cellular flow* observed during transition in vertical enclosures is the visible consequence of incipient buckling in the vertical wall jets. In the same manner we can calculate the buckling frequency number N_B and understand why the flow could be photographed so clearly: the value of N_B is slightly on the laminar side of $O(1)$; hence, the flow is still laminar.

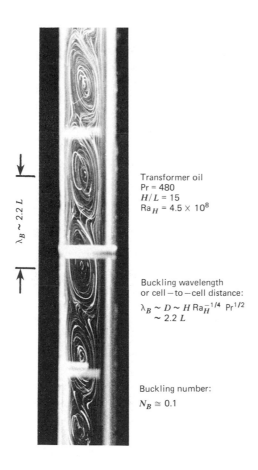

Transformer oil
Pr = 480
$H/L = 15$
$\mathrm{Ra}_H = 4.5 \times 10^8$

Buckling wavelength
or cell – to – cell distance:
$$\lambda_B \sim D \sim H \, \mathrm{Ra}_H^{-1/4} \, \mathrm{Pr}^{1/2}$$
$$\sim 2.2 \, L$$

Buckling number:
$N_B \cong 0.1$

$\lambda_B \sim 2.2 \, L$

Figure 6.7 Transition to turbulent (buckling) flow in a vertical enclosure with different side wall temperatures [44].

According to the same example, the buckling wavelength decreases as the Rayleigh number increases. This means that the observed number of secondary cells that fit in the core must increase as $Ra_H^{1/4}$: This particular scaling is supported convincingly by 27 published photographs and numerical simulations of transition in a vertical enclosure [43].

Very recently, the buckling theory of transition to turbulence in free-jet flow [22] was tested in a special experiment by Pollard [45]. Based on numerous observations, Pollard found that the transition Reynolds number based on nozzle diameter and mean velocity is equal to 26.36 ± 0.26, in very good agreement with the value 25 derived using buckling theory and scale analysis [22]. Pollard's experiments also serve to refine the rough observations reported years ago by A. J. Reynolds [4] (see Table 6.1).

SYMBOLS

A	stream cross-sectional area
c_P	specific heat at constant pressure
C	compressive impulse or reaction [eq. (7)]
D	stream transversal length scale
g	gravitational acceleration
Gr	Grashof number (Table 6.1)
Gr_*	Grashof number based on heat flux (Table 6.1)
I	area moment of inertia of the stream cross-section
k	wave number and thermal conductivity
m	function [eq. (27)]
M	bending moment
N_B	buckling number [eq. (14)]
Pr	Prandtl number
q	heat source strength [W]
Re	Reynolds number (Table 6.1)
Ra_q	Rayleigh number based on source strength [eq. (6)]
t_B	buckling time or time of eddy formation [eq. (13)]
t_v	transversal viscous communication time [eq. (12)]
ΔT	temperature difference
u, v	velocity components (Fig. 6.5)
u', v'	velocity disturbances [eq. (19)]
$\hat{u}, \hat{v}$	disturbance amplitudes [eq. (21)]
U	longitudinal base flow [eq. (19)]
v	stream velocity scale [eq. (7)]
x, y	cartesian coordinates (Fig. 6.5)
y_{tr}	transition height
Y	equilibrium shape of a nearly straight stream [eq. (8)]
α	thermal diffusivity
β	coefficient of thermal expansion

γ	function [eq. (27)]
Γ	condensate mass flowrate (Table 6.1)
Δ	discriminant [eq. (28)]
λ	wavelength
λ_B	buckling wavelength
μ	viscosity
ν	kinematic viscosity
ρ	density
σ	disturbance growth rate [eq. (21)]
ζ	vorticity function [eq. (18)]

REFERENCES

1. H. Schlichting, *Boundary Layer Theory*, 4th ed., McGraw-Hill, New York, 1960, pp. 376–379.

2. J. W. Elder, An experimental investigation of turbulent spots and breakdown to turbulence, *J. Fluid Mech.*, Vol. 9, 1960, pp. 235–246.

3. R. A. Granger, *Fluid Mechanics*, Holt, Rinehart, and Winston, New York, 1984.

4. A. J. Reynolds, Observations of a liquid-into-liquid jet, *J. Fluid Mech.*, Vol. 14, 1962, pp. 552–556.

5. F. Homann, *Einfluss grosser Zähigkeit bei Strömung um Zylinder*, *Forsch. Ingenieurwes.*, Vol. 7, 1936, pp. 1–10; see also Ref. 1, p. 17.

6. R. L. Mahajan and B. Gebhart, An experimental determination of transition limits in a vertical natural convection flow adjacent to a surface, *J. Fluid Mech.*, Vol. 91, 1979, pp. 131–154.

7. A. A. Szewczyk, Stability and transition of the free convection layer along a vertical flat plate, *Int. J. Heat Mass Transfer*, Vol. 5, 1962, pp. 903–914.

8. R. Godaux and B. Gebhart, An experimental study of the transition of natural convection flow adjacent to a vertical surface, *Int. J. Heat Mass Transfer*, Vol. 17, 1974, pp. 93–107.

9. S. Kimura and A. Bejan, Mechanism for transition to turbulence in buoyant plume flow, *Int. J. Heat Mass Transfer*, Vol. 26, 1983, pp. 1515–1532.

10. W. M. Rohsenow and H. Y. Choi, *Heat, Mass and Momentum Transfer*, Prentice-Hall, Englewood Cliffs, NJ, 1961, p. 243.

11. W. M. Kays and M. E. Crawford, *Convective Heat and Mass Transfer*, 2nd ed., McGraw-Hill, New York, 1980, p. 163.

12. E. Reshotko, Boundary layer stability and transition, *Ann. Rev. Fluid Mech.*, Vol. 8, 1976, pp. 311–349.

13. B. Gebhart and R. Mahajan, Characteristic disturbance frequency in vertical natural convection, *Int. J. Heat Mass Transfer*, Vol. 18, 1975, pp. 1143–1148.

14. G. K. Batchelor and A. E. Gill, Analysis of the stability of axisymmetric jets, *J. Fluid Mech.*, Vol. 14, 1962, pp. 529–551.

15. P. G. Drazin and L. N. Howard, Hydrodynamic stability of parallel flow of inviscid fluid, *Adv. Appl. Mech.*, Vol. 9, 1966, pp. 1–89.

16. J. L. Lopez and U. H. Kurzweg, Amplification of helical disturbances in a round jet, *Phys. Fluids*, Vol. 20, 1977, pp. 860–861.

17. S. C. Crow and F. H. Champagne, Orderly structure in jet turbulence, *J. Fluid Mech.*, Vol. 48, 1971, pp. 547–593.

18. J. W. Hoyt and J. J. Taylor, Waves on water jets, *J. Fluid Mech.*, Vol. 83, 1977, pp. 119–127.

19. A. Bejan, *Entropy Generation through Heat and Fluid Flow*, Wiley, New York, 1982, Chapter 4.

20. J. H. Lienhard, *A Heat Transfer Textbook*, Prentice-Hall, Englewood Cliffs, NJ, 1981, pp. 326–328.

21. J. P. Richter, *The Notebooks of Leonardo da Vinci*, Vol. II, Dover, New York, 1970, p. 295.

22. A. Bejan, On the buckling property of inviscid jets and the origin of turbulence, *Lett. Heat Mass Transfer*, Vol. 8, 1981, pp. 187–194.

23. J. P. Den Hartog, *Strength of Materials*, McGraw-Hill, New York, 1949, p. 39.

24. E. G. Cravalho and J. L. Smith, Jr., *Engineering Thermodynamics*, Pitman, Boston, MA, 1981, Chapters 10 and 11.

25. L. Prandtl, *Essentials of Fluid Dynamics*, Blackie & Son, London, 1969, p. 7.

26. N. E. Kotsovinos, Plane turbulent jets. Part 2: Turbulence structure, *J. Fluid Mech.*, Vol. 81, 1977, pp. 45–62.

27. O. M. Griffin, Vortex streets and patterns, *Mech. Eng.*, Vol. 104, No. 3, March 1982, pp. 56–61.

28. L. B. Leopold and M. G. Wolman, River meandering, *Bull. Geol. Soc. Am.*, Vol. 71, 1960, pp. 769–794.

29. A. Bejan, Theoretical explanation for the incipient formation of meanders in straight rivers, *Geophys. Res. Lett.*, Vol. 9, 1982, pp. 831–834.

30. A. Bejan, The meandering fall of paper ribbons, *Phys. Fluids*, Vol. 25, 1982, pp. 741–742.

31. M. G. Stockman and A. Bejan, The nonaxisymmetric (buckling) flow regime of fast capillary jets, *Phys. Fluids*, Vol. 25, 1982, pp. 1506–1511.

32. A. Bejan, Theory of instantaneous sinuous structure in turbulent buoyant plumes, *Wärme Stoffübertrag.*, Vol. 16, 1982, pp. 237–242.

33. S. Kimura and A. Bejan, The buckling of a vertical liquid column, *J. Fluids Eng.*, Vol. 105, 1983, pp. 469–473.

34. H. Schlichting, *op. cit.*, p. 72.

35. P. Feyerabend, *Against Method*, Verso, London, 1978.

36. H. Helmholtz, Über diskontinuierliche Flüssigkeitsbewegungen, *Monatsber. Königl. Preuss. Akad. Wiss. Berlin*, 1868, pp. 215–228; translation by F. Guthrie, On discontinuous movements of fluids, *Philos. Mag.* (4)36, 1868, pp. 337–346.

37. W. Kelvin, The influence of wind on waves in water supposed frictionless, *Philos. Mag.* (4)42, 1871, pp. 368–374.

38. J. W. S. Rayleigh, On the stability, or instability of certain fluid motions, *Proc. London Math. Soc.*, Vol. XI, 1880, pp. 57–70.

39. C. S. Yih, *Fluid Mechanics*, McGraw-Hill, New York, 1969, Chapter 9.

40. S. A. Maslowe, Shear flow instabilities and transition, in H. L. Swinney and J. P. Gollub eds., *Hydrodynamic Instabilities and Transition to Turbulence*, Springer, Berlin, 1981, Chapter 7.

41. H. Lamb, *Hydrodynamics*, 6th ed., Dover, New York, 1945, pp. 670–672.

42. R. Anderson, Ph.D. Thesis, Department of Mechanical Engineering, University of Colorado, Boulder, May 1983.

43. A. Bejan and G. R. Cunnington, Theoretical considerations of transition to turbulence in natural convection near a vertical wall, *Int. J. Heat Fluid Flow*, Vol. 4, No. 3, 1983, pp. 131–139.

44. N. Seki, S. Fukusako, and I. Inaba, Visual observation of natural convective flow in a narrow vertical cavity, *J. Fluid Mech.*, Vol. 84, 1978, pp. 695–704.

45. A. Pollard, Private communication (April 11, 1983), Department of Mechanical Engineering, Queen's University, Kingston, Ontario, Canada.

PROBLEMS

1.   Show that the transition condition listed for boundary layer flow in Table 6.1 corresponds to a critical Reynolds number of order $O(10^2)$, if this critical Reynolds number is based on either the displacement or momentum thickness of the boundary layer.

2.   Show that for the vertical natural convection boundary layer flow, the local Reynolds number based on vertical velocity scale and velocity boundary layer thickness (Table 4.1) is of order $Pr^{-1/2} Ra_y^{1/4}$ if $Pr > 1$. Then, by examining Table 6.1, prove that during transition this local Reynolds number is of order $O(10^2)$.

3.   Prove that in the case of the cigarette-smoke plume of Figs. 6.1 and 6.2, the local Reynolds number based on scales (2, 3) is of the same order as $Ra_q^{1/4}$ during transition. Based on the physical observations compiled in Fig. 6.2, show that this locally defined Reynolds number is of order $O(10^2)$.

4.   Consider the straight inviscid stream (ρ, V, P_0, D, A) shown in Fig. 6.3, and subject this stream to a bending test of prescribed radius of curvature R_∞. Show that in the limit of vanishingly small curvature, $D/R_\infty \to 0$, the resistive bending moment integrated over the stream cross-section is

$$M = \iint_A \left[\rho v(z)^2 + P(z) \right] z \, dA = \frac{\rho V^2 I}{R_\infty}$$

where z is measured radially away from the stream centerline and towards the center of curvature. Hint: Invoke the Bernoulli equation along a streamline and a force balance in the radial direction to derive analytical expressions for the velocity and pressure profiles $v(z)$ and $P(z)$ in the stream cross-section.

5.   Consider the transient flow generated in the vicinity of a flat wall: at times $t < 0$ both the fluid and the wall are motionless, while for $t > 0$ the wall moves along itself with a constant velocity U. Recognizing that the wall is infinitely long (compared with the thickness of the viscous boundary layer forming along the wall), show that the fluid is entrained in laminar flow according to

$$u = U \operatorname{erfc} \left(\frac{y}{2(\nu t)^{1/2}} \right)$$

where u is the fluid velocity in the direction parallel to the wall and y is the distance measured away from the wall. Show that the knee in the above velocity profile resides at $y/2(\nu t)^{1/2} = O(1)$.

6.   Verify that the local Reynolds number of a stream VD/ν is of order $O(10^2)$ during transition, that is, when the buckling frequency number N_B is of order $O(1)$. Hint: Use eqs. (11)–(14).

7.   Determine the range of disturbance wavelengths for which the following inviscid flows are unstable:

(a)   Shear flow: $U = U_0,$ for $y > h$

                                 $U = U_0 y/h,$ for $h > y > -h$

                                 $U = -U_0,$ for $-h > y$

(b)   Wall jet: $U = 0,$ for $y > D$

                                 $U - U_0 \left(2 \quad \dfrac{y}{D/2} \right),$ for $D > y > D/2$

                                 $U = U_0 \dfrac{y}{D/2},$ for $\dfrac{D}{2} > y > 0$

In each case verify that the neutral wavelength scales with the transversal length scale of the flow. Hint: Follow the analytical course traced between eqs. (22) and (30) in the text.

8.   A two-dimensional jet discharges freely into a reservoir that contains the same fluid as the jet. The jet nozzle is a two-dimensional slit of gap size D_0. The Reynolds number based on nozzle size (D_0) and mean velocity through the nozzle (U_0) is $Re_0 = 1$. Consulting Table 6.1 and the scaling laws of transition discussed in this chapter, decide whether the jet is laminar over its entire length, or turbulent, or something else (review the scaling laws revealed by the solution to Problem 10 of Chapter 2).

7

Wall Turbulence

Looking back at the ground covered so far, we began this treatment by formulating the fundamental principles that govern all convection phenomena (Chapter 1), and then we applied these principles to the task of predicting friction and heat transfer in laminar flow (Chapters 2–5). The limitations of the laminar flow description formed the subject of Chapter 6, in which we identified a universal scaling criterion for predicting the transition to turbulent flow. It is natural to continue with a discussion of momentum and heat transfer in turbulent flow: this very important topic is the subject of the next three chapters.

Turbulent heat transfer represents an old field of research that has already generated a large volume of empirical information on how turbulent flows behave and, of importance to heat transfer engineers, how turbulent flows transport heat and mass. It is not the object of this course to review all that has been reported on turbulent heat transfer: this would be an impossible task considering the staggering amount of information available, and, very useful reviews and monographs on turbulence already exist and are highly recommended to the student (see, e.g., Refs. 1–3). The objective of the present treatment is to review some of the concepts, that is, some of the basic ideas behind the contemporary treatment of turbulent heat transfer. This objective is both interesting and feasible because most of the turbulent heat transfer language in use today was coined almost a century ago by people like Reynolds, Boussinesq, and Prandtl, and because during all this time relatively little has been done to question the ideas that produced this language.

It goes without saying that a turbulent flow and temperature field is complicated,[†] as witnessed by the irregularity and randomness in the electrical

---

[†]After all, this is why this type of flow was even named *turbulent* (note the original Latin meaning of this terminology: words like *turba*, *turbidus*, *turbulentus*, etc., were used to describe tumult, uproar, commotion in a crowd of people, confusion, the mentally deranged, muddy waters; see Ref. 2 of Chapter 1).

output from velocity and temperature probes inserted in the flow. It is not difficult to relate this irregularity to the agitated *eddy* motion exhibited so clearly and unpredictably by many flows that surround us (cumulus clouds, muddy rivers, or the post-transition upper section of the cigarette-smoke plume of Fig. 6.1). Thus, there is a great temptation to define turbulence as an irregular and random fluid motion, as is done by the established monographs [1–3] and numerous heat transfer textbooks. But to do this would mean disregarding an important philosophical development, a renaissance in turbulence research, namely, the recognition that turbulent flows possess an *orderly structure* [4, 5]. For this reason we begin the discussion with Figs. 7.1 and 7.2, which show two strikingly similar turbulent flows photographed under strikingly different circumstances by two strikingly different individuals. Perhaps, the only touch of similarity between the two photographs is that both photographers were *not* necessarily looking to photograph the large-scale structure of turbulent flow.

Figure 7.1 shows the turbulent boundary layer downstream from the leading edge of a flat plate; the flow is visualized by boiling, that is, by the entrainment of vapor bubbles formed near the solid surface. Figure 7.2 offers a glimpse of the turbulent boundary layer over the surface of the South Atlantic; the visualization in this case is made possible by flames (fed by an oil slick) and thick black smoke. Of course, one may argue that in both photographs the flow visualization agents are buoyant and interfere with the actual boundary layer flow. This interpretation is correct only for regions situated *sufficiently far* downstream, where buoyancy has had enough time to accelerate the agent to a vertical velocity comparable with the horizontal velocity of unmarked fluid. The visualization method works best, that is, is closest to revealing the truth near the leading edge of each flow. It is near the leading edge that both flows exhibit the same *large-scale structure*. Note the sharpness of the interface between boundary layer and free stream, and note the waviness of this interface. Large eddies protrude through the interface, and their diameters appear to scale with both the local thickness of the boundary layer and the local distance between two consecutive large eddies. This scaling is consistent with the $\lambda_B/D =$ constant scaling law recommended by the buckling theory of inviscid flow as well as by the hydrodynamic stability theory of inviscid flow (Chapter 6). In fact, the theoretical λ_B/D can be used to predict that the angle between the wavy interface and the solid surface is $20°$ [6], which is the angle visible in Figs. 7.1 and 7.2, as well as in photographs of the same flow taken in the laboratory [7].

In this chapter we focus on the mechanism by which turbulent flows transport energy between a stream and a solid wall, as might be the case in Fig. 7.1 in the absence of boiling. The following presentation is constructed along the same lines as classical turbulent heat transfer methodology, which is based on the time-averaged description of the convection phenomenon, and must rely heavily on empirical information in order to solve the so-called *closure problem*

Figure 7.1 Turbulent boundary layer in boiling water flowing from left to right over a flat surface; $U_\infty = 0.52$ m/s, $q_0'' = 4.8 \times 10^5$ W/m^2 (Courtesy of John H. Lienhard; photograph reproduced from J. H. Lienhard, *A Heat Transfer Textbook*, Prentice Hall, Inc., Englewood Cliffs, NJ, 1981, p. 417. Reprinted by permission from Prentice-Hall, Inc., Englewood Cliffs, New Jersey).

Figure 7.2 Smoke visualization of the large-scale structure of the turbulent boundary layer over the ocean (*Wide World Photos*).

(i.e., the fact that the time-averaged governing equations are fewer than the number of unknowns). However, wherever possible, we shall invoke the theoretical arguments of Chapter 6 in order to minimize the empirical content of the time-averaged analysis.

THE TIME-AVERAGED EQUATIONS

Man has found it impossible to determine the turbulent flow solution at any point in space and time by applying the mass, momentum, and energy conservation equations in the form reported in Chapter 1. Mathematically, it has been difficult enough to determine the smooth and time-independent velocity and temperature profiles of simple laminar flows (Chapters 2–5); imagine, then, trying to determine analytically a turbulent profile, for example, the longitudinal velocity profile in turbulent boundary layer flow (when visualized, this wiggly profile fluctuates in time much in the same way as an electric arc). Confronted with this difficulty, Reynolds [8] thought that some of the complications of instantaneous turbulent flow could be removed if one considers not the instantaneous behavior, but the mean behavior averaged over a long enough time period. In terms of mean velocities, pressure, and temperature, the time-averaged flow field imagined by Reynolds is a simpler one, a field without unpredictable fluctuations (eddies). It is important to recognize early that the time-averaged flow behavior is not a simpler *flow* (because only the real turbulent flow exists), rather it is a simpler *way to think* about turbulent flows. Unfortunately, this is also an effective method of simplifying the student's mind, inviting it to see smoothness in turbulence rather than coarseness and orderly structure.

The derivation of the conservation laws for time-averaged flow begins with the transformation

$$u = \bar{u} + u' \qquad P = \bar{P} + P'$$

$$v = \bar{v} + v' \qquad T = \bar{T} + T' \tag{1}$$

$$w = \bar{w} + w'$$

where, by definition, the quantities denoted with an overbar $(\bar{\ })$ represent the mean values obtained by time-averaging over a long enough period, for example,

$$\bar{u} = \frac{1}{\text{period}} \int_0^{\text{period}} u\, d(\text{time}) \tag{2}$$

Combining eqs. (1) and (2), we recognize that, by definition, the fluctuating components denoted with a prime $(\ )'$ average to zero over time, for example,

$$\int_0^{\text{period}} u'd(\text{time}) = 0 \tag{3}$$

Definitions (2) and (3) are the foundation of a special kind of algebra that emerges in the process of substituting the $(\bar{\ }) + (\ )'$ decomposition (1) into the mass, momentum, and energy equations, and then time-averaging these equa-

tions according to definition (2). Some of the rules (theorems) of this algebra
are [9]

$$\overline{u + v} = \bar{u} + \bar{v} \tag{4}$$

$$\overline{\bar{u}u'} = 0 \tag{5}$$

$$\overline{uv} = \overline{\bar{u}\bar{v}} + \overline{u'v'} \tag{6}$$

$$\overline{u^2} = \bar{u}^2 + \overline{u'^2} \tag{7}$$

$$\overline{\left(\frac{\partial u}{\partial x}\right)} = \frac{\partial \bar{u}}{\partial x} \tag{8}$$

$$\frac{\partial \bar{u}}{\partial t} = 0 \tag{9}$$

$$\overline{\left(\frac{\partial u}{\partial t}\right)} = 0 \tag{10}$$

Consider first the transformation of the mass conservation equation (8) of
Chapter 1.

$$\frac{\partial \bar{u}}{\partial x} + \frac{\partial u'}{\partial x} + \frac{\partial \bar{v}}{\partial y} + \frac{\partial v'}{\partial y} + \frac{\partial \bar{w}}{\partial z} + \frac{\partial w'}{\partial z} = 0 \tag{11}$$

Integrating this equation term-by-term over time, and applying rules (3) and
(8) yields

$$\frac{\partial \bar{u}}{\partial x} + \frac{\partial \bar{v}}{\partial y} + \frac{\partial \bar{w}}{\partial z} = 0 \tag{12}$$

which is analytically identical to the original equation [eq. (8) of Chapter 1].
Equation (12) represents the condition for conservation of mass in the imaginary flow described by time-averaged quantities.

Consider next the x momentum equation listed as eq. (19a) in Chapter 1: it
is easy to show that this equation may be rewritten as

$$\frac{\partial u}{\partial t} + \frac{\partial}{\partial x}(u^2) + \frac{\partial}{\partial y}(uv) + \frac{\partial}{\partial z}(uw) = -\frac{1}{\rho}\frac{\partial P}{\partial x} + \nu\nabla^2 u \tag{13}$$

Averaging each term over time and applying rules (8)–(10) yields

$$\frac{\partial}{\partial x}(\overline{u^2}) + \frac{\partial}{\partial y}(\overline{uv}) + \frac{\partial}{\partial z}(\overline{uw}) = -\frac{1}{\rho}\frac{\partial \bar{P}}{\partial x} + \nu\nabla^2\bar{u} \tag{14}$$

Applying now the product averaging rules (6) and (7), we obtain

$$\frac{\partial}{\partial x}\left(\overline{u}^2\right) + \frac{\partial}{\partial y}\left(\overline{u}\,\overline{v}\right) + \frac{\partial}{\partial z}\left(\overline{u}\,\overline{w}\right)$$

$$= -\frac{1}{\rho}\frac{\partial \overline{P}}{\partial x} + \nu\nabla^2\overline{u} - \frac{\partial}{\partial x}\left(\overline{u'^2}\right) - \frac{\partial}{\partial y}\left(\overline{u'v'}\right) - \frac{\partial}{\partial z}\left(\overline{u'w'}\right) \qquad (15)$$

Making use of the mass continuity equation (12), the left-hand side of eq. (15) can be simplified to read

$$\overline{u}\frac{\partial \overline{u}}{\partial x} + \overline{v}\frac{\partial \overline{u}}{\partial y} + \overline{w}\frac{\partial \overline{u}}{\partial z} = -\frac{1}{\rho}\frac{\partial \overline{P}}{\partial x} + \nu\nabla^2\overline{u} - \frac{\partial}{\partial x}\left(\overline{u'^2}\right)$$

$$- \frac{\partial}{\partial y}\left(\overline{u'v'}\right) - \frac{\partial}{\partial z}\left(\overline{u'w'}\right) \qquad (16a)$$

The corresponding time-averaged forms of the momentum equations in the y and z direction are

$$\overline{u}\frac{\partial \overline{v}}{\partial x} + \overline{v}\frac{\partial \overline{v}}{\partial y} + \overline{w}\frac{\partial \overline{v}}{\partial z} = -\frac{1}{\rho}\frac{\partial \overline{P}}{\partial y} + \nu\nabla^2\overline{v} - \frac{\partial}{\partial x}\left(\overline{u'v'}\right) - \frac{\partial}{\partial y}\left(\overline{v'^2}\right) - \frac{\partial}{\partial z}\left(\overline{v'w'}\right)$$

$$(16b)$$

$$\overline{u}\frac{\partial \overline{w}}{\partial x} + \overline{v}\frac{\partial \overline{w}}{\partial y} + \overline{w}\frac{\partial \overline{w}}{\partial z} = -\frac{1}{\rho}\frac{\partial \overline{P}}{\partial z} + \nu\nabla^2\overline{w} - \frac{\partial}{\partial x}\left(\overline{u'w'}\right) - \frac{\partial}{\partial y}\left(\overline{v'w'}\right) - \frac{\partial}{\partial z}\left(\overline{w'^2}\right)$$

$$(16c)$$

Finally, the energy equation expressed as eq. (42) in Chapter 1 yields after a similar time-averaging procedure

$$\overline{u}\frac{\partial \overline{T}}{\partial x} + \overline{v}\frac{\partial \overline{T}}{\partial y} + \overline{w}\frac{\partial \overline{T}}{\partial z} = \alpha\nabla^2\overline{T} - \frac{\partial}{\partial x}\left(\overline{u'T'}\right) - \frac{\partial}{\partial y}\left(\overline{v'T'}\right) - \frac{\partial}{\partial z}\left(\overline{w'T'}\right) \quad (17)$$

The derivation of eq. (17) follows in the steps contained between eqs. (13) and (16a) in the derivation of the time-averaged x momentum equation.

To summarize, the time-averaged conservation laws for constant-property flow are represented by eqs. (12), (16a, b, c) and (17). These laws represent 5 equations for 17 unknowns [the unknowns are $\overline{u}, \overline{v}, \overline{w}, \overline{P}, \overline{T}$, and the 12 terms of type $\partial\,(\overline{u'v'})/\partial y$ appearing in eqs. (16a, b, c) and (17)]; hence, the *closure problem* discussed in the introduction to this chapter. The difference between the number of equations and the number of unknowns has its origin in the original transformation [eqs. (1)], which doubled the number of unknowns; the final number of unknowns ballooned to 17 due to the various product combinations of fluctuating quantities that survive the time-averaging process. Fortunately, the gap between equations and unknowns is not nearly as menacing if we consider especially simple flow configurations such as boundary layers and fully developed flows through ducts.

THE BOUNDARY LAYER EQUATIONS

Consider turbulent flow near a wall parallel to a free stream U_∞, T_∞ oriented in the positive x direction, as in Fig. 2.1 Although the actual turbulent flow is *three-dimensional* regardless of how simple the flow boundaries, from symmetry, the terms representing the $\partial/\partial z$ derivative of time-averaged quantities in eqs. (16a) and (17) must vanish. Furthermore, if we think of u' and v' as the velocity fluctuations caused by an eddy (a rotating fluid blob) as it rides along with the mean flow, then u' and v' are of comparable orders of magnitude. This means that in the boundary layer we can neglect $(\partial/\partial x)\overline{(u'^2)}$ relative to $(\partial/\partial y)\overline{(u'v')}$ in eq. (16a), and $(\partial/\partial x)\overline{(u'T')}$ relative to $(\partial/\partial y)\overline{(v'T')}$ in eq. (17). Applying the other simplifications that result from boundary layer theory (Chapter 2), the x momentum and energy equations reduce to

$$\bar{u}\frac{\partial\bar{u}}{\partial x} + \bar{v}\frac{\partial\bar{u}}{\partial y} = -\frac{1}{\rho}\frac{d\bar{P}}{dx} + \nu\frac{\partial^2\bar{u}}{\partial y^2} - \frac{\partial}{\partial y}\left(\overline{u'v'}\right) \tag{18}$$

$$\bar{u}\frac{\partial\bar{T}}{\partial x} + \bar{v}\frac{\partial\bar{T}}{\partial y} = \alpha\frac{\partial^2\bar{T}}{\partial y^2} - \frac{\partial}{\partial y}\left(\overline{v'T'}\right) \tag{19}$$

Note that the use of $d\bar{P}/dx$ instead of $\partial\bar{P}/\partial x$ in eq. (18) is the result of having taken into account the boundary layer momentum equation in the y direction (which says that $\bar{P}$ in the boundary layer is a function of x only; see Chapter 2). Equations (18) and (19), in conjunction with the two-dimensional mass continuity equation

$$\frac{\partial\bar{u}}{\partial x} + \frac{\partial\bar{v}}{\partial y} = 0 \tag{20}$$

represent the time-averaged conservation laws of the boundary layer region. These equations look very much like their counterparts in laminar flow [eqs. (26), (27), and (7) of Chapter 2] with one important difference, namely, the $(\partial/\partial y)\overline{(u'v')}$, $(\partial/\partial y)\overline{(v'T')}$ terms appearing in eqs. (18) and (19). These terms represent two additional unknowns and account for the closure problem in the two-dimensional boundary-layer geometry.

It is instructive to rewrite eqs. (18) and (19) as

$$\bar{u}\frac{\partial \bar{u}}{\partial x} + \bar{v}\frac{\partial \bar{u}}{\partial y} = -\frac{1}{\rho}\frac{d\bar{P}}{dx} + \frac{1}{\rho}\frac{\partial}{\partial y}\left(\mu\frac{\partial \bar{u}}{\partial y} - \rho\overline{u'v'}\right) \tag{21}$$

$$\bar{u}\frac{\partial \bar{T}}{\partial x} + \bar{v}\frac{\partial \bar{T}}{\partial y} = \frac{1}{\rho c_P}\frac{\partial}{\partial y}\left(k\frac{\partial \bar{T}}{\partial y} - \rho c_P\overline{v'T'}\right) \tag{22}$$

and to ask the question: If the products $\overline{u'v'}$ and $\overline{v'T'}$ survive the time-averaging process, that is, if they are nonzero, then are they negative or positive? The answer is easy to see in Fig. 7.3, which shows two possibilities in the evolution of the instantaneous u velocity at some point N in the boundary layer. The left side of Fig. 7.3 presumes the existence of an eddy that causes a downward velocity through point N, $v' < 0$; if the instantaneous u profile is such that the fluid moves faster if situated farther from the wall, then the short-time effect of $v' < 0$ is to increase the longitudinal velocity at point N, in other words, to induce a positive fluctuation $u' > 0$. The reverse of this scenario is illustrated on the right side of Fig. 7.3 and, what is most interesting, the product $u'v'$ emerges as a negative quantity regardless of the sign of u' and v'. Based on this argument (which applies unchanged to figuring out the sign of $v'T'$), we suspect that the time-averaged products $\overline{u'v'}$ and $\overline{v'T'}$ are both negative. Since the size of the u' and T' fluctuations with which the flow responds to the postulated v' appears to depend on the steepness of the average $\bar{u}$ and $\bar{T}$

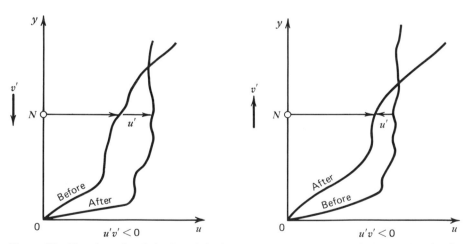

Figure 7.3 The short-time behavior of the instantaneous longitudinal velocity profile, showing how the product $u'v'$ survives the time-averaging of the x momentum equation.

profiles (Fig. 7.3), it makes sense to introduce the notation [10]

$$-\rho\overline{u'v'} = \rho\varepsilon_M \frac{\partial\bar{u}}{\partial y}, \quad \text{eddy shear stress}$$

$$-\rho c_P\overline{v'T'} = \rho c_P\varepsilon_H \frac{\partial\bar{T}}{\partial y}, \quad \text{eddy heat flux} \quad (23)$$

That $-\rho\overline{u'v'}$ and $-\rho c_P\overline{v'T'}$ represent shear stress and heat flux can be seen from the last terms of eqs. (21) and (22): for example, in the momentum equation (21) the molecular diffusion shear stress $\mu(\partial\bar{u}/\partial y)$ is augmented by the time-averaged eddy shear stress $(-\rho\overline{u'v'})$. Comparing the momentum and energy equations for turbulent boundary layer flow [eqs. (21) and (22)] with the laminar boundary layer equations (26) and (27) of Chapter 2, we see that the role played by shear stress and heat flux in turbulent flow is played by augmented expressions.[†]

$$\tau_{\text{app}} = \mu\frac{\partial\bar{u}}{\partial y} - \rho\overline{u'v'} = \rho(\nu + \varepsilon_M)\frac{\partial\bar{u}}{\partial y}, \quad \text{apparent shear stress}$$

$$-q''_{\text{app}} = k\frac{\partial\bar{T}}{\partial y} - \rho c_P\overline{v'T'} = \rho c_P(\alpha + \varepsilon_H)\frac{\partial\bar{T}}{\partial j}, \quad \text{apparent heat flux} \quad (24)$$

Substituting the new notation (23) into the boundary layer equations (21) and (22) yields

$$\bar{u}\frac{\partial\bar{u}}{\partial x} + \bar{v}\frac{\partial\bar{u}}{\partial y} = -\frac{1}{\rho}\frac{d\bar{P}}{dx} + \frac{\partial}{\partial y}\left[(\nu + \varepsilon_M)\frac{\partial\bar{u}}{\partial y}\right] \quad (25)$$

$$\bar{u}\frac{\partial\bar{T}}{\partial x} + \bar{v}\frac{\partial\bar{T}}{\partial y} = \frac{\partial}{\partial y}\left[(\alpha + \varepsilon_H)\frac{\partial\bar{T}}{\partial y}\right] \quad (26)$$

where ε_M and ε_H are two *empirical* functions known as *momentum eddy diffusivity* and *thermal eddy diffusivity*, respectively. Note that ε_M and ε_H are *flow properties*, not fluid properties. Although eqs. (25) and (26) look even more like their correspondents in laminar flow, no real improvement has taken place in our chances of solving the problem theoretically, that is, without relying on experiment. The problem consists of three equations [eq. (20), (25), and (26) outnumbered by five unknowns ($\bar{u}$, $\bar{v}$, $\bar{T}$, ε_M, and ε_H). The search for additional information to close the problem is the objective of the classical activity recognized as *turbulence modeling*. This empirical activity consists of

---

[†]Note that, as in Chapter 2, q'' is defined as positive when heat is transferred from the wall to the fluid.

closely scrutinizing the available body of experimental data in order to identify possible trends and flow characteristics that might lead us to generally applicable expressions (models) for ε_M and ε_H. The state-of-the-art in this area is being reviewed periodically by the *Turbulent Shear Flow Symposiums*, the proceedings of which are recommended to the student [11]. In the present treatment we discuss first the simplest and oldest models that lead to concise engineering formulas for friction and heat transfer; contemporary directions in turbulence modeling are reviewed for the interested student at the end of this chapter.

THE MIXING LENGTH MODEL

The order of magnitude of ε_M can be discussed based on the following scaling argument due to Prandtl [12]. In the boundary layer schematic of Fig. 7.3, imagine a ball of fluid that at some point in time is situated at a distance y where the mean longitudinal velocity is $\bar{u}(y)$. Imagine, next, that this ball migrates toward the wall to the new location $y - l$ where the mean velocity is $\bar{u}(y - l)$; the distance l is the *mixing length* along which the ball of fluid maintains its identity. Assuming that from (y) to $(y - l)$ the ball does not lose its longitudinal momentum, it is clear that the u' fluctuation produced by it at the new level $(y - l)$ is of the order of $\bar{u}(y) - \bar{u}(y - l)$, in other words,

$$O(u') = l\frac{\partial \bar{u}}{\partial y} \tag{27}$$

As argued earlier, in the motion of an eddy superimposed on the time-averaged motion, v' is of the same order of magnitude as u'; hence

$$O(v') = l\frac{\partial \bar{u}}{\partial y} \tag{28}$$

Recalling the message of Fig. 7.3, namely, the negative sign of $\overline{u'v'}$, we write

$$-\overline{u'v'} = l^2\left(\frac{\partial \bar{u}}{\partial y}\right)^2 \tag{29}$$

where l is the length scale associated with travel normal to the wall. Therefore from the definition of momentum eddy diffusivity [eqs. (23)]

$$\varepsilon_M = l^2\left|\frac{\partial \bar{u}}{\partial y}\right| \tag{30}$$

There is no general rule for estimating the mixing length l, since it clearly must vary from one type of flow to another. In a turbulent boundary layer, however, an upper bound for l must be the distance to the wall, therefore, we set

$$l = \kappa y \tag{31}$$

where κ is an empirical constant of order $O(1)$. As is shown later, a suitable value for *von Karman's constant* κ turns out to be 0.4, thus reconfirming the view that y is an upper bound for the mixing length of individual eddies. Combining eqs. (30) and (31) we conclude that Prandtl's mixing-length model for momentum eddy diffusivity is

$$\varepsilon_M = \kappa^2 y^2 \left| \frac{\partial \bar{u}}{\partial y} \right| \tag{32}$$

with the value of κ and the goodness of this model to be decided based on experiment.

Using eq. (32) or an equivalent model for ε_M, the mass and momentum conservation equations (20) and (25) could be integrated numerically to determine the flow field $\bar{u}, \bar{v}$. However, as shown below, some analytical progress can be made based on additional scaling arguments.

THE VELOCITY DISTRIBUTION

If we are primarily interested in how the turbulent boundary layer rubs against the wall and how it carries heat away from the wall, then we can imagine an *inner region* situated close enough to the wall such that the left-hand side of eq. (25) is sufficiently small. If the longitudinal pressure gradient dP/dx is zero, as in the case of uniform flow parallel to a flat wall, then the inner region is also characterized by an apparent shear stress $(\nu + \varepsilon_M)(\partial \bar{u}/\partial y)$ that does not vary with y [see eq. (25)]. Therefore, we can write after Prandtl (see Schlichting [13])

$$\left(\nu + \varepsilon_M \right) \frac{\partial \bar{u}}{\partial y} = \frac{\tau_0}{\rho} \tag{33}$$

in which τ_0 is the actual wall shear stress, that is, the value of τ_{app} at $y = 0$ where the Reynolds stress $-\rho \overline{u'v'}$ vanishes. Keep in mind that τ_0 is the engineering objective of the entire analysis; however, since the dimensions of $(\tau_0/\rho)^{1/2}$ are those of velocity, $(\tau_0/\rho)^{1/2}$ is recognized in the turbulence literature as the *friction velocity*[†]

$$u_* = \left(\frac{\tau_0}{\rho} \right)^{1/2} \tag{34}$$

[†] The fact that u_* has the dimensions of m/s does not mean that u_* is the appropriate *scale* of $\bar{u}$ in the boundary layer (the appropriate scale of $\bar{u}$ is U_∞). That the friction velocity is not the appropriate velocity scale is illustrated by the numerical values of u^+, which are very much different than $O(1)$ (see Fig. 7.4).

and is used in the following nondimensionalization of the flow problem,

$$u^+ = \frac{\bar{u}}{u_*}, \qquad v^+ = \frac{\bar{v}}{u_*}$$

$$x^+ = \frac{xu_*}{\nu}, \qquad y^+ = \frac{yu_*}{\nu} \tag{35}$$

In the above *wall coordinates*, the constant τ_{app} assumption (33) becomes

$$\left(1 + \frac{\varepsilon_M}{\nu}\right)\frac{du^+}{dy^+} = 1 \tag{36}$$

where it should be noted that u^+ is a function of y^+ only (hence, the d/dy^+ derivative sign), and that the x dependence is accounted for in the friction velocity u_* used in definitions (35).

The velocity distribution near the wall $u^+(y^+)$ results from integrating eq. (36) in conjunction with a suitable ε_M model such as eq. (32). The integration is considerably simpler and more instructive if we recognize that the $u^+(y^+)$ function produced by eq. (36) must have two distinct limiting behaviors depending on the relative size of ε_M and ν. That the ratio ε_M/ν must vary with the distance measured away from the wall is argued by the mixing-length model outlined in the preceding section. In the inner layer defined by the constant τ_{app} assumption (33), we can visualize two sublayers:

  (i)   The viscous sublayer, where $\nu \gg \varepsilon_M$.

  (ii) The fully turbulent sublayer (or the *turbulent core*), where $\varepsilon_M \gg \nu$.

These two sublayers mesh at some value of y^+, say, y^+_{VSL}, where ν and ε_M are of the same order of magnitude. Neglecting the term ε_M/ν in eq. (36) and integrating from the wall condition $u^+(0) = 0$, we learn that in the viscous sublayer the velocity profile is linear.

$$u^+ = y^+ \tag{37}$$

In the fully turbulent sublayer, the eddy diffusivity dominates, and eq. (36) reduces to

$$\frac{\varepsilon_M}{\nu}\frac{du^+}{dy^+} = 1 \tag{38}$$

or, using Prandtl's mixing-length model (32),

$$\kappa^2 (y^+)^2 \left(\frac{du^+}{dy^+}\right)^2 = 1 \tag{39}$$

Integrating this equation from the sublayer interface y^+_{VSL} (where, according to eq. (37), $u^+ = y^+_{VSL}$) to any y^+ in the fully turbulent sublayer yields

$$u^+ = \frac{1}{\kappa} \ln y^+ + y^+_{VSL} - \frac{1}{\kappa} \ln y^+_{VSL} \tag{40}$$

in other words,

$$u^+ = A \ln y^+ + B \tag{41}$$

where A and B are two empirical constants. Equation (41) is referred to as the *law of the wall* [14] and is attributed to both Prandtl and Taylor [15]. Fitting the logarithmic expression (41) to experimental measurements, it is found that the constants are approximately[†]

$$A \cong 2.5 \quad \text{and} \quad B \cong 5.5 \tag{42}$$

By examining eq. (40), note that the above constants imply also

$$\kappa \cong 0.4 \quad \text{and} \quad y^+_{VSL} \cong 11.6. \tag{43}$$

Equations (37) and (41) are one way (perhaps, the simplest) to empirically fit the $u^+(y^+)$ measurements that pertain to the inner region. The success of fitting the data with the law of the wall (41) (see Fig. 8 in Ref. 15) demonstrates the goodness of Prandtl's constant τ_{app} assumption (33) and mixing-length model (32). In fact, experimental measurements indicate that the law of the wall holds even in situations with finite pressure gradient [note the dP/dx assumption that preceded eq. (33)]. As expected, eqs. (37) and (41) fail to agree with experiment in the vicinity of $y^+ = y^+_{VSL}$ where neither ν nor ε_M can be neglected (see Fig. 7.4). A number of improved models have been proposed to smooth out the transition between the two limiting behaviors: these have been reviewed in Ref. 15 and are summarized in Table 7.1. It is not difficult to verify that the analytical expressions for $u^+(y^+)$ listed in Table 7.1 fall on the area covered by experimental measurements in the Purtell et al. [22] data reproduced in Fig. 7.4.

The experimental observation that the transition from the viscous sublayer to the fully turbulent sublayer takes place around $y^+_{VSL} = O(10)$ is an interesting example of the general applicability of the transition criterion discussed in Chapter 6. The nature and very existence of a viscous sublayer free of eddy motion has been the subject of debate because, for a long time, it was

[†] The numerical values of these constants vary from one report to another. For example, Coles [14] suggested $A = 2.5$ and $B = 5.1$, while Kays and Crawford [9] and Purtell et al. [22] (Fig. 7.4) use $A = 2.44$ and $B = 5$. In view of all the assumptions and all the handwaving that preceded the law of the wall (41), these slight adjustments in A and B are not crucial. In the present treatment we use the constants listed as eq. (42), as they happen to also correlate the $u^+(y^+)$ measurements in fully developed pipe flow: this topic is treated later in this chapter.

impossible to probe the velocity field so close to the wall. The development of new techniques during the past two decades (e.g., Laser–Doppler Anemometers) has made it possible to observe the evolution of the instantaneous velocity near the wall. The viscous sublayer picture that emerges is one of a laminar shear layer that develops (grows) until it becomes unstable and breaks down (see Ref. 1, pp. 656–684). The breakdown of the laminar layer is punctuated by a burst in which the slow fluid is ejected into the faster moving turbulent core. Kays and Crawford ([9], p. 165) associate the periodic breakdown of the viscous sublayer with the local Reynolds number becoming *supercritical*; in a time-averaged sense, the viscous sublayer maintains the same thickness $y_{VSL}^+ = O(10)$, regardless of the overall thickness of the boundary layer. As is shown in Fig. 7.4, as the overall boundary layer thickness increases, the viscous sublayer occupies a smaller fraction of the boundary layer.

To predict the viscous sublayer scale based on the time-scale argument of Chapter 6, imagine one event that is likely to repeat itself many times near the solid wall. Imagine the high-momentum fluid ball envisioned in the mixing-length argument that led to eq. (32). If this fluid packet runs into the wall,

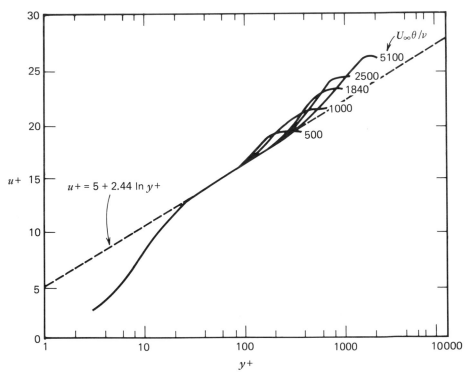

Figure 7.4 Example of $u^+(y^+)$ velocity measurements in turbulent boundary layer flow without longitudinal pressure gradient (after Ref. 22). Note the use of $A \cong 2.44$ and $B \cong 5$ to fit the data.

Table 7.1 Summary of Longitudinal Velocity Expressions for the Inner Region of a Turbulent Boundary Layer (after Kestin and Richardson [15])

$u^+(y^+)$	Range	References
$u^+ = y^+$ $u^+ = 2.5 \ln y^+ + 5.5$	$0 < y^+ < 11.6$ $y^+ > 11.6$	Prandtl and Taylor [13]
$u^+ = y^+$ $u^+ = 5 \ln y^+ - 3.05$ $u^+ = 2.5 \ln y^+ + 5.5$	$0 < y^+ < 5$ $5 < y^+ < 30$ $y^+ > 30$	von Kármán [16]
$u^+ = 14.53 \tanh(y^+/14.53)$ $u^+ = 2.5 \ln y^+ + 5.5$	$0 < y^+ < 27.5$ $y^+ > 27.5$	Rannie [17]
$\dfrac{du^+}{dy^+} = \dfrac{2}{1 + \left\{1 + 4\kappa^2 y^{+2}[1 - \exp(-y^+/A^+)]^2\right\}^{1/2}}$ $\kappa = 0.4 \quad A^+ = 26$	all y^+	van Driest [18]
$u^+ = 2.5 \ln(1 + 0.4 y^+)$ $+ 7.8\,[1 - \exp(-y^+/11)$ $- (y^+/11)\exp(-0.33 y^+)]$	all y^+	Reichardt [19]
$\dfrac{du^+}{dy^+} = \dfrac{1}{1 + n^2 u^+ y^+\left[1-\exp(-n^2 u^- y^+)\right]}$ $n = 0.124$ $u^+ = 2.78 \ln y^+ + 3.8$	$0 < y^+ < 26$	Deissler [20]
$y^+ = u^+ + A[\exp Bu^+ - 1 - Bu^+ - \tfrac{1}{2}(Bu^+)^2$ $- \tfrac{1}{6}(Bu^+)^3 - \tfrac{1}{24}(Bu^+)^4]$ (last term in u^{+4} may be omitted)	all y^+ $A = 0.1108$ $B = 0.4$	Spalding [21]

then, due to its high longitudinal momentum, it will give birth to a shear layer adjacent to the wall (Fig. 7.5). Initially the shear layer will be *laminar* because it is very thin and its transversal viscous communication time is very short. The laminar shear layer grows until the local Reynolds number based on local thickness becomes of order 10^2; hence

$$\frac{y_{\text{VSL}} U_\infty}{\nu} = O(10^2) \tag{44}$$

It was assumed that the velocity scale outside the laminar shear layer is U_∞, which is a reasonable upper bound for the longitudinal velocity of the fluid ball that ran into the wall. Under the same conditions, the wall shear stress scales as $\mu U_\infty / y_{\text{VSL}}$. The scale of y_{VSL}^+ at transition can be calculated as

$$y_{\text{VSL}}^+ = \frac{y_{\text{VSL}}}{\nu} \left(\frac{\tau_0}{\rho} \right)^{1/2} = \left(\frac{y_{\text{VSL}} U_\infty}{\nu} \right)^{1/2} \tag{45}$$

or using the transition criterion (44),

$$y_{\text{VSL}}^+ = O(10) \tag{46}$$

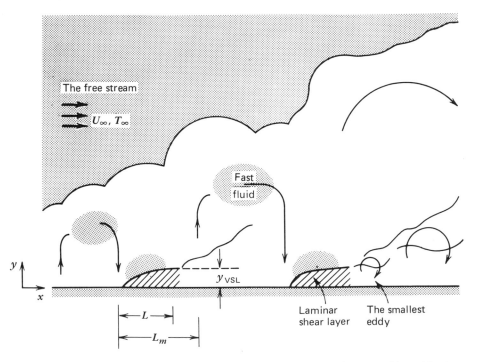

Figure 7.5 The formation of the viscous sublayer as the time-averaged superposition of laminar shear layers with thickness-based Reynolds numbers no greater than $O(10^2)$.

Since $O(10)$ is the measured scale of y^+_{VSL}, the scaling argument that led to eq. (46) implies that the time-averaged viscous sublayer is the superposition of many laminar shear layers terminated by buckling when $N_B = O(1)$ [6].

In conclusion, the time-scale arguments of Chapter 6 provide a purely theoretical basis for the *existence* of a viscous sublayer and for predicting the thickness of such a sublayer. They also provide a means for estimating the B constant in eq. (41) *without* relying on experiment [note the definition of B in terms of y^+_{VSL} and κ in eq. (40)]. An important implication of the $O(y^+_{VSL}) = 10$ analysis presented above is that the *first eddy* that forms immediately after the laminar shear layer reaches a local Reynolds number value of order 10^2 is also characterized by a local Reynolds number of order 10^2. According to the scenario sketched in Fig. 7.5, the first eddy is also the *smallest eddy* in the eddy population resulting from the repeated buckling and rolling-up of the shear layer. Thus, the smallest eddy Reynolds number (based on peripheral velocity and eddy diameter) is of order 10^2 [6]; this conclusion contradicts the often-quoted statement that the small-scale eddy motion is characterized by a Reynolds number equal to one (e.g., Ref. 2, p. 20).

Another interesting observation with which to punctuate this section is to answer the question: Why does the mixing-length-generated velocity profile (41) depend on *two* empirical constants (A and B, or κ and y^+_{VSL}), when in fact the mixing-length model (32) introduces only one such constant (κ)? It is important, I believe, to always question the origin and the proliferation of empirical constants in time-averaged analysis of turbulent flow. To find the answer to the question formulated above, let us derive the $u^+(y^+)$ profile as if we were unaware of the derivation taught by most textbooks [eqs. (37)–(41)]. Combining the mixing-length model (32) with the near-wall momentum equation (36) yields

$$\left[1 + \kappa^2(y^+)^2 \frac{du^+}{dy^+}\right] \frac{du^+}{dy^+} = 1 \qquad (47)$$

Note that this equation was integrated earlier in two extremes, each time by neglecting one of the terms appearing in the square brackets. But, since this time we are "unbiased", we proceed with eq. (47) alone and find that it can actually be integrated in closed form: integrating away from the wall, where the boundary condition is $u^+(0) = 0$, we obtain

$$\kappa u^+ = \frac{\cos\alpha - 1}{\sin\alpha} + \ln\left[\tan\left(\frac{\pi}{4} + \frac{\alpha}{2}\right)\right] \qquad (48)$$

where

$$\alpha = \arctan(2\kappa y^+) \qquad (49)$$

The reader can easily verify that, regardless of the value of κ, the above solution is not a good curvefit for the data presented in Fig. 7.4. For example, in the two limits discussed earlier in this section, the κ-dependent solution, (48) and (49), yields

$$u^+ \to y^+, \quad \text{as } y^+ \to 0$$

$$u^+ \to \frac{1}{\kappa}\ln y^+ + \frac{2\ln 2 + \ln \kappa - 1}{\kappa}, \quad \text{as } y^+ \to \infty \tag{50}$$

Furthermore, if $\kappa \cong 0.4$, the law of the wall recommended by this solution is

$$u^+ = 2.5 \ln y^+ - 1.325 \tag{51}$$

Comparing this result with the curve that agrees with the data [eqs. (41) and (42)], we conclude that the single-constant profile (48) fails in the high y^+ limit. To insure a reasonable curvefit in the high y^+ limit is the function of an *additional* constant, B [eq. (41)]. The additional constant is the result of a true sleight-of-hand, namely, the breaking-up of the integration of eq. (47) into the two cases (i) and (ii). The additional constant B represents the newly created degree of freedom associated with *joining* the two solutions of some intermediate y^+.

As shown in Fig. 7.4, the two-constant curvefitting of the velocity data works for sufficiently low values of y^+, say, $y^+ < 100$. This means that the constant τ_{app} assumption that led to the law of the wall (41) breaks down as we leave the wall region [the departure of $u^+(y^+)$ measurement away from the law of the wall is vividly demonstrated by the right side of Fig. 7.4]. The turbulent boundary layer region emerges as the sandwiching of two distinct zones—an inner zone[†] where the $\tau_{app} = $ constant assumption is fairly good and an outer zone where that assumption fails. Recognizing the logarithmic scale employed on the abscissa of Fig. 7.4, we note that the outer zone is generally much thicker than the constant τ_{app} layer. Furthermore, the thickness of the outer zone increases relative to that of the inner zone as the momentum thickness Reynolds number $U_\infty \theta / \nu$ increases (in other words, the outer zone becomes relatively thicker in the downstream x direction, as θ is expected to increase monotonically in x).

If the outer zone is a region where the $\tau_{app} = $ constant assumption fails, then according to the complete momentum boundary layer equation (25), the outer zone is ruled by a balance between inertia and changes in τ_{app}. It is shown in Chapter 8 that such a balance is characteristic of all turbulent shear flows in regions situated sufficiently far from solid walls (jets, wakes, plumes). For this reason and due to the similar appearance of turbulent wakes and the

---

[†] The inner zone was divided earlier into a viscous sublayer and a fully turbulent sublayer, in the discussion immediately above eq. (37).

outer regions of turbulent boundary layers, the turbulence literature often refers to the outer region of a boundary layer as *the wake region*. (Note that the outer region is the one visible in Figs. 7.1 and 7.2, and note the similarity between the large-scale meandering of this region and the meanders displayed by turbulent jets, wakes, and plumes.)

WALL FRICTION IN BOUNDARY LAYER FLOW

Of interest to engineers is the time-averaged friction force exerted by the turbulent boundary layer on the wall. The wall shear stress τ_0 or, in dimensionless form, the local skin-friction coefficient

$$C_{f,x} = \frac{\tau_0}{\frac{1}{2}\rho U_\infty^2} \tag{52}$$

can be derived from the longitudinal velocity measurements plotted in Fig. 7.4 (Note that τ_0 was already used to nondimensionalize both the abscissa and the ordinate in Fig. 7.4.) Let the function $f_u(y^+)$ be an appropriate curvefit for the velocity data of Fig. 7.4,

$$u^+ \cong f_u(y^+) \tag{53}$$

Examples of f_u expressions can be found in the left column of Table 7.1. Now, assuming that f_u fits the measurements sufficiently well near the high y^+ extremity of the profile (i.e., in the wake region), we can use the curvefit (53) to define an outer boundary layer thickness δ, such that the time-averaged velocity $\bar{u}$ calculated with eq. (53) equals U_∞ when y equals δ. Analytically, this definition amounts to applying eq. (53) at the point of $\bar{u} = U_\infty$ and $y = \delta$:

$$\frac{U_\infty}{(\tau_0/\rho)^{1/2}} \cong f_u\left[\frac{\delta}{\nu}\left(\frac{\tau_0}{\rho}\right)^{1/2}\right] \tag{54}$$

Equation (54) is the source of a particular formula for τ_0 or $C_{f,x}$, a formula that depends on the particular expression chosen for f_u. However, in order to derive this formula, we must determine the outer boundary layer thickness δ. To do this, we first recognize that the thickness δ and its x variation are intimately tied to the behavior of the wake region. In other words, the thickening of the boundary layer must be due to the progressive slowing down of outer layers of the free stream as the flow proceeds in the x direction. To account for this phenomenon, we must consider the complete form of the momentum equation for the boundary layer [eq. (25)]: integrating this equation across the boundary layer, and keeping in mind that in the present case

$d\bar{P}/dx = 0$ yields

$$\frac{d}{dx} \int_0^\infty \bar{u}(U_\infty - \bar{u})\, dy = \frac{\tau_0}{\rho}$$ (55)

The derivation of the more general momentum integral, for finite $d\bar{P}/dx$, is the object of Problem 6 at the end of this chapter.

Equations (54) and (55) are sufficient for determining both $\delta(x)$ and $\tau_0(x)$. For example, using Prandtl's 1/7th power law as the curvefit for the $u^+(y^+)$ data,

$$f_u = 8.7(y^+)^{1/7}$$ (56)

we obtain the following results [23, 24]

$$\frac{\tau_0}{\rho U_\infty^2} = 0.0225 \left(\frac{U_\infty \delta}{\nu} \right)^{-1/4}$$ (57)

$$\frac{\delta}{x} = 0.37 \left(\frac{U_\infty x}{\nu} \right)^{-1/5}$$ (58)

$$\delta = 8\delta^* = \frac{72}{7}\theta$$ (59)

Combining eqs. (57) and (58) yields the desired engineering result, namely, the skin-friction coefficient

$$\frac{\tau_0}{\rho U_\infty^2} = \tfrac{1}{2}C_{f,x} = 0.0296 \left(\frac{U_\infty x}{\nu} \right)^{-1/5}$$ (60)

This formula is shown plotted in Fig. 7.6 next to Schultz–Grunow's empirical correlation [25]

$$C_{f,x} = 0.37 \left[\log_{10} \left(\frac{U_\infty x}{\nu} \right) \right]^{-2.584}$$ (61)

The figure shows that the agreement between formula (60) and measurements deteriorates above Reynolds numbers $U_\infty x/\nu$ of order 10^7–10^8; the imperfect character of expression (60) can be traced back to the approximate character of the velocity profile curvefit f_u chosen as the pedestal for the whole analysis that produced eq. (60). Other curvefits f_u will certainly lead to friction formulas that differ from eq. (60) and to boundary-layer thickness formulas that differ from eq. (58). This last observation is particularly relevant to understanding the not-so-fundamental character of the notion that the turbulent boundary-layer thickness δ always varies as $x^{4/5}$, as might be memorized from eq. (58) or

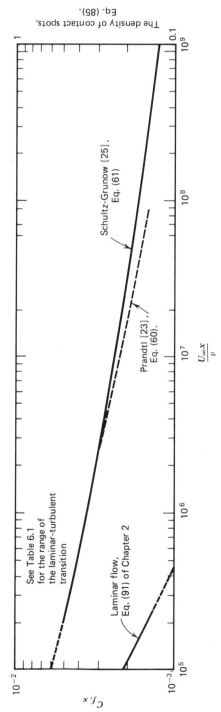

Figure 7.6 Local skin-friction coefficient for turbulent boundary layer flow over a flat wall.

247

a handbook such as Ref. 13. What varies as $x^{4/5}$ is the distance normal to the wall, calculated by intersecting $\bar{u} = U_\infty$ with an arbitrarily chosen curvefit of the velocity profile [eq. (56)]. The distance defined in such an arbitrary manner is certainly not the distance from the wall to the stepped interface so evident in the turbulent boundary layer photographs shown as Figs. 7.1 and 7.2.

HEAT TRANSFER IN BOUNDARY LAYER FLOW

For the heat transfer part of the turbulent boundary layer problem, we focus on the time-averaged energy equation (26), and make the assumption that sufficiently close to the solid wall, the left-hand side of eq. (26) becomes negligible. This move is analogous to the constant τ_{app} assumption made earlier in connection with the velocity profile. Thus, the energy equation reduces to the statement that sufficiently close to the wall the apparent heat flux q''_{app} does not depend on y,

$$(\alpha + \varepsilon_H)\frac{\partial \bar{T}}{\partial y} = \left[(\alpha + \varepsilon_H)\frac{\partial \bar{T}}{\partial y}\right]_{y=0} \tag{62}$$

In other words,

$$(\alpha + \varepsilon_H)\frac{\partial \bar{T}}{\partial y} = \frac{-q''_0}{\rho c_P} \tag{63}$$

As we shall demonstrate shortly, the assumed constancy of q''_{app} leads in relatively few steps to an analytical expression for the temperature distribution in the close vicinity of the wall. Introducing the wall coordinate notation (35), the constant q''_{app} statement (63) becomes

$$\frac{\rho c_P u_*}{-q''_0}\frac{\partial}{\partial y^+}\left(\bar{T} - T_0\right) = \frac{1}{\alpha/\nu + \varepsilon_H/\nu} \tag{64}$$

This form is the basis for defining the following temperature variable in wall coordinates

$$T^+(x^+, y^+) = \left(T_0 - \bar{T}\right)\frac{\rho c_P u_*}{q''_0} \tag{65}$$

In this notation, the integral of eq. (64) reads

$$T^+ = \int_0^{y^+} \frac{dy^+}{1/\mathrm{Pr} + (1/\mathrm{Pr}_t)(\varepsilon_M/\nu)} \tag{66}$$

We see that the temperature profile in the q''_{app} = constant region is governed by the Prandtl number, the turbulent Prandtl number ($Pr_t = \varepsilon_M/\varepsilon_H$), and, via ε_M/ν, the velocity distribution in the same region [see eq. (36)]. The integral appearing on the right-hand side of eq. (66) can be evaluated in closed form based on further assumptions regarding Pr, Pr_t, and ε_M/ν. To begin with, we note that in the fully turbulent region of the constant τ_{app} layer, eqs. (36) and (41) yield

$$\frac{\varepsilon_M}{\nu} = \frac{dy^+}{du^+} = \kappa y^+ \tag{67}$$

Therefore, according to the mixing-length model, the second term in the denominator of the integrand in (66) increases steadily as y^+ increases. Considering the range of values taken by y^+ (Fig. 7.4) and assuming that both Pr and Pr_t do not depart too drastically from $O(1)$, chances are good that in the integrand of eq. (66) the term $(\varepsilon_M/\nu)/Pr_t$ will outweigh the term $1/Pr$ if y^+ is sufficiently large. This observation suggests the following two-step integration of (66):

$$T^+ = \int_0^{y^+_{CSL}} \frac{dy^+}{\dfrac{1}{Pr} + \left(\begin{array}{c} \text{negligible} \\ \text{term} \end{array}\right)} + \int_{y^+_{CSL}}^{y^+} \frac{dy^+}{\left(\begin{array}{c} \text{negligible} \\ \text{term} \end{array}\right) + \dfrac{1}{Pr_t}\dfrac{\varepsilon_M}{\nu}} \tag{68}$$

where y^+_{CSL} is the dimensionless thickness of a conduction sublayer in which the molecular mechanism outweighs the eddy transport of heat. Combining eqs. (68) and (67) and regarding Pr_t as y independent, we obtain a broken line expression for the temperature profile.

$$T^+ = \begin{cases} Pr\, y^+, & \text{if } y^+ < y^+_{CSL} \\ Pr\, y^+_{CSL} + \dfrac{Pr_t}{\kappa}\ln\dfrac{y^+}{y^+_{CSL}}, & \text{if } y^+ > y^+_{CSL} \end{cases} \tag{69}$$

This result depends on three empirical constants, Pr_t, κ, and y^+_{CSL}. According to Ref. 26, good agreement with temperature measurements is achieved if

$$Pr_t \cong 0.9, \qquad \kappa \cong 0.41, \qquad y^+_{CSL} \cong 13.2 \tag{70}$$

provided that the Prandtl number of the fluid is in the range 0.5–5. Substituting these values into the $y^+ > y^+_{CSL}$ portion of the T^+ profile (69) yields [26]

$$T^+ = 2.195\ln y^+ + 13.2\, Pr - 5.66 \tag{71}$$

We conclude that in the fully turbulent region of the layer where q''_{app} is constant, the temperature profile T^+ is analytically the same as the law of the wall encountered earlier [eq. (41)].

A wall heat flux formula consistent with the temperature distribution developed above can be derived by making the assumption that eq. (69) holds well enough near the outer edge of the boundary layer (i.e., at the edge of the wake region). This assumption is, of course, made for analytical convenience, since the wake region is neither one of constant q''_{app} nor one of constant τ_{app}. However, this assumption turns out to be a fairly good one, and the heat-transfer coefficient formula facilitated by it turns out to be very instructive. Setting $\bar{T} = T_\infty$ at $y = \delta$ in eq. (69), we obtain

$$\rho c_p u_* \frac{T_0 - T_\infty}{q''_0} = \Pr y^+_{\text{CSL}} + \frac{\Pr_t}{\kappa} \ln\left(\frac{\delta u_*/\nu}{y^+_{\text{CSL}}}\right) \tag{72}$$

Note that the heat-transfer coefficient $h = q''_0/(T_0 - T_\infty)$ appears explicitly and that δ is an unknown. In order to determine δ we might be tempted to rely on eq. (58): this would be inappropriate, because Prandtl's formula for $\delta(x)$ is based on the 1/7th power law velocity profile, while the temperature profile (69) is based on the law of the wall. Therefore, to be consistent, we apply the law of the wall (41) at the outermost edge of the turbulent boundary layer,

$$\frac{U_\infty}{u_*} = \frac{1}{\kappa} \ln\left(\frac{\delta u_*}{\nu}\right) + B \tag{73}$$

Eliminating $(\delta u_*/\nu)$ between eqs. (72) and (73) and keeping in mind that by definition

$$\frac{U_\infty}{u_*} = \left(\frac{2}{C_{f,x}}\right)^{1/2} \tag{74}$$

we obtain the desired heat transfer result,

$$\frac{h}{\rho c_p U_\infty} = \frac{\frac{1}{2}C_{f,x}}{\Pr_t + \left(\frac{1}{2}C_{f,x}\right)^{1/2}\left(\Pr y^+_{\text{CSL}} - B\Pr_t - \frac{\Pr_t}{\kappa}\ln y^+_{\text{CSL}}\right)} \tag{75}$$

The left-hand side of this equation is a dimensionless way of reporting heat-transfer coefficient in turbulent flow, namely, the local *Stanton number*.

$$\text{St}_x = \frac{h}{\rho c_p U_\infty} \tag{76}$$

The right-hand side of eq. (75) can be refined by using appropriate values for the empirical constants: for example, using the constants listed as eq. (70) and taking $B \cong 5.1$, the heat transfer rate formula becomes [26]

$$\text{St}_x = \frac{\frac{1}{2}C_{f,x}}{0.9 + \left(\frac{1}{2}C_{f,x}\right)^{1/2}(13.2\Pr - 10.25)} \tag{77}$$

This result is worth thinking about before proceeding further. Looking at the already weak relationship between $C_{f,x}$ and Re_x displayed by Fig. 7.6, we get the feeling that the denominator in expression (77) is not very sensitive to changes in the Reynolds number. Also, the Prandtl number Pr was already assumed to be in the range 0.5–5, which means that the denominator is of order $O(1)$ and Pr-dependent. We reach the interesting conclusion that at any x along the wall, the Stanton number and the skin-friction coefficient are proportional and of the same order of magnitude, and that the proportionality factor is a function of the Prandtl number. Indeed, heat transfer measurements over a wider Pr range ($0.6 < Pr < 60$) satisfy a famous *empirical formula* suggested by Colburn [27, 28]

$$St_x Pr^{2/3} = \tfrac{1}{2} C_{f,x} \tag{78}$$

The analysis outlined here between eqs. (63) and (77) is a simplified version of what von Karman conceived based on a smoother, three-region, integration of eq. (66) [16]. The three y^+ regions considered by von Karman are listed in Table 7.1. If $Pr_t = 1$, the local Stanton number expression produced by this analysis is

$$St_x = \frac{\tfrac{1}{2} C_{f,x}}{1 + 5\left(\tfrac{1}{2} C_{f,x}\right)^{1/2} \left\{ Pr - 1 + \ln\left[1 + \tfrac{5}{6}(Pr - 1)\right] \right\}} \tag{79}$$

It is instructive to plug numerical values for Re_x and Pr into formulas (77) and (79) to discover that they both predict practically the same heat transfer rate as Colburn's empirical correlation (78). The success of an extremely compact analytical expression to consistently convey the same message as fancier formulas produced by increasingly fancier analyses suggests that Colburn's formula is the carrier of fundamental information regarding the very nature of turbulent heat transfer near a solid wall in both boundary layer flow and duct flow. I use this opportunity to present a theory [29] that predicts[†] the scaling law documented by Colburn's formula (78).

Of pedagogic interest is the fact that the Pr = 1 equivalent of Colburn's formula can be derived rather easily based on an argument recognized as the *Reynolds analogy* between heat transfer and wall friction in turbulent flow [30],

$$St_x = \tfrac{1}{2} C_{f,x}, \qquad (Pr = 1) \tag{80}$$

The same result follows from setting Pr = 1 in von Karman's expression (79). To derive the Reynolds analogy from the basic premises of this chapter, recall

---

[†] Colburn did not offer any theoretical basis for the $Pr^{2/3}$ factor appearing in eq. (78). A number of authors before him proposed similar empirical formulas with the Pr exponent ranging from 0.6 to 0.7: Colburn chose the $0.66 = 2/3$ exponent "because it is more or less an average value" [27].

that the near-wall region is characterized by both constant τ_{app} and constant q''_{app},

$$\tau_0 = (\mu + \rho\varepsilon_M)\frac{d\bar{u}}{dy} \tag{81}$$

$$-q''_0 = (k + \rho c_P \varepsilon_H)\frac{d\bar{T}}{dy} \tag{82}$$

If Pr = 1 and $\varepsilon_M = \varepsilon_H$, then dividing eqs. (81) and (82) yields

$$\frac{\tau_0}{-q''_0} = \frac{1}{c_P}\frac{d\bar{u}}{d\bar{T}} \tag{83}$$

Integrating this from the wall ($\bar{u} = 0, \bar{T} = T_0$) to the free stream ($\bar{u} = U_\infty, \bar{T} = T_\infty$; note the reality-bending fault of this boundary condition) leads to

$$\frac{\tau_0}{-q''_0} = \frac{U_\infty}{c_P(T_\infty - T_0)} \tag{84}$$

Recalling the definitions of the skin-friction coefficient [eq. (52)] and the Stanton number [eq. (76)], we conclude that eq. (84) is the same as eq. (80). Considering the $Pr_t = 1$ assumption that led to eqs. (80) and (84), the Reynolds analogy is nothing but shorthand for the view that each eddy has the same propensity to convect heat as it has to transfer momentum in the direction normal to the wall. Although the Reynolds analogy works for Pr $\cong$ 1 fluids such as the common gases, it must be distinguished from Colburn's formula (78) because any time we try to relax the Pr = Pr_t = 1 assumptions in time-averaged analysis we end up with St_x expressions that *never* reproduce the Prandtl number scaling envisioned by Colburn [e.g., eqs. (77) and (79)].

The above observation is in fact a very good clue for the path we might follow towards deriving the Colburn scaling on a purely theoretical basis. I can think that if time-averaged analysis never leads to Colburn's correlation of so many experimental results, then perhaps the opposite sort of analysis does. To try the opposite analysis (in this case, an analysis based on the view that the turbulent boundary layer is a coarse structure pulsating in time, not a smooth picture that does not change in time) is to proceed *against method* [31]. It is worth noting that by recognizing the fluctuating character of the boundary layer, we have already been able to predict the existence of a viscous sublayer as well as its time-averaged thickness [see Fig. 7.5 and eqs. (44)–(46)]. Relying on the same point of view, we conclude that the instantaneous distribution of shear stress $\tau_0(t)$ and heat flux $q''_0(t)$ along the wall of Fig. 7.5 must be as is qualitatively shown in Fig. 7.7. Each spot on the wall that is instantaneously in contact with U_∞-fast and T_∞-cold fluid, is instantaneously characterized by maximum shear stress and heat flux. The in-between regions, being covered by slow and already hot fluid that is ejected from the wall, are regions of substantially lower shear stress and heat flux. Instantaneous pictures of $\tau_0(t)$ and $q_0''(t)$ (of the type sketched in Fig. 7.7) dance along the $x \rightarrow x'$ wall section as time passes. Therefore, in an order of magnitude sense, the time-averaged quantities $\tau_0(x)$ and $q_0''(x)$ are both related to the local *density*

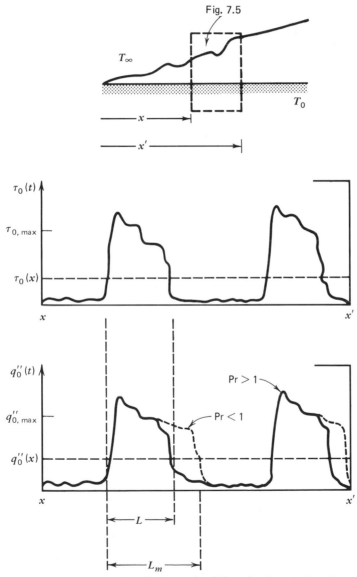

Figure 7.7 The analogy between skin friction and wall heat flux in turbulent flow near a solid wall.

of contact spots defined as

$$\eta(x) = \frac{\text{cumulative length of direct contact spots}}{\text{total length of sample wall section } (x \rightarrow x')} \tag{85}$$

That relationship is

$$\frac{\tau_0}{\tau_{0,\max}} \sim \eta \sim \frac{q_0''}{q_{0,\max}''} \tag{86}$$

Focusing for the time being on only the extreme ends of this relation and writing the time-averaged τ_0 and q_0'' in terms of the local coefficients $C_{f,x}$ and St_x, we obtain

$$\frac{\mathrm{St}_x}{\frac{1}{2}C_{f,x}} \sim \frac{q_{0,\max}''}{T_0 - T_\infty} \frac{U_\infty}{c_P \tau_{0,\max}} \tag{87}$$

Finally, since the direct contact spots are covered with laminar shear flow, and since outside these laminar layers the flow temperature conditions are described by U_∞ and T_∞, the scales of $\tau_{0,\max}$ and $q_{0,\max}''$ are (see Chapter 2)

$$\tau_{0,\max} \sim \rho U_\infty^2 \left(\frac{U_\infty L}{\nu}\right)^{-1/2} \tag{88}$$

$$\frac{q_{0,\max}''}{T_0 - T_\infty} \sim \frac{k}{L}\mathrm{Pr}^{1/3}\left(\frac{U_\infty L}{\nu}\right)^{1/2}, \qquad (\mathrm{Pr} > 1) \tag{89}$$

In eqs. (88) and (89), L is the longitudinal length scale of each laminar shear layer. Combining eqs. (87)–(89), we obtain the following scaling law

$$\frac{\mathrm{St}_x}{\frac{1}{2}C_{f,x}} \sim \mathrm{Pr}^{-2/3}, \qquad (\mathrm{Pr} > 1) \tag{90}$$

This is the same as the scaling demonstrated by Colburn's formula (78). Therefore, experimental measurements of both wall friction and wall heat flux support the scaling laws of Chapter 6 which, after all, were the basis for the intermittent-contact scenario illustrated in Figs. 7.5 and 7.7.

There is more to the above theory than the ability to derive Colburn's formula. Regarding the geometric concept of *density of contact spots* [eq. (85)], we can easily estimate the order of magnitude of η,

$$\eta \sim \frac{\tau_0}{\tau_{0,\max}} \sim \frac{C_{f,x}}{(U_\infty L/\nu)^{-1/2}} \tag{91}$$

Recalling that the growth of each laminar layer is terminated when the thickness-referenced Reynolds number is of order 10^2 (Chapter 6) and that in laminar shear flow the thickness varies as $L^{1/2}$, we conclude that the Reynolds number based on spot length scale L has a characteristic order of magnitude,[†]

$$\frac{U_\infty L}{\nu} \sim 10^4 \qquad (92)$$

Substituting this result into the density formula (91), we find

$$\eta \sim 10^2 C_{f,x} \qquad (93)$$

This conclusion is very interesting. Looking at Fig. 7.6, which shows that $C_{f,x}$ is consistently of order 10^{-3}–10^{-2}, we learn that the density of contact spots is of order 0.1–1; in other words, the laminar contact regions cover a significant percentage of the wall {this is why the laminar shear layer thickness alone predicted successfully the time-averaged viscous sublayer thickness [eqs. (44)–(46)]}. Most important, however, is the proportionality between η and $C_{f,x}$: this proportionality gives us a geometric interpretation for the meaning of the skin-friction coefficient in turbulent boundary layer flow and, in addition, eq. (93) explains why the $C_{f,x}$ values are so much smaller than unity.

The turbulent boundary layer emerges as a sequence of laminar spots terminated when their buckling number N_B exceeds $O(1)$ (Chapter 6), a population whose density decreases very gradually as the outer thickness of the boundary layer increases in the downstream direction.

The scale analysis that led to eq. (90) demonstrates that the so-called Colburn analogy (78) is valid in the $\Pr > 1$ range. What, then, is the corresponding scaling law for liquid-metal near-wall turbulence? To answer this question, we must keep in mind that in $\Pr < 1$ fluids the thermal diffusivity is such that, given the same fluid layer thickness, the thermal communication time across the layer is shorter than the viscous communication time. Thus, although the laminar spot breaks down when it reaches a length L, the direct thermal contact between wall (T_0) and outer fluid (T_∞) persists over a longer length L_m. In other words, if the Prandtl number is sufficiently small, the wall will communicate via thermal diffusion even across the first (smallest) eddies formed in the wake of laminar spot breakdown. As in the case of transition to turbulence (Chapter 6), the maximum eddy diameter D_T across which the thermal diffusion effect will still have time to travel can be estimated by equating the thermal diffusion time $D_T^2/(16\alpha)$ with the rolling period $D_T/(U_\infty/2)$; hence

$$\frac{D_T U_\infty}{\alpha} = O(10^2) \qquad (94)$$

[†]Note that this prediction is consistent also with Elder's observations of transition in a laminar boundary layer in the presence of disturbances (see Table 6.1).

From the study of laminar thermal boundary layers in liquid metals (Chapter 2), we recall that

$$\frac{D_T}{L_m} \sim \mathrm{Pr}^{-1/2}\left(\frac{U_\infty L_m}{\nu}\right)^{-1/2} \tag{95}$$

Combining eqs. (94) and (95), we find

$$\frac{L_m U_\infty}{\nu} \sim 10^4 \mathrm{Pr}^{-1} \tag{96}$$

or, using eq. (92),

$$\frac{L_m}{L} \sim \frac{1}{\mathrm{Pr}} > 1 \tag{97}$$

We conclude that for liquid metals, which occupy the $O(\mathrm{Pr})$ range $10^{-1} - 10^{-3}$, the ratio L_m/L evaluated above would be of order $10 - 10^3$. However, since L_m cannot be greater than L/η [Fig. 7.7 and eq. (85)], and since η is of order 10^{-1} (Fig. 7.6), we learn that the ratio L_m/L cannot exceed $O(10)$. This means that in liquid metals the intermittent thermal contact between free stream and wall is not terminated by criterion (94) but by the next inrush of free-stream fluid (note that the longitudinal length scale between two consecutive "inrush" events is of order L/η). The thermal contact between wall and free-stream is characterized therefore by the front portion of a thermal boundary layer of length L/η in the $\mathrm{Pr} \to 0$ limit (Chapter 2):

$$q_0'' \sim \frac{k(T_0 - T_\infty)}{L/\eta} \mathrm{Pr}^{1/2}\left(\frac{U_\infty L/\eta}{\nu}\right)^{1/2} \tag{98}$$

Combining this estimate with eqs. (92, 93) we anticipate the following scaling law between St_x and $C_{f,x}$ for liquid metals

$$\frac{St_x}{\left(1/2 C_{f,x}\right)^{1/2}} \sim 10^{-1} \mathrm{Pr}^{-1/2} \tag{99}$$

The validity of eq. (99) cannot be determined directly against experiment because of the absence of heat transfer measurements for liquid metal turbulent boundary layer flow. The absence of measurements is explained by the absence of practical applications that might demand this sort of information. However, liquid metal heat transfer data for duct flow exist and, as shown in the next section, those data are consistent with eq. (99).

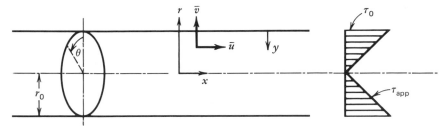

Figure 7.8 Round tube geometry and the distribution of apparent shear stress in fully developed turbulent flow.

DUCT FLOW

The time-averaged analysis of friction and heat transfer in turbulent duct flow involves many of the assumptions already made in the presentation of the boundary layer problem. Since differences between the two analyses exist, and in order to avoid repetition, we will highlight only the differences. Consider turbulent flow through the round tube geometry sketched in Fig. 7.8. The time-averaged equations for mass, momentum, and energy are

$$\frac{\partial \bar{u}}{\partial x} + \frac{1}{r}\frac{\partial}{\partial r}(r\bar{v}) = 0 \tag{100}$$

$$\bar{u}\frac{\partial \bar{u}}{\partial x} + \bar{v}\frac{\partial \bar{u}}{\partial r} = -\frac{1}{\rho}\frac{d\bar{P}}{dx} + \frac{1}{r}\frac{\partial}{\partial r}\left[r(\nu + \varepsilon_M)\frac{\partial \bar{u}}{\partial r}\right] \tag{101}$$

$$\bar{u}\frac{\partial \bar{T}}{\partial x} + \bar{v}\frac{\partial \bar{T}}{\partial r} = \frac{1}{r}\frac{\partial}{\partial r}\left[r(\alpha + \varepsilon_H)\frac{\partial \bar{T}}{\partial r}\right] \tag{102}$$

Note that eqs. (101) and (102) have already been simplified based on boundary layer arguments of the type employed earlier in this chapter [see eqs. (25) and (26)].

Friction Factor

The first feature that distinguishes the pipe flow from the boundary layer flow is that if the pipe flow is fully developed[†] the inertia terms vanish from the left-hand side of eq. (101). This means that in pipe flow there is no room for a wake region, not even in the close vicinity of the centerline. Integrating the momentum equation (101) in conjunction with the control volume force balance

$$-\frac{d\bar{P}}{dx} = 2\frac{\tau_0}{r_0} \tag{103}$$

[†]Review the beginning of Chapter 3 for the scaling implications of the concept of fully developed flow.

yields

$$\frac{\tau_{app}}{\tau_0} = 1 - \frac{y}{r_0} \qquad (104)$$

where $y = r_0 - r$ is the transversal coordinate measured away from the wall (Fig. 7.8), and

$$\tau_{app} = \rho(v + \varepsilon_M)\frac{\partial \bar{u}}{\partial y} \qquad (105)$$

According to the fully developed flow momentum equation (104), the apparent shear stress τ_{app} decreases linearly to the pipe centerline. Close enough to the wall (i.e., such that $y \ll r_0$), the turbulent flow is one of constant τ_{app}. For this reason, the mixing-length analysis that produced the law of the wall (41) for boundary layer flow could be applied here as well, with the understanding that it necessarily breaks down near the pipe centerline (where τ_{app} is no longer approximately equal to τ_0). It is found that the broken line velocity distribution represented by eqs. (37) and (41) (with $A \cong 2.5$ and $B \cong 5.5$) fits measurements sufficiently well. The major drawback of the $\tau_{app} \cong \tau_0$ approximation of eq. (104) is that it produces a velocity profile with a finite slope at the pipe centerline,

$$\left(\frac{du^+}{dy^+}\right)_{y=r_0} = 2.5\frac{v}{r_0(\tau_0/\rho)^{1/2}} \qquad (106)$$

An empirical velocity profile with zero slope at the centerline was proposed by Reichardt [19],

$$u^+ = 2.5\ln\left[\frac{3(1 + r/r_0)}{2[1 + 2(r/r_0)^2]}y^+\right] + 5.5 \qquad (107)$$

This profile becomes identical to the law of the wall (41) in the limit $y^+ \to 0$.

To each analytical expression that fits the measured velocity distribution corresponds a certain formula for τ_0 or its dimensionless counterpart, the friction factor (Chapter 3)

$$f = \frac{\tau_0}{\frac{1}{2}\rho U^2} \qquad (108)$$

As in all duct flow problems, the role of velocity scale U is played by the velocity averaged over the duct cross-section; in the geometry of Fig. 7.8, the average velocity is

$$U = \frac{1}{\pi r_0^2}\int_0^{2\pi}\int_0^{r_0}\bar{u}\,r\,dr\,d\theta \qquad (109)$$

To illustrate the derivation of a friction factor formula, consider Prandtl's

1/7th power law (56) under the assumption that it holds all the way to the pipe centerline ($\bar{u} = U_c$, $y = r_0$)

$$\frac{U_c}{(\tau_0/\rho)^{1/2}} \cong 8.7 \left[\frac{r_0(\tau_0/\rho)^{1/2}}{\nu} \right]^{1/7} \tag{110}$$

Noting that the f definition (108) implies $(\tau_0/\rho)^{1/2} = U(f/2)^{1/2}$, and that eq. (109) provides the needed relationship between the centerline velocity U_c and the average velocity U, the friction factor formula derived from eq. (110) is (Problem 9)

$$f \cong 0.078 \, \text{Re}_D^{-1/4} \tag{111}$$

In this notation, Re_D is the Reynolds number based on average velocity and pipe diameter, $D = 2r_0$. Formula (111) agrees well with measurements for Re_D values up to 80,000 [12]; this formula is nearly identical to the one proposed by Blasius in 1913, $f \cong 0.0791 \, \text{Re}_D^{-1/4}$ [32].

An alternative formula that has wider applicability is obtained working with the law of the wall, $u^+ = 2.5 \ln y^+ + 5.5$, [eq. (41)], instead of the 1/7th power law. The result is [12]

$$\frac{1}{f^{1/2}} = 1.737 \ln\left[\text{Re}_D f^{1/2} \right] - 0.396 \tag{112}$$

which agrees with measurements for Re_D values of up to $O(10^6)$. The heat transfer literature refers to eq. (112) as the Karman–Nikuradse relation [33]: this relation is displayed as the lowest curve in Fig. 7.9. The figure shows that regardless of the Reynolds number, the friction factor in turbulent flow is considerably greater than that in laminar flow, in the *hypothetical case* that the laminar regime can exist at such large Reynolds numbers. The same observation can be made in connection with the local skin-friction coefficient $C_{f,x}$ displayed in Fig. 7.6. The fact that for the same Reynolds number, the values of f and $C_{f,x}$ in turbulent flow are greater than their counterparts in imaginary laminar flow, might tempt the student to think that turbulent flows are more viscid than laminar flows. This thought would be in grave error: as pointed out in Chapter 6, if there exists a viscid flow, then that can only be the laminar flow (i.e., the flow that is thoroughly penetrated by the viscous information emitted by its confining walls). The friction factor is greater in turbulent flow than in the imaginary laminar case, because in turbulent flow the shear flow is a sequence of laminar shear layers (Fig. 7.5) whose thicknesses are much smaller than the pipe radius (i.e., much smaller than the shear layer thickness in the imaginary laminar flow).

The jump exhibited by both f and $C_{f,x}$ across the laminar–turbulent transition [Figs. 7.6 and 7.9] says something about the nature of the turbulent

Surface condition	k_s [mm]
Riveted steel	0.9-9
Concrete	0.3-3
Wood stave	0.18-0.9
Cast iron	0.26
Galvanized iron	0.15
Asphalted cast iron	0.12
Commercial steel or Wrought iron	0.05
Drawn tubing	0.0015

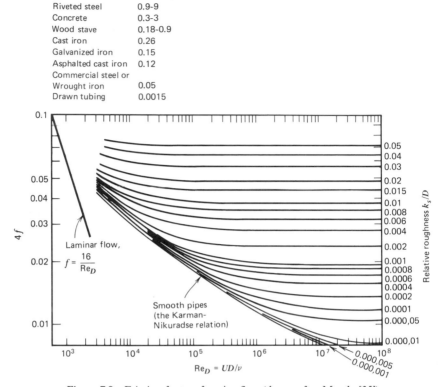

Figure 7.9 Friction factors for pipe flow (drawn after Moody [35]).

flow in the vicinity of each laminar contact region. If the duct flow is in the Re_D range that makes the transition possible, and if turbulence occurs in one isolated region on the wall, then the flow will redistribute itself so as to go around the turbulent spot. This effect is due to a greater resistance to flow through the turbulent region than through the remaining laminar region, as both regions are subjected to the same driving pressure gradient. The same flow redistribution effect is a well-known problem that causes the clogging of parallel heat-exchanger tubes (the more fouled a tube, the slower the flow through it; hence, the greater its fouling tendency; the opposite feedback mechanism works in the parallel tube that has to make up the flowrate lost in the fouled tube). Closer to home, the imbalance of flow through two parallel ducts is why we breather through one nostril most of the time. Thus, each spot of direct contact between wall and outer fluid (Fig. 7.5) is a region of *three-dimensional* flow, as the surrounding fluid must go around what it perceives as a region of relatively higher resistance. These words are illustrated with great clarity by A. Anand's photograph of transition to turbulence in a water rivulet sandwiched between two glass plates [38] (Fig. 7.10). The laminar

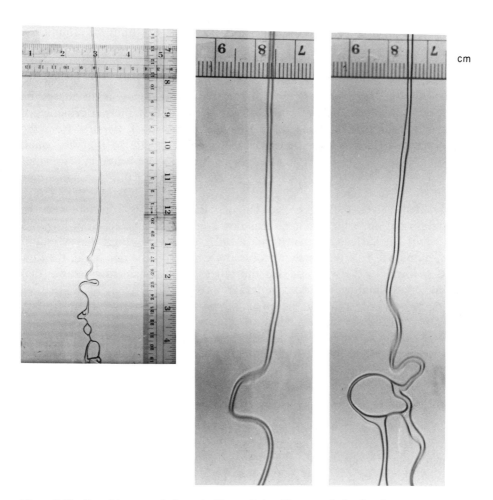

cm

Figure 7.10 Transition to turbulence in Hagen–Poiseuille water rivulet flow between two transparent parallel plates [38]; plate-to-plate spacing = 1.45 mm; volumetric flowrate = 1.35 ml/s, transition Reynolds number based on mean velocity and plate-to-plate spacing ≅ 1950. Left side = overall view. Right side = close-ups of the transition zone.

straight rivulet is a strand of the Hagen–Poiseuille flow that would have filled the entire parallel-plate channel: this strand shows that whenever a turbulent region develops, the flow seeks ways to go around this region (note that the Reynolds number based on gap thickness is about 2000, in agreement with Table 6.1).

For fully developed flow through ducts with cross-sections other than round, the Karman–Nikuradse relation (112) still holds if Re_D is replaced by the Reynolds number based on hydraulic diameter, Re_{D_h}. Note that for a duct of noncircular cross-section, the time-averaged τ_0 is not uniform around the periphery of the cross-section; hence, in definition (108), τ_0 is the perimeter-averaged wall shear stress.

From an engineering standpoint, an issue that must be taken into account in the calculation of pressure drop in turbulent flow through a straight duct is the effect of wall roughness. Experimentally it is found that the performance of commercial surfaces that do not feel rough to the touch departs from the performance of well-polished surfaces. This effect is due to the very small thickness acquired by the laminar sublayer in many applications [e.g., since $U_\infty y_{VSL}/\nu$ is of order $O(10^2)$, i.e., $y_{VSL}^+ = O(10)$, in water flow through a pipe with $U \sim 10$ m/s and $\nu \sim 0.01$ cm^2/s, y_{VSL} is approximately 0.01 mm!). Consequently, even slight imperfections of the surface may interfere with the natural formation of the laminar-shear-flow contact spots sketched in Figs. 7.5 and 7.7. If the surface irregularities are taller than y_{VSL}, then these "mountains" alone will rule the friction process by which, on the one hand, the flow exerts a force on the wall and, on the other, the wall generates eddies that partially slow the flow down in a boundary layer.

Nikuradse [34] measured the effect of surface roughness on the friction factor by coating the inside surface of pipes with sand of definite grain size glued as tightly as possible to the wall. If k_s is the grain size in Nikuradse's sand roughness, then the friction factor *fully rough limit* is given by

$$f \cong \left[1.74 \ln\left(\frac{D}{k_s}\right) + 2.28 \right]^{-2} \qquad (113)$$

The fully rough limit is that regime where the roughness size exceeds the order of magnitude of what would have been the laminar sublayer in time-averaged turbulent flow over a smooth surface,

$$k_s^+ = \frac{k_s(\tau_0/\rho)^{1/2}}{\nu} > O(10) \qquad (114)$$

The roughness effect described by Nikuradse is illustrated by the upper curves on the Moody chart (Fig. 7.9).

Moody's chart [35] is reproduced in Fig. 7.9 for two reasons. First, it represents a very useful tool for calculating pressure drop in many applications involving diverse flow regimes, duct cross-sections, and roughness conditions. Second, Moody's chart shines as an example of the importance of investing creativity into the graphic reporting of research results. Moody is not the one who discovered the duct friction information projected on the chart that bears his name: these discoveries are mainly the work of Nikuradse. Moody compiled what was known in the 1940s (e.g., Nikuradse's experiments [34] and the analyses triggered by it [36, 37]) and displayed this information on a single chart, in terms of easy-to-use dimensionless groups. The graphic presentation of this information eliminated much of the difficulty associated with handling implicit friction factor formulas on the slide rule. In addition to being very useful, Moody's original chart is also a high-quality drawing: this, I feel, is partly responsible for the so-frequent reproduction of this drawing in its original form, in some cases, without even mentioning Nikuradse.

Stanton Number

The heat transfer potential of turbulent pipe flow may be deduced from the friction factor information discussed above by adopting further simplifying assumptions. As a start, it is instructive to rewrite the energy equation (102) for fully developed flow

$$\rho c_p \bar{u} \frac{\partial \bar{T}}{\partial x} = \frac{1}{r} \frac{\partial}{\partial r} \left(r q_{app}'' \right) \tag{115}$$

and then to integrate this form twice

$$\int_0^r \rho c_p \bar{u} \frac{\partial \bar{T}}{\partial x} r \, dr = r q_{app}'' \tag{116}$$

$$\int_0^{r_0} \rho c_p \bar{u} \frac{\partial \bar{T}}{\partial x} r \, dr = r_0 q_0'' \tag{117}$$

Dividing eqs. (116) and (117) side-by-side and using the distribution of τ_{app} (104) as a guide, we obtain [39]

$$\frac{q_{app}''}{q_0''} = M \left(1 - \frac{y}{r_0} \right) \tag{118}$$

where

$$M = \frac{\dfrac{2}{r^2} \displaystyle\int_0^r \bar{u} \frac{\partial \bar{T}}{\partial x} r \, dr}{\dfrac{2}{r_0^2} \displaystyle\int_0^{r_0} \bar{u} \frac{\partial \bar{T}}{\partial x} r \, dr} \tag{119}$$

In particular, if the pipe is heated with x-independent heat flux, the longitudinal temperature gradient $\partial \bar{T}/\partial x$ is independent of r: hence

$$M = \frac{\dfrac{2}{r^2} \displaystyle\int_0^r \bar{u} r \, dr}{\dfrac{2}{r_0^2} \displaystyle\int_0^{r_0} \bar{u} r \, dr} \tag{120}$$

Function M is therefore the ratio of $\bar{u}$ averaged over a round cross-section of radius r divided by $\bar{u}$ averaged over the entire pipe cross-section. Since the pipe velocity profile $\bar{u}$ resembles the slug profile (because the laminar sublayer is one or two orders of magnitude thinner than the pipe radius), in many practical cases, M is approximately equal to one, while not a strong function of r. Therefore, an acceptable approximation for the distribution of apparent heat flux over the pipe cross-section is

$$\frac{q_{app}''}{q_0''} \cong 1 - \frac{y}{r_0} \tag{121}$$

In conclusion, the apparent heat flux follows a distribution that is practically the same as the one followed by the apparent shear stress. This observation is the starting point in an analysis that ties the Stanton number to the friction factor information presented earlier in this section. The following analysis was first reported by Prandtl in 1910 (more detailed accounts of this analysis may be found in Ref. 12, pp. 403–407 and Ref. 13, pp. 494–496). Dividing eqs. (104) and (121) and recognizing the definitions of q_{app}'' and τ_{app} yields

$$\frac{\nu + \varepsilon_M}{\tau_0} d\bar{u} = \frac{c_P(\alpha + \varepsilon_H)}{q_0''} d\bar{T} \tag{122}$$

Now imagine that the pipe cross-section is composed of two distinct regions—an annular region near the wall $(0 < y < y_1)$ where $\nu \gg \varepsilon_M$ and $\alpha \gg \varepsilon_H$ and a disk-shaped region in the center $(y_1 < y < r_0)$ where $\nu \ll \varepsilon_M$ and $\alpha \ll \varepsilon_H$. Integrating eq. (122) from $y = 0$ to $y = y_1$, neglecting both ε_M and ε_H, we obtain

$$\frac{\nu}{\tau_0} \bar{u}_1 = \frac{c_P \alpha}{-q_0''} (\bar{T}_1 - \bar{T}_0) \tag{123}$$

where $\bar{u}_1$ and $\bar{T}_1$ are the time-averaged quantities at $y = y_1$. Next, we integrate eq. (122) from $y = y_1$ to $y = y_2$, where y_2 is chosen in such a way that $\bar{T}(y_2)$ is equal to the *mean* temperature $\bar{T}_m$ (see Chapter 3) and $\bar{u}(y_2)$ is *approximately* the same as the mean velocity U; the result of this second integration is

$$\frac{\varepsilon_M}{\tau_0} (U - \bar{u}_1) = \frac{c_P \varepsilon_H}{-q_0''} (\bar{T}_m - \bar{T}_1) \tag{124}$$

Eliminating $\bar{T}_1$ between eqs. (123) and (124) and using the definitions of friction factor $\tau_0/(\frac{1}{2}\rho U^2)$ and the Stanton number $h/(\rho c_p U)$, we find

$$St = \frac{f/2}{Pr_t + (\bar{u}_1/U)(Pr - Pr_t)} \tag{125}$$

As in the case of turbulent boundary layers, the Stanton numbers turns out to be proportional to the dimensionless wall shear stress, $f/2$ in this case. The proportionality factor in this relation is a function of the Prandtl number and two more parameters $(Pr_t, \bar{u}_1/U)$ that have to be adjusted empirically. Prandtl's formula (125) agrees with the measurements involving fluids with Prandtl numbers greater than 0.5, if $Pr_t = \varepsilon_M/\varepsilon_H$ is taken as unity and $\bar{u}_1/U$ is replaced by Hofmann's empirical correlation $\bar{u}_1/U \cong 1.5\,Re_D^{-1/8}Pr^{-1/6}$ [40].

Much more consistent agreement with measurements is registered by Colburn's [27] more compact correlation.

$$St\,Pr^{2/3} \cong \frac{f}{2} \tag{126}$$

This formula is analytically the same as the one encountered in boundary layer flow. Its success further supports the validity of the theoretical argument used to deduce eqs. (85)–(90). The Dittus–Boelter equation [41] is more popular among heat transfer engineers:

$$\frac{hD}{k} = 0.0243\,Pr^{0.4}Re_D^{0.8} \tag{127}$$

Its accuracy is recognized in the range $0.7 < Pr < 160$, and $Re_D > 10^4$ (see Ref. 28, p. 406). Note that the Colburn correlation (126) is nearly identical to the Dittus–Boelter correlation (127) if

$$f \cong 0.046\,Re_D^{-0.2} \tag{128}$$

which is a good power-law approximation for the smooth-pipe friction factor curve drawn on Moody's chart (Fig. 7.9).

For liquid metals $(0.003 < Pr < 0.05)$, the Nusselt number for smooth circular tubes with constant heat flux has been correlated as [42]

$$\frac{hD}{k} \cong 6.3 + 0.0167\,Re_D^{0.85}Pr^{0.93} \tag{129}$$

In the high-Re_D limit, this correlation and eq. (128) can be combined to yield

$$\frac{St}{(f/2)^{1/2}} \cong (0.11\,Re_D^{-0.05})Pr^{-0.07} \tag{130}$$

In the Re_D range 10^4–10^6, the right-hand side of this equation is equal to a number of order $O(10)$ times a function that decreases with increasing Pr. This result confirms most of the scaling law proposed on theoretical grounds in eq. (99). The fact that the Pr exponent in eq. (130) is not as large as in eq. (99) does not necessarily spell "disagreement", considering the narrow Pr range on which the experimental correlation (129) is based.

NATURAL CONVECTION ALONG VERTICAL WALLS

We studied the behavior of laminar boundary layers in natural convection in Chapters 4 and 5, and then featured this phenomenon in order to identify the scaling laws of transition to turbulence (Chapter 6). In this section we focus on what happens beyond transition, in other words, we seek to predict the heat-transfer coefficient in a natural convection boundary layer where the Rayleigh number is large enough so that the local Reynolds number based on stream width is considerably greater than $O(10^2)$. The Rayleigh number range in which the laminar–turbulent transition can take place can be determined by combining the local Reynolds number criterion $O(10^2)$ with the laminar scales assembled in Table 4.1 (see Ref. 43 and Table 6.1).

Although a number of analytical attempts have been made to predict the Nusselt number in turbulent natural convection over a vertical wall, some of the trends revealed by experiments have not been accounted for satisfactorily. For this reason we begin with the empirical correlation reported by Churchill and Chu [44] for the (isothermal wall)–(isothermal reservoir) geometry sketched in Fig. 7.11, for both the laminar and the turbulent ranges known experimentally ($10^5 < \mathrm{Ra}_H < 10^{12}$).

$$\frac{h_{0-H}H}{k} \cong \left\{ 0.825 + \frac{0.387\,\mathrm{Ra}_H^{1/6}}{\left[1 + (0.492/\mathrm{Pr})^{9/16}\right]^{8/27}} \right\}^2 \tag{131}$$

In this expression h_{0-H} is the heat-transfer coefficient averaged over the entire wall of height H and Ra_H is the Rayleigh number based on H and $(T_0 - T_\infty)$. Of special interest are the asymptotes suggested by the Churchill–Chu correlation in the turbulent flow limit (large Ra_H).

$$\frac{h_{0-H}H}{k} \cong \begin{cases} 0.15\,\mathrm{Ra}_H^{1/3}, & \text{if } \mathrm{Pr} \gg 1 \\ 0.19\,(\mathrm{Ra}_H\mathrm{Pr})^{1/3}, & \text{if } \mathrm{Pr} \ll 1 \end{cases} \tag{132}$$

In the lower expression, the 0.19 coefficient is replaced by 0.198 if the wall condition is of uniform heat flux. The heat transfer results summarized in eqs. (132) should not be seen as exact or definitive answers to the heat transfer question; nevertheless, they do offer a bird's-eye view of the trends followed by heat transfer data.

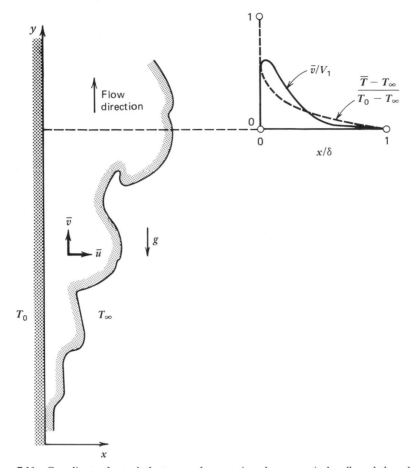

Figure 7.11 Coordinates for turbulent natural convection along a vertical wall, and the velocity and temperature profiles used by Eckert and Johnson [45].

On the analytical side, the continuing effort of explaining the above trends has been influenced by the first analysis reported by Eckert and Jackson [45]. Below, I outline an *abbreviated version* of the Eckert–Jackson analysis, a version that is better suited for the classroom because it accomplishes as much as Eckert and Jackson's analysis [45] with less algebra. Relative to the time-averaged turbulent flow analysis presented earlier in this chapter, the new element in vertical boundary layer convection is the momentum equation with the body force term. In the frame of Fig. 7.11, the time-averaged boundary layer momentum equation is

$$\bar{u}\frac{\partial \bar{v}}{\partial x} + \bar{v}\frac{\partial \bar{v}}{\partial y} = \frac{\partial}{\partial x}\left[(\nu + \varepsilon_M)\frac{\partial \bar{v}}{\partial x}\right] + g\beta(\bar{T} - T_\infty) \qquad (133)$$

Note the use of the Boussinesq incompressible flow model already seen in the treatment of laminar flow (Chapter 4). Integrating eq. (133) across the boundary layer region yields

$$\frac{d}{dy}\int_0^\infty \bar{v}^2\, dx = -\frac{\tau_0}{\rho} + g\beta \int_0^\infty (\bar{T} - T_\infty)\, dx \tag{134}$$

Eckert and Jackson used this momentum integral in conjunction with the following velocity and temperature profiles

$$\bar{v} = V_1\left(1 - \frac{x}{\delta}\right)^4 \left(\frac{x}{\delta}\right)^{1/7}$$

$$\bar{T} - T_\infty = (T_0 - T_\infty)\left[1 - \left(\frac{x}{\delta}\right)^{1/7}\right] \tag{135}$$

that have been sketched in Fig. 7.11. If *we now neglect the inertia term* in eq. (134), we obtain

$$\frac{\tau_0}{\rho} = \frac{\delta}{8} g\beta (T_0 - T_\infty) \tag{136}$$

The energy equation for the boundary layer flow of Fig. 7.11 is

$$\bar{u}\frac{\partial \bar{T}}{\partial x} + \bar{v}\frac{\partial \bar{T}}{\partial y} = \frac{\partial}{\partial x}\left[(\alpha + \varepsilon_H)\frac{\partial \bar{T}}{\partial x}\right] \tag{137}$$

Integrating this equation across the boundary layer,

$$\frac{d}{dy}\int_0^\infty \bar{v}(\bar{T} - T_\infty)\, dx = \frac{q_0''}{\rho c_P} \tag{138}$$

and making use of the assumed profiles [eqs. (135)] yields

$$0.0366\,(T_0 - T_\infty)\frac{d}{dy}(V_1\delta) = \frac{q_0''}{\rho c_P} \tag{139}$$

At this stage, we have two equations for four unknowns ($\tau_0, q_0'', \delta, V_1$) [eqs. (136) and (139)]. In order to close the problem, Eckert and Jackson made two additional assumptions:

(i)   The wall shear stress τ_0 depends on the velocity scale V_1 and on the outer boundary layer thickness δ in the same manner as in *forced convection*,

$$\tau_0 \cong 0.0225\rho V_1^2\left(\frac{V_1\delta}{\nu}\right)^{-1/4} \tag{140}$$

(ii) The local shear stress and heat flux are related through a scaling law of the Colburn type,

$$\frac{q_0''}{(T_0 - T_\infty)\rho c_p V_1} Pr^{2/3} = \frac{\tau_0}{\rho V_1^2} \tag{141}$$

Solving the problem represented by eqs. (136) and (139)–(141), we obtain the local Nusselt number (Problem 14),

$$Nu_y = \frac{q_0'' y}{(T_0 - T_\infty)k} = 0.039 \, Pr^{1/5} Ra_y^{2/5} \tag{142}$$

If the inertia term of eq. (134) is not neglected as in the analysis sketched above, the local Nusselt number takes the seemingly more general form derived by Eckert and Jackson [45].

$$Nu_y = 0.0295 \frac{Pr^{1/15}}{(1 + 0.494 \, Pr^{2/3})^{2/5}} Ra_y^{2/5} \tag{143}$$

Equation (143) becomes the same as eq. (142) as $Pr \to \infty$, leaving the impression that by taking the inertia term into account it is valid for low Prandtl numbers as well. In reality, the Eckert and Jackson formula (143) can only refer to fluids with Pr of order $O(1)$ or greater,[†] because the $Pr > 1$ range is a constraint brought into the analysis through the adoption of the scaling law (141) (see the scaling analysis built around Fig. 7.7). Therefore, the retention of the inertia term in eq. (134) only complicates the analysis, because it is inconsistent with a simplifying assumption inserted later on: this inconsistency is corrected in Problem 15, where the retention of the inertia term is combined with a more appropriate assumption for $Pr < 1$, that is, with eq. (99) in place of the Colburn form (141).

Although the predictions based on the abbreviated analysis agree numerically with the heat transfer rates measured in air and water at Rayleigh numbers Ra_H of order 10^9–10^{12} (Problem 16), the analytical trend (142) is not the same as the one revealed by experiments [eq. (132)]. This discrepancy casts some doubt on assumptions (140) and (141), which were brought in to close the problem; in view of scaling arguments advanced in connection with Fig. 7.7, more questionable seems the first assumption [worded as (i)]. A number of alternative solutions to the turbulent natural convection problem have been reviewed by Gebhart [46] and, in greater detail, by Jaluria [47]: the lack of complete agreement between these analyses and eq. (132) persists.

The problem of turbulent heat transfer in a rectangular enclosure heated from the side is even more challenging. So far, this problem has been the

[†]As eq. (142) does after considerably less algebra.

subject of experimental studies designed to measure the overall heat transfer rate [48] and document the flow and temperature fields inside the fluid layer [49]. Jakob [48] correlated the heat transfer data produced by a number of experimental reports; for air-filled cavities he recommended

$$\mathrm{Nu}_L \cong 0.065 \, \mathrm{Gr}_L^{1/3} \left(\frac{H}{L} \right)^{-1/9} \tag{144}$$

for the Gr_L range $(2)10^5$–$(1.1)10^7$ and the height/spacing ratio range $H/L > 3$. The notation in eq. (144) is $\mathrm{Nu}_L = q''_{0-H}/(k\,\Delta T/L)$, and $\mathrm{Gr}_L = g\beta\Delta T L^3/\nu^2$; using H as the length scale, the Jakob correlation (144) reads

$$\frac{q''_{0-H}}{\Delta T} \frac{H}{k} \cong 0.065 \, \mathrm{Gr}_H^{1/3} \left(\frac{H}{L} \right)^{-1/9} \tag{145}$$

Noting that the heat transfer must overcome two vertical boundary layers in a parallel-plate enclosure heated from the side, it is found that the Jakob correlation is highly consistent with the vertical single-wall measurements summarized in eq. (132) (see Problem 17).

MORE REFINED TURBULENCE MODELS

The fundamental feature that distinguishes the analysis of time-averaged turbulent flow from other analyses (laminar flow, Chapters 2–5, 9; porous media, Chapters 9–11) is the closure problem. Even in relatively simple time-averaged turbulent flows such as the boundary layer near a flat wall or the fully developed flow through a round tube, the number of unknowns exceeded the number of equations. To proceed with the analysis, the difference between these numbers was made up by introducing additional equations whose validity is supported by a combination of intuitive reasoning and laboratory measurements. Thus, in the case of the flow part of the turbulent convection problem, we relied on Prandtl's mixing-length model [eq. (32)] to evaluate the unknown called momentum eddy diffusivity ε_M. Likewise, for the heat transfer part of the convection problem, we assumed a constant value for the turbulent Prandtl number [eq. (70)], in order to be able to evaluate the thermal eddy diffusivity ε_H. In both cases, ε_M and ε_H resulted from the assumed algebraic expressions, that is, from *algebraic models*. Alternatives to the mixing-length model as an algebraic model for ε_M are summarized in Table 7.1. Alternatives to writing $\mathrm{Pr}_t \cong 0.9$, constant, as a means of evaluating ε_H are reviewed by A. J. Reynolds [50].

The main shortcoming of these simple models is their proven lack of universal applicability. For example, a model that works near the wall in turbulent pipe flow breaks down near the pipe centerline [eq. (106)]. Also, as is shown in Chapter 8, turbulent flow regions situated sufficiently far from solid walls demand eddy diffusivity models that differ from eq. (32) of this chapter. The need for a universally applicable turbulence model is obvious; however, the idea that such a model could be invented has been met with varying degrees of skepticism by proponents and users of turbulence models. Nevertheless, due to technological advances in high-speed computers, during the past two decades the field of heat transfer has witnessed the emergence of a new generation of more powerful turbulence models—a generation that most certainly takes us closer to the objective of simulating a new time-averaged turbulent flow on the computer. The progress in this area of research has been reviewed in articles and monographs that are highly recommended [51–56]. Below, we illustrate this relatively new approach to turbulence modeling by means of the $k–\varepsilon$ model [57], which has become the most popular.

The starting point in the $k–\varepsilon$ model and other nonalgebraic (one-equation and two-equation) models is the analogy that can be drawn between the motion of a fluid packet in turbulent flow and the random motion of a molecule in an ideal gas. A classical result in the kinetic theory of gases (a result derived first by Maxwell in 1860 [58]) is that the kinematic viscosity of a gaseous substance may be calculated with the formula $v = \frac{1}{3}\bar{a}\lambda$, where $\bar{a}$ is the mean speed of the molecule and λ is the mean free-path length. In the case of a fluid packet in turbulent flow, the mean speed scale is $k^{1/2}$, where k is the so-called *turbulence energy*

$$k = \tfrac{1}{2}\left[\overline{(u')^2} + \overline{(v')^2} + \overline{(w')^2}\right] \tag{146}$$

Therefore, the momentum eddy diffusivity ε_M may be modeled as

$$\varepsilon_M = C_\mu k^{1/2} L \tag{147}$$

where C_μ is a dimensionless empirical constant, and for the time being, L is an unknown length scale that plays the same role for the fluid packet as the mean free-path length plays for the molecule of a gas. The eddy diffusivity model represented by eq. (147) was proposed independently by Kolmogorov [59] and Prandtl [60]: this model says that in order to calculate ε_M, that is, in order to close the flow part of the problem, we must determine two more local quantities, k and L. As shown below, these two quantities follow from two more equations—the k-equation and the ε-equation.

The k-equation may be derived from the complete momentum equations. The procedure consists of: multiplying the x momentum equation (13) by u', the y momentum equation by v', and the z momentum equation by w'; then time-averaging these three equations; then adding them term by term. From the resulting equation we subtract the equation obtained by first multiplying the x momentum equation by $\bar{u}$, the y momentum equation by $\bar{v}$, and the z momentum equation by $\bar{w}$, and then time-averaging and adding these equations term by term. A more direct approach is to imagine a control volume in a *slender flow region* [e.g., boundary layer (Fig. 7.5)] and to argue that the convection of k into the control volume (Dk/Dt) equals the eddy diffusion of k in the transversal (y) direction plus the rate of k generation minus the rate of k destruction [55]. The rate of k diffusion in the transversal direction may be written as (assuming $\varepsilon_M \gg \nu$)

$$\frac{\partial}{\partial y}\left(\frac{\varepsilon_M}{\sigma_k}\frac{\partial k}{\partial y}\right) \tag{148}$$

where σ_k is a dimensionless empirical constant. The rate of k production can be evaluated by multiplying the eddy shear stress ($\varepsilon_M \partial \bar{u}/\partial y$) by the time-averaged velocity gradient ($\partial \bar{u}/\partial y$),

$$\varepsilon_M\left(\frac{\partial \bar{u}}{\partial y}\right)^2 = C_\mu k^{1/2} L\left(\frac{\partial \bar{u}}{\partial y}\right)^2 \tag{149}$$

Finally, the rate of k destruction or the *dissipation rate* ε may be evaluated by imagining a fluid packet of diameter L, oscillating with velocity $k^{1/2}$ in turbulent flow field. The drag force on this fluid packet is of order $C_D \rho L^2 (k^{1/2})^2$, where C_D is a *drag coefficient* approximately equal to 1 [55]; the mechanical power dissipated per unit mass is $k^{1/2}C_D\rho L^2(k^{1/2})^2/(\rho L^3) \sim C_D(k^{1/2})^3/L$, in other words,

$$\varepsilon = C_D \frac{k^{3/2}}{L} \tag{150}$$

In conclusion, the k-equation for a boundary layer-type flow region is

$$\underbrace{\frac{Dk}{Dt}}_{\text{Convection}} = \underbrace{\frac{\partial}{\partial y}\left(\frac{\varepsilon_M}{\sigma_k}\frac{\partial k}{\partial y}\right)}_{\text{Diffusion}} + \underbrace{\varepsilon_M\left(\frac{\partial \bar{u}}{\partial y}\right)^2}_{\text{Generation}} - \underbrace{\varepsilon}_{\text{Destruction}} \tag{151}$$

where the dissipation rate ε should not be confused with the momentum eddy diffusivity ε_M.

The ε-equation for a boundary layer-type region may be constructed in similar manner

$$\frac{D\varepsilon}{Dt} = \frac{\partial}{\partial y}\left(\frac{\varepsilon_M}{\sigma_\varepsilon}\frac{\partial \varepsilon}{\partial y}\right) + C_1\varepsilon_M\left(\frac{\partial \bar{u}}{\partial y}\right)^2\frac{\varepsilon}{k} - C_2\frac{\varepsilon^2}{k} \tag{152}$$

where σ_ε, C_1, and C_2 are three more dimensionless empirical constants. Finally, setting $C_D = 1$ and eliminating the unknown length scale L between eqs. (147) and (150) yields

$$\varepsilon_M = C_\mu \frac{k^2}{\varepsilon} \tag{153}$$

The three equations (151)–(153) are sufficient for determining the three un-knowns $(k, \varepsilon, \varepsilon_M)$. The recommended values of the five empirical constants appearing in these equations are [53]

$$C_\mu = 0.09, \qquad C_1 = 1.44, \qquad C_2 = 1.92$$

$$\sigma_k = 1, \qquad \sigma_\varepsilon = 1.3 \tag{154}$$

These values have been found to be appropriate for the plane jets and plane shear layers discussed in Chapter 8. A nearly identical set of constants works for turbulent boundary layers; hence, it is reasonable to expect the above constants to adequately serve the numerical simulation of boundary layers as well.

In order to solve the heat transfer part of the problem, the most common approach is to combine the above ε_M calculation with the statement that the turbulent Prandtl number $\varepsilon_M/\varepsilon_H$ is equal to 0.9.

It is instructive to look back at the mixing-length model used throughout this chapter, and to see if there is any overlap between that simple model and the k–ε model of eqs. (151)–(153). According to the mixing-length model [eq. (30)], we can write

$$\left(\frac{\partial \bar{u}}{\partial y} \right)^2 = \frac{\varepsilon_M^2}{l^4} \tag{155}$$

where l is the mixing length. This means that the k-generation term in the k-equation (151) assumes the form

$$\varepsilon_M \left(\frac{\partial \bar{u}}{\partial y} \right)^2 = \frac{\varepsilon_M^3}{l^4} = C_\mu^3 k^{3/2} \frac{L^3}{l^4} = \frac{C_\mu^3}{C_D} \left(\frac{L}{l} \right)^4 \varepsilon \tag{156}$$

which shows that the k-generation term equals the k-destruction term ε if the length scale L is taken as

$$L = l \left(\frac{C_D}{C_\mu^3} \right)^{1/4} \tag{157}$$

We conclude that in the inner region of a turbulent boundary layer (i.e., in the layer where the mixing-length model works) the generation of k is balanced by the dissipation of k. For this reason, the inner layer is usually referred to as the *equilibrium layer*. In general, a given turbulent flow region is the ballpark for competition between four effects: convection, diffusion, generation and de-struction, as is indicated by eq. (151).

The k–ε model outlined above is valid only in those flow regions that are strongly turbulent, that is, in regions where the eddy diffusivity ε_M overwhelms the molecular diffusivity ν. [This limitation is amply illustrated by the construction of the diffusion terms in eqs. (151) and (152).] In the case of turbulent boundary layer flow or fully developed turbulent flow through a duct, eqs. (151)–(153) do not apply in the viscous sublayer. One way to bridge the viscous sublayer and to impose the solid-wall boundary condition on the k–ε simulation of the strongly turbulent regions is the so-called *wall-function* method. Let y_c be the physical distance to the solid wall from a point situated just outside the viscous sublayer, where the logarithmic law of the wall holds (Fig. 7.4). The wall-function approach consists of the assumption that at $y = y_c$ the velocity component parallel to the wall obeys the logarithmic law of the wall [eq. (40)],

$$\frac{\bar{u}_c}{u_*} = \frac{1}{\kappa} \ln\left(\frac{u_* y_c}{\nu}\right) + B \tag{158}$$

and that at the same location the generation of k is in equilibrium with the destruction of k. If k_c and ε_c are the turbulence energy and dissipation rate at $y = y_c$, then it can be shown that the conditions imposed on eqs. (151) and (152) in lieu of solid-wall boundary conditions are (Problem 20) [54]

$$k_c = \frac{u_*^2}{C_\mu^{1/2}}, \qquad \varepsilon_c = \frac{u_*^3}{\kappa y_c} \tag{159}$$

For further information on the numerical implementation of the k–ε model, the reader is directed to Refs. 51–54.

SYMBOLS

A, B	constants in the logarithmic law of the wall [eqs. (41) and (42)]
c_P	specific heat at constant pressure
C_D	dimensionless coefficient [eq. (150)]
$C_{f,x}$	local skin friction coefficient [eq. (52)]
C_μ, C_1, C_2	constants [eq. (154)]
D_T	distance of maximum thermal penetration in the y direction, in the vicinity of a direct contact spot [eq. (94)]
f	friction factor [eq. (108)]
f_u	curvefit for the velocity profile [eq. (53)]
h	heat transfer coefficient
H	height of enclosure [eq. (144)]
k	turbulence energy [eq. (146)]
k_s	sand grain size [eq. (113)]
l	mixing length [eq. (27)]
L	length of direct viscous contact [eq. (92)]

L	horizontal dimension of enclosure [eq. (144)]
L	length scale [eq. (147)]
L_m	length of direct thermal contact [eq. (95)]
M	function [eq. (119)]
Nu_y	local Nusselt number [eq. (142)]
P	pressure
Pr	Prandtl number
Pr_t	turbulent Prandtl number [eq. (66)]
q''_{app}	apparent heat flux [eq. (24)]
q''_0	wall heat flux [cq. (63)]
$q''_{0,\,\mathrm{max}}$	maximum heat flux, under a direct thermal contact spot [eq. (86)]
r	radial position
r_0	tube radius
Re_{D_h}	Reynolds number based on hydraulic diameter, D_h
Ra_H	Rayleigh number based on enclosure height, H
St_x	local Stanton number [eq. (76)]
St	Stanton number for fully developed turbulent flow [eq. (125)]
T	temperature
T_0	wall temperature
T_∞	free-stream temperature
u, v, w	velocity components in cartesian coordinates
u, v	velocity components in cylindrical coordinates (Fig. 7.8)
u_*	friction velocity [eq. (34)]
U	duct-averaged velocity [eq. (109)]
U_c	centerline velocity [eq. (110)]
U_∞	free-stream velocity
x, y, z	cartesian coordinates
α	thermal diffusivity
β	coefficient of thermal expansion
δ	outer boundary layer thickness
δ^*	displacement thickness [eq. (59)]
ε	turbulence dissipation function [eq. (150)]
ε_H	thermal eddy diffusivity [eq. (23)]
ε_M	momentum eddy diffusivity [eq. (23)]
η	density of contact spots [eq. (85)]
θ	momentum thickness [eq. (59)]
κ	von Karman's constant [eq. (31)]
μ	viscosity
ν	kinematic viscosity
ρ	density
$\sigma_k, \sigma_\epsilon$	constants [eq. (154)]
τ_{app}	apparent shear stress [eq. (24)]

τ_0 wall shear stress [eq. (33)]

$\tau_{0,\max}$ maximum shear stress, under a direct viscous contact
                            spot [eq. (86)]

$(\ )_c$ properties at the point where the wall condition is imposed
                            on the k–ε model [eqs. (158) and (159)]

$(\ )_{CSL}$ conduction sublayer

$(\ )_{VSL}$ viscous sublayer

$\overline{(\ )}$ time-averaged part

$(\ )'$ fluctuating part

$(\ )^+$ wall coordinates and wall variables [eq. (35)]

$(\ )_{0-H}$ averaged from $y = 0$ to $y = H$

REFERENCES

1. J. O. Hinze, *Turbulence*, 2nd ed., McGraw-Hill, New York, 1975.

2. H. Tennekes and J. L. Lumley, *A First Course in Turbulence*, The MIT Press, Cambridge, MA, 1972.

3. A. A. Townsend, *The Structure of Turbulent Shear Flow*, 2nd ed., Cambridge University Press, Cambridge, England, 1976.

4. A. Roshko, Structure of turbulent shear flows: A new look. Dryden Research Lecture, *AIAA J.*, Vol. 14, 1976, pp. 1349–1357.

5. B. J. Cantwell, Organized motion in turbulent flow, *Ann. Rev. Fluid Mech.*, Vol. 13, 1981, pp. 457–515.

6. A. Bejan, *Entropy Generation through Heat and Fluid Flow*, Wiley, New York, 1982, pp. 88, 89.

7. P. Bandyopadhyay, Large structure with a characteristic upstream interface in turbulent boundary layers, *Phys. Fluids*, Vol. 23, 1980, pp. 2326, 2327.

8. O. Reynolds, On the dynamical theory of incompressible viscous fluids and the determination of the criterion, *Philos. Trans. R. Soc. London Ser. A*, Vol. 186, 1895, pp. 123–164.

9. W. M. Kays and M. E. Crawford, *Convective Heat and Mass Transfer*, McGraw-Hill, New York, 1980, p. 40.

10. J. Boussinesq, Théorie de l'écoulement tourbillonant, *Mem. Pre. par. div. Sav.* XXIII, Paris, 1877.

11. F. Durst, B. E. Launder, F. W. Schmidt, L. J. S. Bradbury, J. W. Whitelaw, *Turbulent Shear Flows*, Vols. I–III, Springer-Verlag, New York, 1979, 1980, 1983.

12. L. Prandtl, *Essentials of Fluid Dynamics*, Blackie & Son, London, 1969, p. 117; in German, *Führer durch die Strömungslehre*, Vieweg, Braunschweig, 1949.

13. H. Schlichting, *Boundary Layer Theory*, 4th ed., McGraw-Hill, New York, 1960, p. 489.

14. D. Coles, The law of the wall in turbulent shear flow, *50 Jahre Grenzschichtforschung*, H. Görtler and W. Tollmien, eds., Vieweg, Braunschweig, 1955, pp. 153–163.

15. J. Kestin and P. D. Richardson, Heat transfer across turbulent, incompressible boundary layers, *Int. J. Heat Mass Transfer*, Vol. 6, 1963, pp. 147–189.

16. T. von Karman, The analogy between fluid friction and heat transfer, *Trans. ASME*, Vol. 61, 1939, pp. 705–710.

17. W. D. Rannie, Heat transfer in turbulent shear flow, *J. Aero Sci.*, Vol. 23, 1956, pp. 485–489.

18. E. R. van Driest, On turbulent flow near a wall, *J. Aero. Sci.*, Vol. 23, 1956, pp. 1007–1011.

19. H. Reichardt, Die Grundlagen des turbulenten Wärmeüberganges, *Arch. Gesamte Waermetech.*, Vol. 2, 1951, pp. 129–142.

20. R. G. Deissler, Analysis of turbulent heat transfer, mass transfer and friction in smooth tubes at high Prandtl and Schmidt number, NACA TN 3145, 1954.

21. D. B. Spalding, A single formula for the "law of the wall", *J. Appl. Mech.*, Vol. 28, 1961, pp. 455–457.

22. L. P. Purtell, P. S. Klebanoff, and F. T. Buckley, Turbulent boundary layer at low Reynolds number, *Phys. Fluids*, Vol. 24, May 1981, pp. 802–811.

23. L. Prandtl, *op. Cit.*, p. 127.

24. H. Schlichting, *op. Cit.*, p. 536.

25. F. Schultz–Grunow, Neues Widerstandsgesetz für glatte Platten. *Luftfahrtforschung*, Vol. 17, 1940, p. 239; also NACA TM-986, 1941.

26. W. M. Kays and M. E. Crawford, *op. Cit.*, p. 211.

27. A. P. Colburn, A method of correlating forced convection heat transfer data and a comparison with fluid friction, *Trans. Am. Inst. Chem. Eng.*, Vol. 29, 1933, pp. 174–210.

28. F. P. Incropera and D. P. De Witt, *Fundamentals of Heat Transfer*, Wiley, New York, 1981, p. 302.

29. A. Bejan, Analytical prediction of turbulent heat transfer parameters, the second annual report, CUMER 82-6, December 1982, Department of Mechanical Engineering, University of Colorado, Boulder.

30. O. Reynolds, On the extent and action of the heating surface for steam boilers, *Proc. Manchester Lit. Phil. Soc.*, Vol. 14, 1874, pp. 7–12.

31. P. Feyerabend, *Against Method*, Verso, London, 1978.

32. H. Blasius, *Forschungsarbeiten des Ver. Deutsch. Ing.*, No. 131, 1913; see also Ref. 12, p. 163.

33. W. M. Kays and H. C. Perkins, Forced convection, internal flow in ducts, in W. M. Rohsenow and J., P. Hartnett, eds., *Handbook of Heat Transfer*, Section 7, McGraw-Hill, New York, 1973.

34. J. Nikuradse, Strömungsgesetze in rauhen Rohren, *VDI-Forschungsh.*, Vol. 361, 1933, pp. 1–22.

35. L. F. Moody, Friction factors for pipe flow, *Trans. ASME*, Vol. 66, 1944, pp. 671–684.

36. T. von Karman, Mechanische Ähnlichkeit und Turbulenz, *Nachr. Ges. Wiss. Göttingen Math. Phys. Kl. Fachgruppe 1*, 1930 No. 5, pp. 58–76; also "Mechanical similitude and turbulence", NACA TM No. 611, 1931.

37. L. Prandtl, Neuere Ergebnisse der Turbulenzforschung, *Z. Ver. Dtsch. Ing.*, Vol. 77, 1933, pp. 105–114.

38. A. Anand, M. S. Thesis, Department of Mechanical Engineering, University of Colorado, Boulder, August 1983.

39. W. M. Rohsenow and H. Y. Choi, *Heat, Mass and Momentum Transfer*, Prentice-Hall, Englewood Cliffs, NJ, 1961, p. 183.

40. E. Hofmann, Die Wärmeübertragung bei der Strömung im Rohr, *Z. Gesamte Kälte-Ind.*, Vol. 44, 1937, pp. 99–107.

41. L. M. K. Boelter, V. H. Cherry, H. A. Johnson, and R. C. Martinelli, *Heat Transfer Notes*, McGraw-Hill, New York, 1965, p. 552.

42. C. A. Sleicher and M. W. Rouse, A convenient correlation for heat transfer to constant and variable property fluids in turbulent pipe flow, *Int. J. Heat Mass Transfer*, Vol. 18, 1975, pp. 677–683.

43. A. Bejan and G. R. Cunnington, Theoretical considerations of transition to turbulence near a vertical wall, *Int. J. Heat Fluid Flow*, Vol. 4, No. 3, 1983, pp. 131–139.

44. S. W. Churchill and H. H. S. Chu, Correlating equations for laminar and turbulent free convection from a vertical plate, *Int. J. Heat Mass Transfer*, Vol. 18, 1975, pp. 1323–1329.

45. E. R. G. Eckert and T. W. Jackson, Analysis of turbulent free convection boundary layer on a flat plate, NACA Report 1015, 1951.

46. B. Gebhart, *Heat Transfer*, 2nd ed., McGraw-Hill, New York, 1971, pp. 368–369.

47. Y. Jaluria, *Natural Convection Heat and Mass Transfer*, Pergamon, Oxford 1980, pp. 252–270.

48. M. Jakob, *Heat Transfer*, Vol. I, Wiley, New York, 1949.

49. J. W. Elder, Turbulent free convection in a vertical slot, *J. Fluid Mech.*, Vol. 23, 1965, pp. 99–111.

50. A. J. Reynolds, The prediction of turbulent Prandtl and Schmidt numbers, *Int. J. Heat Mass Transfer*, Vol. 18, 1975, pp. 1055–1069.

51. D. Anderson, J. C. Tannehill, and R. H. Pletcher, *Computational Fluid Mechanics and Heat Transfer*, Hemisphere, Washington, D.C., 1984.

52. B. E. Launder, and D. B. Spalding, *Mathematical Models of Turbulence*, Academic Press, New York, 1972.

53. B. E. Launder and D. B. Spalding, The numerical computation of turbulent flows, *Comput. Methods Appl. Mech. Eng.*, Vol. 3, 1974, pp. 269–289.

54. W. Rodi, Examples of turbulence models for incompressible flows, *AIAA J.*, Vol. 20, 1982, pp. 872–879.

55. E. R. G. Eckert and R. M. Drake, Jr., *Analysis of Heat and Mass Transfer*, McGraw-Hill, New York, 1972, pp. 364–368.

56. B. E. Launder, Modelling of turbulent flow in gas turbine blading: achievements and prospects, *Int. J. Heat Fluid Flow*, Vol. 3, 1982, pp. 171–184.

57. F. H. Harlow and P. Nakayama, Transport of turbulence energy decay rate, Los Alamos Science Lab., University of California, Report LA-3854 (1968).

58. R. D. Present, *Kinetic Theory of Gases*, McGraw-Hill, New York, 1958, p. 41.

59. A. N. Kolmogorov, Equations of turbulent motion of an incompressible turublent fluid, *Izv. Akad, Nauk SSSR, Ser. Fiz.*, VI, No. 1–2, 1942, pp. 56–58.

60. L. Prandtl, Über ein neues Formelsystem für die ausgebildete Turbulenz, *Nachr. Akad. Wiss. Göttingen*, 1945.

PROBLEMS

1. Prove the validity of the time-averaging rules expressed as eqs. (4)–(10) in Chapter 7.

2. Derive the time-averaged form of the energy equation [eq. (17)] by starting with eq. (42) of Chapter 1 and applying the rules of time-averaging [eqs. (4)–(10)]. Use the derivation of eq. (16a) as a guide.

3. One way to improve the mixing-length model, that is, to achieve a smoother overlap between measurements and the empirically adjusted curves suggested by the mixing-length model. (Fig. 7.4) is to do away with the assumption that the eddy diffusivity ε_M is zero in a layer of finite thickness y_{VSL}. Instead, as proposed by van Driest [18], assume that ε_M decays rapidly as

y decreases and becomes zero strictly at the wall. Starting with the new mixing-length model

$$l = \kappa y \left(1 - e^{-y^+/A^+} \right)$$

and assuming that it is valid throughout the inner region defined by the constant shear stress postulate (33), develop the analytical means for calculating u^+ as a function of y^+, κ, and A^+. Setting $\kappa = 0.4$, show that the new u^+ calculation fits the data of Fig. 7.4 smoothly if the empirical constant A^+ is approximately 25.

4. Integrate the constant τ_{app} equation (47), using the wall condition $u^+(0) = 0$. The necessary hint for performing the integration is hidden in eq. (49). Show graphically that in the $y^+ \to \infty$ limit the $u^+(\kappa, y^+)$ formula (48) does not agree with the data plotted in Fig. 7.4. Show also that the disagreement persists, even if the value of κ is lowered below 0.4.

5. Prove that in a time-averaged turbulent boundary layer, the total flowrate through the viscous sublayer, $\int_0^{y_{\text{VSL}}} \bar{u}\,dy$ is independent of the longitudinal position x.

6. Derive the integral momentum equation for a boundary layer with a finite longitudinal pressure gradient. Start with eq. (25) and integrate it across the boundary layer; show that the resulting momentum equation is

$$\frac{d\theta}{dx} + (H + 2)\frac{\theta}{U_\infty}\frac{dU_\infty}{dx} = \frac{\tau_0}{\rho U_\infty^2}$$

where θ is the momentum thickness [see eq. (86) of Chapter 2]. Parameter H is the so-called *shape factor*

$$H = \frac{\delta^*}{\theta}$$

where δ^* is the displacement thickness [see eq. (85) of Chapter 2]. Is the above momentum integral valid only for time-averaged turbulent flow?

7. Derive the skin-friction coefficient formula recommended by Prandtl's 1/7th power law velocity profile [eq. (56)]. Note that in this case eq. (54) implies $\bar{u}/U_\infty = (y/\delta)^{1/7}$. Step by step, compare your findings with those listed as eqs. (57)–(60).

8. The x variation of the boundary layer thickness δ [eq. (58)] depends on the choice of the analytical curvefit for the $u^+(y^+)$ data. Prove this statement by deriving general formulas for $\tau_0(x)$ and $\delta(x)$ based on the general curvefit

$$f_u = C(y^+)^{1/m}$$

where C and m are constants. Note that the above expression is a generalization of eq. (56), and that the present problem is a generalization of Problem 7.

9.   Derive the friction factor formula (111) recommended by Prandtl's 1/7th power velocity distribution (56). In the course of this derivation show that the average velocity is only slightly smaller than the centerline velocity, $U = 0.817$ U_c.

10.   Use the scaling laws of transition discussed in Chapter 6 to explain the jump in the value of f as the laminar flow breaks down (Fig. 7.9).

11.   Determine the function M [eq. (120)] for Hagen–Poiseuille flow through a pipe, and compare it with the M function that corresponds to the time-averaged turbulent profile fitted with the 1/7th power law, $\bar{u}/U_c = (y/r_0)^{1/7}$. Comment on the validity of the linear apparent heat flux distribution shown as eq. (121).

12.   Consider the fully developed turbulent flow through a parallel-plate channel with uniform heat flux. Following the procedure outlined in the text for turbulent pipe flow, show that τ_{app} decreases linearly from τ_0 at the wall to zero along the centerline. Derive also the equivalent of eq. (118), and show that the apparent heat flux follows (approximately) the same linear distribution.

13.   Consider the case of uniform heat flux to fully developed turbulent flow in a pipe, and derive a relationship between St, $f/2$, and Pr [i.e., an equivalent of eq. (125)] in the following manner. Start with the energy equation (115) and assume that $\bar{u}$ is practically independent of r and equal to U. Show that in wall coordinates the integral of the energy equation reads

$$T^+ = \int_0^{y^+} \frac{1 - y/r_0}{1/\mathrm{Pr} + \varepsilon_H/\nu}\, dy^+$$

Integrate this result by breaking the $0 < y^+ < r_0^+$ interval into two subintervals, a conduction sublayer $0 < y^+ < y_{CSL}^+$ where $\varepsilon_H/\nu \ll 1/\mathrm{Pr}$, and a core region in which $1/\mathrm{Pr}$ is negligible relative to ε_H/ν. Make the additional assumptions that Pr_t is a constant and the ε_M/ν is adequately represented by the mixing-length model [see eq. (38)]. Derive the relationship St($f/2$, Pr, Pr_t, κ, y_{CSL}^+) by writing $T^+(r_0^+) = T_c^+$ and noting the difference between centerline temperature $\bar{T}_c$ and mean temperature $\bar{T}_m$ in the definition of heat transfer coefficient. Assume the 1/7th power law for the distribution of both $\bar{u}(r)$ and $(T_0 - \bar{T})/(T_0 - \bar{T}_c)$ in order to calculate the ratios $U/\bar{u}_c$ and $(T_0 - \bar{T}_m)/(T_0 - \bar{T}_c)$.

14.   Consider the turbulent natural convection boundary layer shown in Fig. 7.11. Derive the local Nusselt number formula (142) by combining the momentum and energy integrals (136) and (139) with Eckert and Jackson's assumptions (140) and (141). (Hint: Assume power law expressions for velocity and boundary layer thickness, $V_1 = Ay^m, \delta = By^n$). Derive the corresponding formula for the wall shear stress τ_0, and sketch qualitatively the variation of τ_0 with altitude y.

15. As noted in the text, Eckert and Jackson's formula (143) cannot be applied to Pr < 1 fluids. Construct an abbreviated analysis of the type shown in the text between eqs. (135)–(142), this time for the Pr < 1 limit. In the integral momentum equation (134) retain only the buoyancy and inertia terms, and instead of the Colburn analogy (141), use the scaling law for Pr < 1 fluids [eq. (99)]. Compare your result with eqs. (132) and (143) for Pr ≪ 1 fluids.

16. In the cases of air and water, compare numerically the abbreviated analysis result (142) with the Pr > 1 asymptote followed by measurements [eq. (132)]. Determine the Ra_H range in which the agreement between these two formulas is better than 5 percent. (Note that eq. (142) refers to the local Nusselt number.)

17. Estimate the net heat transfer rate across a rectangular cavity of height H and horizontal spacing L, filled with air in the turbulent regime. Do this by modeling the enclosure flow as the "sandwich" of two boundary layers of the kind shown in Fig. 7.11, each boundary layer being driven by $\Delta T/2$, where ΔT is the overall wall-to-wall temperature difference. Using eq. (132) for the heat transfer rate across each boundary layer, derive a simpler substitute for the Jakob correlation (145).

18. Consider the classical view that there exists an inner (near-wall) region in turbulent boundary layer flow where the apparent shear stress τ_{app} and the apparent heat flux q''_{app} are constant, that is, independent of the distance to the wall y. What is then the constant (q''_{app}/τ_{app}) according to the Colburn correlation $St_x Pr^{2/3} = \frac{1}{2} C_{f,x}$?

19. Consider the heat transfer in boundary layer flow from an isothermal wall T_0 to a constant temperature stream (U_∞, T_∞). The leading laminar section of the boundary layer has a length comparable with the length of the trailing turbulent section; consequently, the heat flux averaged over the entire wall length L is influenced by both sections. Derive a formula for the L-averaged Nusselt number, assuming that the laminar–turbulent transition is located at a point x (between $x = 0$ and $x = L$) where $xU_\infty/\nu = 3.5 \times 10^5$. (Table 6.1).

20. Consider the turbulent flow near a solid wall, and let y_c be the distance to the wall from a point situated in the layer where the logarithmic law of the wall (158) holds. Prove that, according to the k–ε model, the turbulence energy k_c and the dissipation rate ε_c at $y = y_c$ are related to the friction velocity u_* by eqs. (159).

8

Free-Stream Turbulence

It is a tradition in the field of convection heat transfer to discuss the transport characteristics of turbulent flow in the context of the wall-friction and wall heat-transfer problems stated in Chapter 2. The wall problems are very important because they constitute the backbone of many applied activities (e.g., heat-exchanger development). For this reason, the near-wall characteristics of turbulent flow were treated first. In the present chapter we depart somewhat from tradition and focus on a different class of turbulent transport problems, namely, the turbulent flow of exergy-carrying fluid through regions situated sufficiently far from solid walls.

To include free-stream turbulent flows in a course on convection heat transfer is to recognize an important development in contemporary engineering. That development is the new emphasis placed by the *affluent* industrial societies on the problem of coexisting with and protecting the environment. Almost without exception, the processes selected by man to effect the interaction between his lifestyle and his environment are turbulent transport processes. The smoke plume swept by the wind in the wake of an industrial area and the water jets discharged by a city into the river are examples of how our presence impacts what surrounds us. These free-stream flows rely on *turbulent mixing* to diffuse away our refuse and, in this way, to minimize the negative effect that high concentrations of such refuse might have on the biosphere. Our almost total reliance on turbulent jets and plumes to disperse fluid pollutants (thermal as well as chemical) is easily explained by the fact that turbulent mixing is the most effective transport mechanism known [1]. To introduce the student to the most basic characteristics of this mechanism is the objective of this chapter.

FREE SHEAR LAYERS

One of the simplest turbulent mixing problems concerning regions far away from solid surfaces is sketched in Fig. 8.1. Consider the time-averaged growth (swelling) of the turbulent interface between a stream and a stagnant fluid reservoir, and keep in mind that the instantaneous picture of the interface (the *shear layer*) is in fact dominated by a chain of large eddies, so large that their diameters define the visual thickness of the shear layer. The instantaneous large-scale structure of a two-dimensional shear layer is shown in Fig. 8.2.: two shear layers of this kind form on both sides of a two-dimensional jet discharging into a reservoir as is shown in Fig. 8.1.

In connection with the time-averaged development of the shear layer, we ask the question of how thick the shear layer will be at a certain distance x downstream from the edge of the nozzle: the answer to this question is relevant to predicting the degree of mixing of (U_0, T_0) fluid with stagnant reservoir fluid (U_∞, T_∞). Considering the flow part of the mixing problem first, we begin with the time-averaged continuity and momentum equations for the (x, y) frame drawn in Fig. 8.1,

$$\frac{\partial \bar{u}}{\partial x} + \frac{\partial \bar{v}}{\partial y} = 0 \tag{1}$$

$$\bar{u}\frac{\partial \bar{u}}{\partial x} + \bar{v}\frac{\partial \bar{u}}{\partial y} = -\frac{1}{\rho}\frac{d\bar{P}}{dx} + \frac{\partial}{\partial y}\left[(v + \varepsilon_M)\frac{\partial \bar{u}}{\partial y}\right] \tag{2}$$

Note that the momentum equation (2) is of the boundary layer type; hence, it applies to the space occupied by the shear layer *only if the shear layer is slender*. The two unknowns to be determined from eqs. (1) and (2) are $\bar{u}$ and $\bar{v}$; however, in order to proceed with an analytical solution it is necessary to invoke additional simplifying assumptions:

(i) The longitudinal pressure gradient in the shear layer $(d\bar{P}/dx)$ is zero, as the static pressure is the same on both sides of the shear layer. This assumption is fairly good for most turbulent free-shear-layer flows, including the two-dimensional jet development sketched in Fig. 8.1. It is worth noting that the pressure inside a developing jet is practically the same as that of the ambient fluid at rest [2].

(ii) The momentum eddy diffusivity ε_M is much greater than the kinematic viscosity v, so that v can be omitted in the right-hand side of eq. (2). This assumption is made without much hesitation in classical treatments of the free shear layer problem (e.g., Refs. 2–4), presumably because the flow region is already assumed to be situated sufficiently far from solid surfaces. However, looking at the tip section of the shear layer of Fig. 8.2, we see that near its origin the shear layer is laminar, in other words, sufficiently close to $x = 0$ and $v \ll \varepsilon_M$ assumption must break down. The fact that the $v \ll \varepsilon_M$ assumption is valid only beyond a certain value of x (hence, beyond a certain shear layer thickness) should be expected based on the transition scaling laws discussed in Chapter 6. Note first that the governing equations (1) and (2) could be solved for laminar flow by setting $\varepsilon_M = 0$ and applying the boundary layer methods

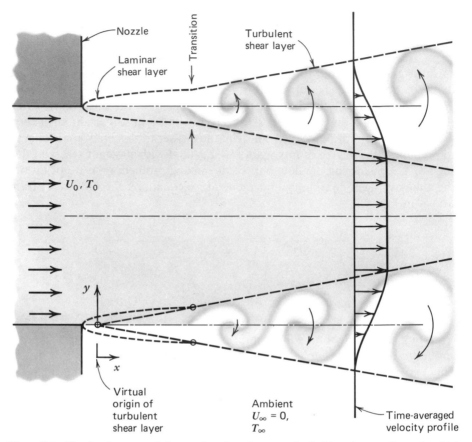

Figure 8.1 The development of the two free shear layers on both sides of a two-dimensional jet.

of Chapter 2. Since there is nothing in the steady state laminar shear layer solution to suggest that it does not hold for all values of x, and since the transition scale criterion (15) of Chapter 6 appears to have universal applicability, the laminar shear layer is destined to break down when it exceeds a certain thickness, in other words, when x exceeds a certain order of magnitude (see Problem 1). The $\nu \ll \varepsilon_M$ assumption is then valid only in the turbulent section of the shear layer, only sufficiently far downstream from the laminar–turbulent transition region, such that the largest eddies are much bigger than the critical thickness of the shear layer during transition (i.e., much bigger than the *smallest eddies* discussed in Chapter 7).

    (iii) The third assumption concerns the eddy diffusivity ε_M: this assumption is absolutely necessary for closing the $(\bar{u}, \bar{v})$ problem described by eqs. (1) and (2). It is very important to understand from the beginning that what is

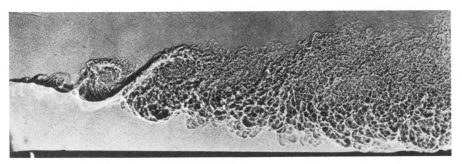

Figure 8.2 The discrete, stepwise growth of a turbulent shear layer (upper side: helium, $U_0 = 9.15$ m/s; lower side: nitrogen, $U_\infty = 1/3$ m/s; pressure = 7 atm). (Photograph reproduced from Ref. 6 with permission from Cambridge University Press.)

usually assumed as an expression for ε_M [namely, eq. (4)] is purely *empirical* in origin: this ε_M model is the result of observing that as Fig. 8.2. suggests, *the shear layer thickness D appears to be proportional to x*. Based on this observation and subject to assumptions (i), (ii) justified above, the momentum equation (2) represents the following balance of scales.

$$\frac{U_0^2}{x} \sim \varepsilon_M \frac{U_0}{D^2} \tag{3}$$

Since the eye sees $D \sim x$, the scaling law (3) requires

$$\varepsilon_M \sim U_0 x \tag{4}$$

The ε_M scale derived above can also be derived by invoking the mixing-length model (Chapter 7) *coupled* with the $D \sim x$ observation, as was done originally by Prandtl [2] (see also Ref. 3). Thus writing

$$\varepsilon_M = l^2 \left| \frac{\partial \bar{u}}{\partial y} \right| \tag{5}$$

and by taking in an order of magnitude sense

$$\frac{\partial \bar{u}}{\partial y} \sim \frac{U_0}{D} \quad \text{and} \quad l \sim D \tag{6}$$

the mixing-length model (5) becomes identical to the scaling law (4). In conclusion, physical observations alone suggest that ε_M is proportional to x and that ε_M is not a function of y. The conclusion that ε_M is independent of y may seem paradoxical considering that, regardless of how large ε_M is inside the shear layer, ε_M must decrease to zero outside the shear layer where, as Fig. 8.2 plainly shows, there are no eddies. In fact, no paradox exists, simply because eq. (4) is the result of scale analysis valid *inside* the slender shear layer region only.

According to assumptions (i) and (ii), the boundary-layer momentum equation (2) reduces to

$$\bar{u}\frac{\partial \bar{u}}{\partial x} + \bar{v}\frac{\partial \bar{u}}{\partial y} = \varepsilon_M \frac{\partial^2 \bar{u}}{\partial y^2} \tag{7}$$

The $D \sim x$ observation that led to the ε_M model (4) suggests the following formulation for seeking a similarity solution to the turbulent free shear layer problem [2–4],

$$\eta = \sigma\frac{y}{x}, \qquad \varepsilon_M = \frac{1}{4\sigma^2}U_0 x$$

$$\bar{u} = \frac{\sigma}{2}U_0 F'(\eta), \qquad \bar{\psi} = \frac{x}{2}U_0 F(\eta) \tag{8}$$

where σ is an empirical constant, η is the similarity variable, and $\bar{\psi}$ is the time-averaged streamfunction defined as $\bar{u} = \partial\bar{\psi}/\partial y, \bar{v} = -\partial\bar{\psi}/\partial x$. Function $F'(\eta)$ represents the shape of the dimensionless velocity profile in the shear layer region. In the notation of eqs. (8), the mass and momentum equations (1) and (7) collapse into

$$F''' + 2\sigma FF'' = 0 \tag{9}$$

subject to the following boundary conditions:

$$\bar{u} = U_0 \text{ as } y \to \infty, \quad \text{or } F' = \frac{2}{\sigma} \text{ as } \eta \to \infty$$

$$\bar{u} = 0 \text{ as } y \to -\infty, \quad \text{or } F' = 0 \text{ as } \eta \to -\infty \tag{10}$$

$$\bar{v} = 0 \text{ at } y = 0, \quad \text{or } F = 0 \text{ at } y = 0$$

The similarity problem (9, 10) can be solved numerically, and the resulting velocity profile resembles the shape already sketched in Fig. 8.1. A closed-form curvefit to this numerical solution is [5]

$$\bar{u} \cong \frac{U_0}{2}\left[1 + \text{erf}\left(\sigma\frac{y}{x}\right)\right] \tag{11}$$

Comparison with measurements indicates that

$$\sigma \cong 13.5 \tag{12}$$

is the appropriate value for the empirical constant that accounts for the linear growth of the shear layer.

The observed large-scale structure of the shear layer (Fig. 8.2) is responsible for the closed-form solution, (11) and (12): geometrically, this solution implies that the shear layer thickness grows linearly in x, the actual growth rate depending on the *definition* of shear layer thickness. For example, if the effective thickness is defined as the knee-to-knee distance D_{k-k} sketched in Fig.

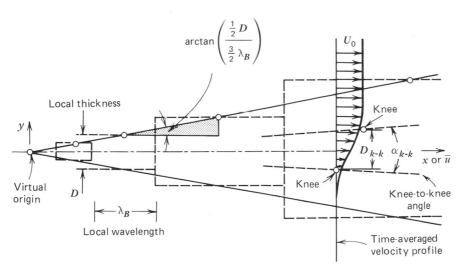

Figure 8.3 The constant-angle growth of a turbulent shear layer, as the repeated manifestation of the $\lambda_B \sim D$ scaling law of Chapter 6.

8.3,[†] then the knee-to-knee growth angle α_{k-k} is

$$\alpha_{k-k} = \arctan\left(\frac{\pi^{1/2}}{\sigma}\right) \cong 7.5° \tag{13}$$

Although the 7.5° growth angle is an important constant to remember in turbulence research, it is worth keeping in mind that the numerical value of α_{k-k} is intimately connected to the definition of knee-to-knee thickness. Figure 8.2, for example, shows very clearly that the visual growth angle is considerably greater (of the order of 20° [6]): a purely theoretical explanation for the photographed growth rate is proposed in Ref. 7 based on the $\lambda_B \sim D$ scaling law discussed in Chapter 6. The same scaling law provides a theoretical basis for the time-averaged linear growth rate of the shear layer, a fact that was accepted empirically in the development of the ε_M model given by eq. (4). (See Ref. 7, pp. 84, 89.)

The scaling argument that explains the constant-angle shape of the shear layer is the following: If the $\lambda_B \sim D$ proportionality is indeed *a property* of the shear layer as a finite-size region, then the λ_B wave will be rolled into a large eddy in time of order $t_B \sim \lambda_B/(U_0/2)$. (Note that from symmetry, $U_0/2$ is the scale of the relative velocity between the shear layer fluid and the fluid situated on either side of the shear layer.) During the same time period, the formation of the eddy effects the stepwise thickening of the shear layer into a new region of thickness $D_{new} > D_{old}$ [the scale analysis of the lateral growth of each elbow of the λ_B wave indicates that $D_{new} \sim 2D_{old}$ (Ref. 7, p. 75)]. Also during a time of order t_B, the shear layer fluid travels downstream to a distance of order λ_B, since the shear layer fluid velocity relative to one of the fluid reservoirs at rest is $U_0/2$. In summary, the shear layer region is one that thickens stepwise in the downstream direction and, as is shown schematically in Fig. 8.3, each step is D-thick and λ_B-long. Furthermore, if the $\lambda_B \sim D$ proportionality is universal, that is, if it applies anywhere along the shear layer, then all the steps are geometrically similar. Averaging in time the parade of large eddies through this sequence of steps, we anticipate that the time-averaged shear layer region must *visibly flare out linearly* with a constant half-angle of order $\arctan[(D/2)/(3\lambda_B/2)]$.

The linear growth of turbulent free shear layers (and turbulent jets and plumes) is amply documented and, through the eddy diffusivity model (4), it forms the backbone of the classical time-averaged analytical description exhibited in this chapter. Because of this ample documentation, the scaling argument constructed in the preceding paragraph could be traveled in the reverse direction to conclude that the universally observed constant-angle geometry of turbulent shear flow validates the theoretical view that *the $\lambda_B \sim D$*

[†]The two knees are the intersections of the line $\bar{u}/y = (d\bar{u}/dy)_{y=0}$ with the two vertical lines $\bar{u} = U_0$ and $\bar{u} = 0$ on Fig. 8.3.

scaling is a fundamental property of the flow. Additional evidence that supports this view has been compiled in Chapter 6 and in Ref. 7.

Turning our attention to the heat transfer part of the problem, namely, the *thermal mixing* between T_∞ and T_0 fluids in the shear layer region, we note that assumptions (ii) and (iii) can also be applied to the eddy thermal diffusivity ε_H. Hence, based on the assumptions that ε_H is much greater than α and that ε_H is not a function of y inside the shear layer region, the time-averaged energy equation assumes the simpler form

$$\bar{u}\frac{\partial \bar{T}}{\partial x} + \bar{v}\frac{\partial \bar{T}}{\partial y} = \varepsilon_H \frac{\partial^2 \bar{T}}{\partial y^2} \tag{14}$$

Making the additional assumption that the turbulent Prandtl number is a constant equal to one

$$\mathrm{Pr}_t = \frac{\varepsilon_M}{\varepsilon_H} \cong 1 \tag{15}$$

the eddy thermal diffusivity ε_H is given by the same expression as that for ε_M in eqs. (8). Finally, comparing the $y \to \pm\infty$ boundary conditions for temperature $\bar{T}$ with those for longitudinal velocity $\bar{u}$ and noting the identical form of the simplified momentum and energy equations [eqs. (7) and (14)], the temperature field problem $\bar{T}(x, y)$ becomes identical to the flow problem $\bar{u}(x, y)$ already discussed (see also Problem 5).

$$\frac{\bar{T} - T_\infty}{T_0 - T_\infty} = \frac{\bar{u}}{U_0} \cong \frac{1}{2}\left[1 + \mathrm{erf}\left(\sigma\frac{y}{x}\right)\right] \tag{16}$$

According to this solution, the temperature gradient $\partial \bar{T}/\partial y$ is maximum at $y = 0$, in other words, the heat transfer rate between the two semiinfinite fluid reservoirs is maximum across the plane of original velocity discontinuity. If we are asked to evaluate the time-averaged heat flux across the $y = 0$ plane, we might be tempted to quickly write $q''_{y=0} = k(\partial \bar{T}/\partial y)_{y=0}$, as in the early chapters of this course: this would be wrong, because the temperature field solution (16) was developed based on the assumption that the eddy heat transport mechanism is much more effective than the molecular mechanism [assumption (ii)]. Therefore, the proper way to evaluate the midplane heat flux is by writing

$$q''_{y=0} = \rho c_P \varepsilon_H \left(\frac{\partial \bar{T}}{\partial y}\right)_{y=0} \tag{17}$$

which combined with the eddy diffusivity model [eq. (8)] and with the Pr_t

assumption (15) yields [4]

$$\frac{q''_{y=0}}{T_0 - T_\infty} = \frac{1}{4\sigma\pi^{1/2}}\rho c_P U_0 \tag{18}$$

We reach the interesting conclusion that the midplane heat-transfer coefficient is x-independent, or that the midplane Stanton number is a constant equal to $(4\sigma\pi^{1/2})^{-1} \cong 0.0104$.

JETS

Two-Dimensional Jets

Another frequent example of turbulent mixing in freestream flow is the jet flow already discussed in connection with Fig. 8.1. In the case of a two-dimensional jet (U_0, T_0) injected through a slit of width D_0 into a stagnant isothermal fluid reservoir (T_∞), the initial mixing between the two fluids is ruled by the free shear layer phenomenon. Since the two shear layers grow linearly in the direction of the flow, they are destined to merge downstream from the nozzle at a characteristic distance that scales with the nozzle dimension D_0. In the

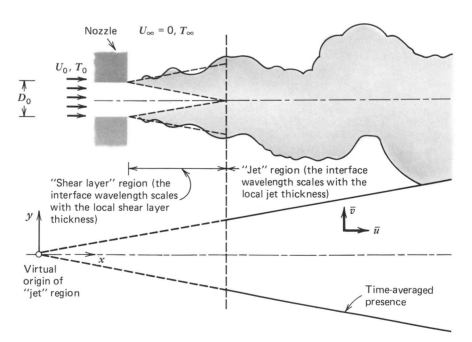

Figure 8.4 The large-scale instantaneous structure of a two-dimensional turbulent jet and the constant-angle shape of the time-averaged flow region.

shear layer section of the flow (Fig. 8.4), the centerline velocity is very close to U_0 and practically independent of x. Beyond the point where the two shear layers merge, the stream proceeds as a *jet* and the centerline velocity $\bar{u}_c$ decreases monotonically with x. The following analysis applies to the jet region of the flow, again, based on the boundary-layer theory that this region is *slender*.

An analytical description of the time-averaged jet profile is possible [5] if the mass and momentum equations (1) and (2) are coupled with assumptions (i)–(iii) identified earlier in this chapter. In particular, assumption (iii) is justified by the empirical observation that the jet region flares out linearly in the x direction. As is shown in the bottom-half of Fig. 8.4, if the jet thickness D is proportional to x, then the point of $x = 0$ represents a fictitious, point-size origin of the jet flow. After assumptions (i)–(iii), the momentum equation assumes the simpler form (7) with the eddy diffusivity obeying the model

$$\varepsilon_M = \frac{1}{4\gamma^2}\bar{u}_c x \tag{19}$$

where γ is the empirical constant accounting for the growth rate of the jet region. Before proceeding with the similarity solution to eqs. (1), (7), and (19), it is instructive to deduce first the relationship between $\bar{u}_c$ and x. Instructed by Problem 10 of Chapter 2, by integrating the x momentum equation from $y = -\infty$ to $y = \infty$, we find

$$\int_{-\infty}^{\infty} \bar{u}^2 dy = U_0^2 D_0, \text{ constant} \tag{20}$$

In an order of magnitude sense, eq. (20) states

$$\bar{u}_c^2 D \sim U_0^2 D_0 \tag{21}$$

or, defining x_0 such that $D/D_0 = x/x_0$, we can now write

$$\frac{\bar{u}_c}{U_0} = \left(\frac{x}{x_0}\right)^{-1/2} \tag{22}$$

In conclusion, in a two-dimensional turbulent jet, the centerline velocity decays as $x^{-1/2}$. This scaling is the basis for the following construction of the similarity solution:

$$\eta(x, y) = \gamma\frac{y}{x}, \qquad \bar{u} = U_0\left(\frac{x}{x_0}\right)^{-1/2}F'(\eta) \tag{23}$$

In terms of the time-averaged streamfunction,

$$\bar{\psi} = \frac{1}{\gamma}U_0 x_0^{1/2} x^{1/2} F(\eta) \tag{24}$$

where $\bar{u} = \partial\bar{\psi}/\partial y, \bar{v} = -\partial\bar{\psi}/\partial x$, the momentum equation (7) reduces to

$$(F')^2 + FF'' + \tfrac{1}{2}F''' = 0 \tag{25}$$

The appropriate boundary conditions on the dimensionless streamfunction profile F are

$$F = 0 \quad \text{and} \quad F' = 1 \quad \text{at } \eta = 0$$

$$F' = 0 \qquad\qquad\qquad \text{at } \eta = \infty \tag{26}$$

The problem represented by eqs. (25) and (26) can be solved in closed-form (see Ref. 3),

$$\bar{u} = U_0\left(\frac{x}{x_0}\right)^{-1/2}(1 - \tanh^2\eta) \tag{27}$$

In order to use the velocity solution (27), we have to first calculate the value of x_0 in terms of the *jet strength* $U_0 D_0$, which is assumed known. Substituting eq. (27) into the jet strength constraint (20) yields

$$\frac{x_0}{\gamma D_0} = \int_{-\infty}^{\infty}(1 - \tanh^2\eta)^2 d\eta = \frac{4}{3} \tag{28}$$

Finally, it is found that the linear growth parameter γ that produces the best agreement between experimental measurements and the similarity profile is [8]

$$\gamma \cong 7.67 \tag{29}$$

This empirical constant is comparable with the one determined for shear layers, stressing once more the universality of constant-angle growth in free-stream turbulent flow.

The temperature distribution in the jet region is closely related to the velocity distribution. The close connection between the two fields is to be expected in view of the large-scale eddy formation process that, as in shear layers (Fig. 8.3), is responsible for the lateral growth of the jet. Indeed, starting with the assumption that ε_H is equal to ε_M, it is not difficult to show that the temperature excess function $(\bar{T} - T_\infty)$ is given by an expression analogous to eq. (27), as the boundary layer momentum and energy equations become identical and the jet strength constraint (20) is replaced by an enthalpy flow constraint (see Problem 11 of Chapter 2). However, experimentally it is found that the temperature field data are fitted better by [9]

$$\frac{\bar{T} - T_\infty}{\bar{T}_c - T_\infty} \cong \left(\frac{\bar{u}}{\bar{u}_c}\right)^{\text{Pr}_t} \tag{30}$$

where $\mathrm{Pr}_t \cong 0.5$ and $(\overline{T}_c - T_\infty)$ is the time-averaged temperature difference between jet centerline and stagnant ambient. The temperature difference $(\overline{T}_c - T_\infty)$ decreases in the downstream direction as $x^{-1/2}$, that is, in the same manner as the centerline velocity $\overline{u}_c$. The curvefit (30) shows that at a fixed longitudinal position x the temperature profile is *broader* than the velocity profile, because the ratio $\overline{u}/\overline{u}_c$ is always less than one and the exponent Pr_t is also less than one.

Round Jets

An even more common free-turbulent flow, more common than the two-dimensional jet of Fig. 8.4, is the round jet formed by discharge from a nozzle into a fluid reservoir. The initial section of this type of jet is also governed by the free shear layer flow sketched in Fig. 8.1; however, this time, the shear layer fills an annular region that surrounds the fluid issuing from the nozzle. The jet section of the flow begins at a distance of approximately $5D_0$ downstream from the nozzle [10], D_0 being the nozzle diameter. An analytical solution for the time-averaged flow in the jet section is again possible based on assumptions (i)–(iii) [11]. In a cylindrical coordinate system (r, x) drawn as in the lower-half of Fig. 8.4 (where r replaces y), the mass continuity and the simplified momentum equations are

$$\frac{\partial \overline{u}}{\partial x} + \frac{1}{r}\frac{\partial}{\partial r}(r\overline{v}) = 0 \tag{31}$$

$$\overline{u}\frac{\partial \overline{u}}{\partial x} + \overline{v}\frac{\partial \overline{u}}{\partial r} = \varepsilon_M \frac{1}{r}\frac{\partial}{\partial r}\left(r\frac{\partial \overline{u}}{\partial r}\right) \tag{32}$$

where ε_M is again modeled as proportional to the product $(x\overline{u}_c)$. In addition, the jet strength defined as

$$K = 2\pi \int_0^\infty \overline{u}^2 r\, dr \tag{33}$$

must have the same value at any distance x downstream from the fictitious point-size origin of the jet (Fig. 8.4). The jet strength constraint (33) implies

$$\overline{u}_c^2 D^2 \sim K, \quad (\text{not a function of } x) \tag{34}$$

in other words,

$$\overline{u}_c \sim \frac{K^{1/2}}{x} \tag{35}$$

since, visually, the jet thickness D is proportional to x. Recognizing once more the eddy diffusivity model, we conclude that

$$\varepsilon_M \sim x\overline{u}_c \sim K^{1/2}, \quad (\text{constant}) \tag{36}$$

The conclusion that in turbulent round jets the eddy diffusivity can be modeled as a constant independent of x or y is very important. Its immediate implication is that, analytically, the turbulent round jet problem is identical to the laminar round jet problem [12], as the constant eddy diffusivity ε_M replaces the kinematic viscosity ν in the boundary layer formulation of the momentum equation [eq. (32)]. Therefore the solution for the velocity profile can be written immediately

$$\bar{u} = \frac{\gamma_0}{2}\left(\frac{3}{\pi}\right)^{1/2}\frac{K^{1/2}}{x}\left(1 + \frac{\eta^2}{4}\right)^{-2} \tag{37}$$

where η is the similarity variable

$$\eta = \gamma_0\frac{r}{x} \tag{38}$$

and γ_0 is the empirical constant related to the angle of the cone filled by the time-averaged round jet. Reichardt's experiments indicate that the value

$$\gamma_0 \cong 15.2 \tag{39}$$

is adequate for matching expression (37) to longitudinal velocity measurements ([3], p. 608).

An alternative purely empirical way to curvefit the time-averaged velocity profile is to use the Gaussian form

$$\bar{u} = \bar{u}_c\exp\left[-\left(\frac{r}{b}\right)^2\right] \tag{40}$$

where b is a characteristic radial dimension proportional to the transversal length scale D. Fischer et al. [1] surveyed the $\bar{u}$ data produced by 15 independent experiments and concluded that

$$b = (0.107 \pm 0.003)x \tag{41}$$

In other words, the Gaussian profile used more often as a substitute for expression (37) is

$$\bar{u} = \bar{u}_c\exp\left[-\left(9.35\frac{r}{x}\right)^2\right] \tag{42}$$

The reader can easily verify that expressions (37) and (42) are essentially equivalent if $\eta < 2$. Furthermore, substituting eq. (42) into the jet strength constraint (33) yields $\bar{u}_c = 7.46K^{1/2}/x$, which is only 0.4 percent higher than the corresponding quantity on the right-hand side of eq. (37).

The temperature distribution is again closely related to the velocity field. Thus, it is found that the transversal scale of the flow region heated by the turbulent jet is proportional to x. Using Gaussian forms to curvefit the temperature profile

$$\overline{T} - T_\infty = (\overline{T}_c - T_\infty)\exp\left[-\left(\frac{r}{b_T}\right)^2\right] \qquad (43)$$

the 15 experiments compared by Fischer et al. [1] showed that the transversal length scale b_T varies little from one report to another,

$$b_T = (0.127 \pm 0.004)x \qquad (44)$$

Formulas (41) and (44) indicate that at a given axial location x the temperature profile is slightly broader than the velocity profile,

$$\frac{b_T}{b} \cong 1.19 \qquad (45)$$

The consistency of this observation, not only here but in two-dimensional jets and free shear layers, is worthy of our curiosity (Problem 5).

PLUMES

Continuing on the road from simple to complex, in this section we analyze the mixing in turbulent jet flows driven not by the strength given to them by the nozzle [eqs. (20) and (33)] but by the effect of buoyancy. Consider the vertical flow of heated fluid above a point heat source or above a round nozzle discharging upward, and attach to this flow a cylindrical coordinate system (y, r) such that the axial point $y = 0$ coincides with the *virtual* origin of the time-averaged plume: as suggested by Fig. 8.5, it is widely observed that at sufficiently large values of y the plume thickness D is proportional to y. In the case of free shear layers and jets we saw that the observed constant-angle growth of the flow region leads in relatively few steps to similarity solutions that represent the time-averaged flow quite adequately. In this section we conduct an integral analysis of the round plume: the chief conceptual focus of this analysis will be to highlight the relationship between the universally observed $D \sim y$ proportionality and a practice described as the *entrainment hypothesis* in the turbulent-jet literature [1, 13].

The mass, momentum, and energy equations applicable to the plume as a *slender* flow region are

$$\frac{1}{r}\frac{\partial}{\partial r}(r\overline{u}) + \frac{\partial \overline{v}}{\partial y} = 0 \qquad (46)$$

$$\frac{\partial}{\partial y}(\overline{v}^2) + \frac{1}{r}\frac{\partial}{\partial r}(r\overline{u}\overline{v}) = \frac{1}{r}\frac{\partial}{\partial r}\left[r(\nu + \varepsilon_M)\frac{\partial \overline{v}}{\partial r}\right] + g\beta(\overline{T} - T_\infty) \qquad (47)$$

$$\frac{\partial}{\partial y}(\overline{v}\overline{T}) + \frac{1}{r}\frac{\partial}{\partial r}(r\overline{u}\overline{T}) = \frac{1}{r}\frac{\partial}{\partial r}\left[r(\alpha + \varepsilon_H)\frac{\partial \overline{T}}{\partial r}\right] \qquad (48)$$

The integral analysis begins with integrating eqs. (46)–(48) over the flow cross-section defined by the plane y = constant. Integrating the mass continuity equation (46) yields

$$(r\bar{u})_\infty - (r\bar{u})_0 + \frac{d}{dy}\int_0^\infty \bar{v}r\,dr = 0 \qquad (49)$$

In this equation, the second term is zero due to symmetry. The first term requires special attention, because it is quite tempting to see a vanishing $\bar{u}$ as r becomes large: we must keep in mind, however, that in this boundary-layer analysis "large r" means "sufficiently larger than the transversal length scale of the plume." Therefore, as we look at the instantaneous picture of the *edge* of the plume (in Fig. 8.5 or in an actual industrial smoke discharge), we conclude that $(r\bar{u})_\infty$ must be finite because at the edge both r and $\bar{u}$ are finite. The actual scale of $(r\bar{u})_\infty$ will become evident later in this section. In conclu-

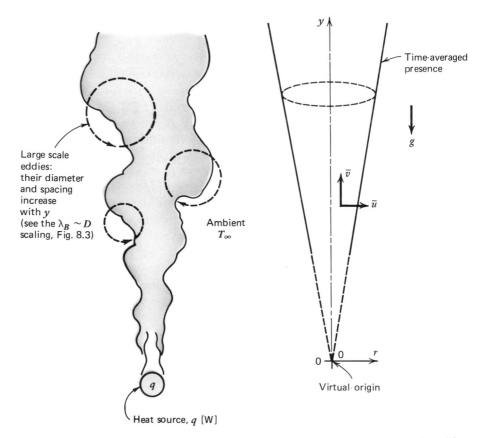

Figure 8.5 The meandering or buckled shape of a turbulent plume above a concentrated heat source and the funnel shape of the time-averaged flow region.

sion, eq. (49) reduces to

$$\frac{d}{dy} \int_0^\infty \bar{v} r \, dr = -(r\bar{u})_\infty \tag{50}$$

Turning our attention to the momentum equation (47), its corresponding area integral is

$$\frac{d}{dy} \int_0^\infty \bar{v}^2 r \, dr + (r\bar{u}\bar{v})_\infty - (r\bar{u}\bar{v})_0 = \left[r(v + \varepsilon_M) \frac{\partial \bar{v}}{\partial r} \right]_\infty$$
$$- \left[r(v + \varepsilon_M) \frac{\partial \bar{v}}{\partial r} \right]_0 + g\beta \int_0^\infty (\bar{T} - T_\infty) r \, dr \tag{51}$$

We can now argue that, although $(r\bar{u})_\infty$ is finite, the second term $(r\bar{u}\bar{v})_\infty$ vanishes because $\bar{v}$ vanishes as we approach the T_∞ reservoir. The third term $(r\bar{u}\bar{v})_0$ is zero due to symmetry. On the right-hand side of eq. (51), the first term vanishes because $\bar{v} = 0$ and the second term is zero, again due to symmetry. Thus the momentum equation (47) reduces to

$$\frac{d}{dy} \int_0^\infty \bar{v}^2 r \, dr = g\beta \int_0^\infty (\bar{T} - T_\infty) r \, dr \tag{52}$$

Finally, the energy equation (48) can be integrated over the cross-section to yield

$$\frac{d}{dy} \int_0^\infty \bar{v} \bar{T} r \, dr + (r\bar{u}\bar{T})_\infty - (r\bar{u}\bar{T})_0 = \left[r(\alpha + \varepsilon_H) \frac{\partial \bar{T}}{\partial r} \right]_\infty - \left[r(\alpha + \varepsilon_H) \frac{\partial \bar{T}}{\partial r} \right]_0 \tag{53}$$

The last three terms in this equation drop out based on arguments similar to those preceding eq. (52). The second term $(r\bar{u}\bar{T})_\infty$ is in fact $(r\bar{u})_\infty T_\infty$, where $(r\bar{u})_\infty$ is given by the mass conservation equation (50). Therefore, combining eqs. (50) and (53) we obtain the expected result that the enthalpy flowrate is conserved in each cross-section,

$$\frac{d}{dy} \int_0^\infty \bar{v} (\bar{T} - T_\infty) r \, dr = 0 \tag{54}$$

A more useful version of eq. (54) is to write that the integral alone is y-independent and proportional to the strength of the heat source q: setting q equal to the enthalpy flowrate through each cross-section yields

$$\int_0^\infty \bar{v} (\bar{T} - T_\infty) r \, dr = \frac{q}{2\pi\rho c_P} \tag{55}$$

The second phase of any integral analysis is to assume known the actual variation of the unknowns in the direction in which the governing equations were integrated. Assuming the Gaussian profiles recommended by the survey of round jet data [1]

$$\bar{v} = \bar{v}_c \exp\left[-\left(\frac{r}{b}\right)^2\right]$$

$$\bar{T} - T_\infty = (\bar{T}_c - T_\infty)\exp\left[-\left(\frac{r}{b_T}\right)^2\right] \tag{56}$$

and keeping in mind that the ratio b_T/b is a constant of order one [eq. (45)], the integral equations (50), (52), and (55) reduce to

$$\frac{d}{dy}\left(\bar{v}_c b^2\right) = -2(r\bar{u})_\infty \tag{57}$$

$$\frac{d}{dy}\left(\bar{v}_c^2 b^2\right) = 2g\beta\left(\bar{T}_c - T_\infty\right)b_T^2 \tag{58}$$

$$\bar{v}_c\left(\bar{T}_c - T_\infty\right) = \frac{q}{\pi\rho c_P}\frac{1 + (b/b_T)^2}{b^2} \tag{59}$$

These three equations are sufficient for determining $\bar{v}_c(y)$, $\bar{T}_c(y)$, and $b(y)$ *provided* the entrainment term $(r\bar{u})_\infty$ is known. The scale of $(r\bar{u})_\infty$ is the direct consequence of *observing* that the time-averaged thickness of the plume is proportional to the plume height,

$$b \sim y \tag{60}$$

From this observation and eq. (57) we deduce the following scaling law

$$\bar{v}_c b \sim (r\bar{u})_\infty \tag{61}$$

In other words, we conclude that near the sharp interface between the plume stream and the ambient, the entrainment velocity must scale with the longitudinal (vertical) velocity. This conclusion is consistent with the eddy formation mechanism recommended by the $\lambda_B \sim D$ property of any inviscid stream: the growth of the stream thickness is effected by D-wide eddies which, rotating like bicycle wheels on a track (on the ambient), bring ambient fluid into the stream with a velocity proportional to the peripheral velocity of the eddy. Noting that in the case of a plume $\bar{v}_c > 0$ and $(r\bar{u})_\infty < 0$, we write

$$(r\bar{u})_\infty = -\hat{\alpha}b\bar{v}_c \tag{62}$$

where $\hat{\alpha}$ is an empirical constant tied ultimately to the actual cone angle of the plume. [This linear growth feature—either accepted empirically or derived theoretically based on the $\lambda_B \sim D$ property—is the source of eq. (62).]

The entrainment model (62) becomes the necessary substitute for the similarity variable $\eta = \gamma_0 r/x$ [eq. (38)], if the similarity formulation of the earlier sections is replaced by the integral formulation presented in this section. The same entrainment hypothesis was used by Morton, Taylor, and Turner [14] in the analysis of buoyant turbulent jets in stratified media. Thus, combining the entrainment assumption (62) with the integral equations (57)–(59) and appropriate starting conditions (at $y = 0$), we have the means to derive the y dependence of $\bar{v}_c$ and $\bar{T}_c$.

The starting conditions for integrating eqs. (57)–(59) require special attention. We can focus on the simplest case—the so-called simple plume [1]—in which we can assume that the strength of the jet $\bar{v}_c^2 b^2$ is zero at $y = 0$. In addition, we can take the flowrate $\bar{v}_c b^2$ to be zero at $y = 0$. These assumptions make the simple plume one that originates from a fictitious point, as is shown on the right-hand side of Fig. 8.5. The integral solution subjected to these starting conditions is [1, 13, 15]:

$$b = \frac{6}{5}\hat{\alpha}y \tag{63}$$

$$\bar{v}_c = \left[\frac{25}{24\pi\hat{\alpha}^2}\frac{qg\beta}{\rho c_P y}\left(1 + \frac{b_T^2}{b^2}\right)\right]^{1/3} \tag{64}$$

$$\bar{T}_c - T_\infty = 0.685\left(1 + \frac{b^2}{b_T^2}\right)\left(1 + \frac{b_T^2}{b^2}\right)^{-1/3}\left(\frac{q}{\pi\rho c_P}\right)^{2/3}\hat{\alpha}^{-4/3}y^{-5/3}(g\beta)^{-1/3} \tag{65}$$

An approximate value for the empirical constant is 0.12 [13].

Historically, the turbulent simple-plume problem was first solved and published in 1941 by Wilhelm Schmidt [16], who used a similarity formulation of the type presented earlier in this chapter (in other words, he used the mixing-length eddy diffusivity model instead of the entrainment hypothesis of integral analysis). Fifteen years later, Morton et al. [14] published an integral solution for the more general problem involving ambient stratification and pointed out that the integral solution for zero stratification [eqs. (63)–(65)], "is of the same form as that of Schmidt" [16]. They also pointed out that the agreement between Schmidt's solution and the integral solution for zero stratification "illustrates the fact that the entrainment assumption is consistent with Schmidt's mixture length assumptions."

It must be stressed that the time-averaged turbulent plume description provided by eqs. (63)–(65) in conjunction with the assumed Gaussian profiles (56) is adequate only for sufficiently large values of y, depending on what particular device acts as heat source. For example, in the case of the smoke plume shown in Fig. 6.1, the section immediately above the heat source is laminar; hence, the $y = 0$ point of Fig. 8.5 does not coincide with the heat source. Another example is the plume generated by a very large heat source (e.g., a barbecue or a campfire): in its initial stages this flow accelerates upwards as an inviscid stream and, over a height of only a few diameters, its thickness *decreases*. Eventually, the formation of large eddies takes over as a plume-thickening mechanism, and the upper section of the plume falls in line with eqs. (63)–(65). However, even near the base the stream exhibits the $\lambda_B \sim D$ scaling law: Fig. 8.6 compares the sinuous shape predicted by buckling theory [17] with the photographed base section of a natural gas fire [7].

The integral equations (57)–(59) can be solved for the more general case where the initial section of the plume is in fact a forced jet, one defined by a nozzle producing a jet of known flowrate $(\bar{v}_c b^2)_{y=0}$ and known strength $(\bar{v}_c^2 b^2)_{y=0}$. Given enough time, that is, above a certain height, the strength of the buoyant jet becomes dominated by the effect of buoyancy [eq. (58)] and the behavior of the buoyant jet becomes similar to that of the simple plume. The height range above which the initial jet becomes a plume can be evaluated by comparing the local strength of the simple plume [$\bar{v}_c^2 b^2$, eqs. (63) and (64)] with the imposed jet strength $(\bar{v}_c^2 b^2)_{y=0}$.

Based on the foregoing treatment of turbulent free shear layers, jets, and plumes, we are in the position to draw the very important conclusion that, in a

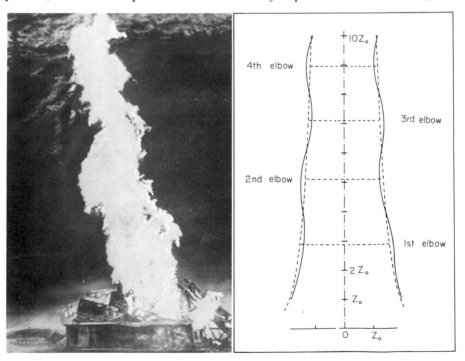

Figure 8.6 The initial buckled shape of a turbulent plume rising from rest; left side: night photograph of a natural gas well on fire [7] (*Wide World Photos*); right side: the large-scale sinuous structure predicted by buckling theory [17].

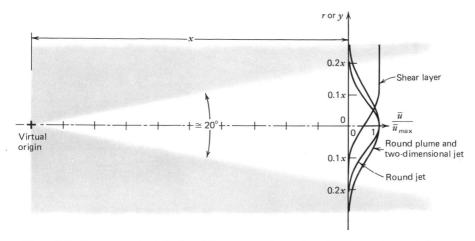

Figure 8.7 The geometric similarity of time-averaged turbulent shear layers, jets and plumes.

time-averaging sense, all these flows and their associated temperature fields are *geometrically similar*. Figure 8.7 shows a drawn-to-scale summary of the mean velocity profiles discussed earlier in this chapter; as such, this drawing is a restatement of the observation that the transversal length scale of shear layers, jets, and plumes is proportional to the distance x measured from a fictitious origin. This physical observation was the basis for the similarity treatment of shear layers and jets, and for the entrainment hypothesis incorporated in the integral analysis of the simple plume. The constant-angle growth appears to be a consequence of the $\lambda_B \sim D$ scaling law of inviscid flow (Fig. 8.3).

THERMAL WAKES BEHIND CONCENTRATED SOURCES

A relatively simple way to model the dispersion of thermal pollution in the wake of a concentrated heat source is shown in Fig. 8.8. Consider a line heat source of strength q' [W/m], positioned normal to a time-averaged uniform stream $(\overline{U}_\infty, \overline{T}_\infty)$ populated throughout by eddies that have the same characteristic size and peripheral speed. This sort of turbulence can be created in the laboratory, immediately behind a turbulence-generating grid installed normal to the flow in a wind tunnel [18]. In nature, *grid-generated turbulence* is no more than an approximate model for the eddy transport capability of large streams that bathe concentrated sources of heat or mass (e.g., the atmospheric boundary layer and the mainstream section of a river). The eddy population in such streams is the result of earlier stream–wall and stream–stream interactions of the type discussed in Chapter 7 and the preceding sections of this chapter.

In order to determine the time-averaged temperature field in the wake of the line source of Fig. 8.8, consider the energy equation applicable to that

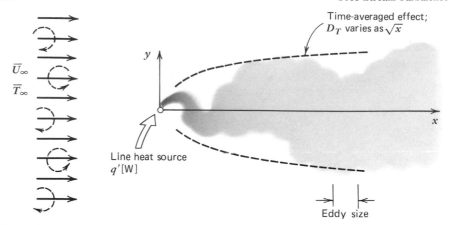

Figure 8.8 The development of a thermal wake behind a line heat source normal to a uniform stream with grid-generated turbulence.

situation:

$$\overline{U}_\infty \frac{\partial \overline{T}}{\partial x} = \varepsilon_H \frac{\partial^2 \overline{T}}{\partial y^2} \tag{66}$$

Equation (66) has already been simplified based on the following assumptions:

The thermal eddy diffusivity is much greater than the molecular diffusivity, $\varepsilon_H \gg \alpha$.

The thermal wake region is slender; hence, eq. (66) is of the boundary layer type.

The eddy diffusivity ε_H is not a function of either y or x; the value of this constant, assumed known, is controlled by the mechanism that generates turbulence in the $\overline{U}_\infty$ stream.

With these assumptions in mind, the scale analysis of the energy equation shows that the thermal wake thickness scales as $(\varepsilon_H x / \overline{U}_\infty)^{1/2}$. Also, the scale analysis of the enthalpy conservation constraint

$$q' = \rho c_P \int_{-\infty}^{\infty} \overline{U}_\infty (\overline{T} - \overline{T}_\infty) \, dy \tag{67}$$

shows that the centerline-to-ambient temperature difference $(\overline{T}_c - \overline{T}_\infty)$ scales as $(q'/\rho c_P)(\overline{U}_\infty \varepsilon_H x)^{-1/2}$. The similarity solution recommended by these scales is

$$\overline{T}(x, y) - \overline{T}_\infty = \frac{q'/\rho c_P}{(\overline{U}_\infty \varepsilon_H x)^{1/2}} \theta(\eta) \tag{68}$$

$$\eta = y \left(\frac{\overline{U}_\infty}{\varepsilon_H x} \right)^{1/2} \tag{69}$$

where the similarity profile θ is given by the solution to the following problem

$$-\tfrac{1}{2}(\theta + \eta\theta') = \theta''; \qquad \theta \to 0 \text{ as } \eta \to \pm\infty, \quad \text{and} \quad \int_{-\infty}^{\infty} \theta \, d\eta = 1 \quad (70)$$

The result is

$$\theta = \frac{1}{2\sqrt{\pi}} \exp\left(-\frac{\eta^2}{4}\right) \qquad (71)$$

In conclusion, the time-averaged temperature field behind the line source has a Gaussian profile whose span increases as $x^{1/2}$. The centerline temperature difference $(\overline{T}(x,0) - \overline{T}_\infty)$ decreases as $1/x^{1/2}$ in the flow direction [eq. (68)]. The temperature field is known if the eddy diffusivity ε_H associated with the uniform turbulence is known; in fact, the above solution can be combined with actual measurements of $\overline{T}(x, y)$ in order to calculate the ε_H value of a certain population of eddies produced by a certain laboratory technique.

The thermal wake behind a point source immersed in grid-generated turbulence can be analyzed according to the same model. The temperature field is given by (Problem 4)

$$\overline{T}(x,r) - \overline{T}_\infty = \frac{q}{4\pi\rho c_P \varepsilon_H x} \exp\left(-\frac{\overline{U}_\infty r^2}{4\varepsilon_H x}\right) \qquad (72)$$

where q [W] is the strength of the point source and r is the radial distance measured away from the wake centerline.

SYMBOLS

b	radial length scale of round velocity jet [eq. (40)]
b_T	radial length scale of round thermal jet [eq. (43)]
c_P	specific heat at constant pressure
D	transversal length scale of flow region
D_{k-k}	knee-to-knee thickness of time-averaged turbulent shear layer (Fig. 8.3)
F	similarity streamfunction profile
g	gravitational acceleration
K	jet strength [eq. (33)]
l	mixing length [eq. (5)]
P	pressure
Pr_t	turbulent Prandtl number
q	point heat source strength [W]
q'	line heat source strength [W/m]
q''	heat flux [W/m²]
r	radial position

t_B buckling time or time of large eddy formation

T temperature

T_0 temperature of nozzle fluid

T_∞ free-stream temperature

u longitudinal velocity component

U_0 velocity of nozzle fluid

U_∞ free-stream velocity

v transversal velocity component

x longitudinal coordinate

x_0 starting length, defined in eqs. (22) and (28)

y transversal coordinate

α thermal diffusivity

$\hat{\alpha}$ empirical constant [eq. (62)]

α_{k-k} knee-to-knee angle of time-averaged turbulent shear layer
              (Fig. 8.3)

β coefficient of thermal expansion

γ empirical constant [eq. (19)]

γ_0 empirical constant [eq. (37)]

ε_H thermal eddy diffusivity

ε_M momentum eddy diffusivity

η similarity variable

θ similarity temperature profile [eq. (68)]

λ_B local buckling wavelength [eq. (83)]

ν kinematic viscosity

ρ density

σ empirical constant [eq. (8)]

ψ streamfunction

$\overline{(\ \ )}$ time-averaged quantity

$(\ \ )_c$ property measured along the stream centerline

REFERENCES

1. H. B. Fischer, E. J. List, R. C. Y. Koh, J. Imberger, N. H. Brooks, *Mixing in Inland and Coastal Waters*, Academic Press, New York, 1979, Chapter 9.

2. L. Prandtl, *Essentials of Fluid Dynamics*, Blackie & Son, London, 1969, p. 122.

3. H. Schlichting, *Boundary Layer Theory*, 4th ed., McGraw-Hill, New York, 1960, p. 596.

4. E. R. G. Eckert and R. M. Drake, Jr., *Analysis of Heat and Mass Transfer*, McGraw-Hill, New York, 1972, p. 386.

5. H. Görtler, Berechnung von Aufgaben der freien Turbulenz auf Grund eines neuen Näherungsansatzes, *Z. Angew. Math. Mech.*, Vol. 22, 1942, pp. 244–254.

6. G. L. Brown and A. Roshko, On density effects and large scale structures in turbulent mixing layers, *J. Fluid Mech.*, Vol. 64, 1974, pp. 775–816.

7. A. Bejan, *Entropy Generation through Heat and Fluid Flow*, Wiley, New York, 1982.

8. H. Reichardt, Gesetzmässigkeiten der freien Turbulenz, *VDI-Foschungsh.*, 1942, p. 414, 2nd ed., 1951; also Ref. 3, p. 607.

9. H. Reichardt, Impuls und Wärmeaustausch in freier Turbulenz, *Z. Angew. Math. Mech.*, Vol. 24, 1944, p. 268.

10. S. C. Crow and F. H. Champagne, Orderly structure in jet turbulence, *J. Fluid Mech.*, Vol. 48, 1971, pp. 547–592.

11. W. Tollmien, Berechnung turbulenter Ausbreitungsvorgänge, *Z. Angew. Math. Mech.*, Vol. 6, 1926, pp. 468–478; also NACA TM 1085, 1945.

12. H. Schlichting, Laminare Strahlausbreitung, *Z. Angew. Math. Mech.*, Vol. 13, 1933, p. 260; also Ref. 3, p. 181.

13. J. S. Turner, *Buoyancy Effects in Fluids*, Cambridge University Press, Cambridge, England, 1973.

14. B. Morton, G. I. Taylor, and J. S. Turner, Turbulent gravitational convection from maintained and instantaneous sources, *Proc. R. Soc. London, Ser. A*, Vol. 234, 1956, pp. 1–23.

15. Y. Jaluria, *Natural Convection Heat and Mass Transfer*, Pergamon, Oxford, 1980, pp. 131, 132.

16. W. Schmidt, Turbulente Ausbreitung eines Stromes erhitzter Luft, *Z. Angew. Math. Mech.*, Vol. 21, 1941, pp. 265–278 and pp. 351–363.

17. A. Bejan, Theory of instantaneous sinuous structure in turbulent buoyant plumes, *Wärme Stoffübertrag.*, Vol. 16, 1982, pp. 237–242.

18. H. Tennekes and J. L. Lumley, *A First Course in Turbulence*, MIT Press, Cambridge, MA, 1972, p. 242.

PROBLEMS

1. Study the slender shear layer region of Fig. 8.1 as a laminar flow problem. Rely on scale analysis to show that the thickness of this laminar region must increase as $(\nu x/U_0)^{1/2}$. Note the difference between the nonlinear growth of the laminar layer and the linear growth of the turbulent case treated in the text. Apply the local Reynolds number criterion for the laminar–turbulent transition [eq. (15), Chapter 6], and determine the length of the laminar portion of the shear layer. Sketch to scale the structure of the shear layer, showing both the laminar and turbulent sections for two cases (a) and (b) such that $(U_0)_a = 2(U_0)_b$. Comment on the change in this structure as U_0 increases.

2. Repeat the scaling argument of p. 288 and Fig. 8.3 for the more general case where both U_0 and U_∞ are finite (note that Fig. 8.2 corresponds to such a case). Show that relative to the observer at rest, the constant angle of the mixing region must scale as

$$\arctan\left[\frac{D(U_0 - U_\infty)}{3\lambda_B(U_0 + U_\infty)}\right]$$

were λ_B/D is a constant of order 2. The implication of this result is that the angle photographed in Fig. 8.2 stands to decrease as U_∞ approaches U_0. Test this conclusion against the comprehensive set of photographs published in Ref. 6 and against Fig. 7 of Ref. 6.

3. Consider the time-averaged development of a two-dimensional turbulent plume above a horizontal line heat source of strength q' [W/m]. In the usual cartesian system (x, y) and $(\bar{u}, \bar{v})$, with y and $\bar{v}$ pointing upward, derive the integral equations for mass, momentum, and energy in each horizontal cut through the plume. Combining the mass continuity equation with the lateral entrainment hypothesis (discussed in the text in conjunction with round plumes), and assuming that the velocity profile thickness D is of the same order as the temperature profile thickness D_T, use scale analysis to derive the scales of D_T, $\bar{v}$, and $(\overline{T}_c - T_\infty)$. In this notation, $(\overline{T}_c - T_\infty)$ is the temperature between a point on the centerplane and the ambient fluid reservoir. Sketch qualitatively the variation of D_T, $\bar{v}$, and $(\overline{T}_c - T_\infty)$ with altitude y. Assuming Gaussian profiles for both $\bar{v}$ and $(\overline{T} - T_\infty)$, derive expressions for the centerline velocity and temperature difference, $\bar{v}_c$ and $(\overline{T}_c - T_\infty)$.

4. Consider the development of a thermal wake behind a point source of strength q (watts) situated in a uniform stream $\overline{U}_\infty$ that contains grid-generated eddies. The eddy thermal diffusivity of this stream ε_H is constant and assumed known. Develop a similarity solution for the time-averaged temperature field behind the point source: Express your result as

$$\overline{T}(r, x) - \overline{T}_\infty = \text{function}(q, \varepsilon_H, U_\infty, r, x)$$

where (r, x) is the cylindrical coordinate system attached to the point source, with the x axis pointing downstream (i.e., in the same direction as U_∞). Note also the analogy between the energy equation simplified for this problem and the transient equation for pure radial conduction away from a line source. Show that your solution for the thermal wake behind a point source is analytically the same as that for transient conduction around a line source releasing instantly a finite amount of energy per unit length.

5. Consider the free shear layer formed between two streams at different temperatures (Fig. 8.1). Show that in Pr > 1 fluids the thickness of the velocity profile, D, must be of the same order as the thickness of the temperature profile, D_T. HINT: based on scale analysis, estimate the thermal penetration distance by which the D-size eddy swells as a temperature field during one rotation. What is the scale of D_T/D in Pr < 1 fluids?

9

Mass Transfer

The convection heat transfer phenomena in Nature are often accompanied by *mass transfer*, that is, by the transport of a certain substance that acts as a component (constituent, species) in the fluid mixture. The circulation of atmospheric air is in many cases driven by differential heating; however, in an industrial area this air flow will also act as a carrier for the many $\dot{m}_{out}$'s put on drawing boards by engineers, and, eventually, by factories into the atmosphere. By the same token, certain ocean currents driven by differential heating also act as freight trains for salt (in the form of saline water). Closer to home, the forced ventilation system in my building lets me know whenever a cigar is lit in somebody else's office.

Beyond these environmental engineering applications, convection mass transfer processes alone (in the absence of heat transfer) constitute the backbone of many operations in the chemical industry. This seems like enough reason to include mass transfer in this convection course. An additional argument in favor of this decision is the *analogy* that exists between convective mass transfer and convective heat transfer. This analogy is pedagogically very important because it gives the student an opportunity to organize his own understanding of heat transfer and to learn mass transfer with the least memorization. The present chapter highlights this analogy: although the focus is on mass transfer, the structure of the presentation constitutes a review of heat transfer. I use this review as an opportunity to extend the analogy between mass transfer and heat transfer to the topics of Chapters 4 and 10: the topics of natural convection mass transfer and mass convection through porous media are usually not included in the textbook treatment of mass transfer.

PROPERTIES OF MIXTURES

The study of convective heat transfer—the object of the first eight chapters of this course—began with a review of the thermodynamics of pure substances. That review was demanded, first, by the role of *mutual relative* played by thermodynamics to both heat transfer and fluid mechanics [1], and, second, by the fact that as an undergraduate the student is likely to have been exposed first to thermodynamics. For the same reasons, we begin the discussion of mass convection with a brief review of the thermodynamics and, especially, the *nomenclature* of mixtures of substances [2, 3].

Consider a batch of fluid of overall volume V and total mass m. Like the air surrounding a fin in natural convection or like the river water sweeping away the discharge from a chemical plant, this fluid batch is a mixture of identifiable *components* (e.g., in the case of air, the components are nitrogen, oxygen, carbon dioxide, inert gases, and impurities). Let m_i be the individual masses of the components that constitute the mixture. Then, by definition, the *concentration* of component i in the mixture (m, V) is

$$C_i = \frac{m_i}{V} \tag{1}$$

with units of kg/m³. Note that *concentration* is another word for the component density ρ_i demanded by the viewpoint that each component (m_i) fills the entire volume V,

$$\rho_i = \frac{m_i}{V} \tag{2}$$

Since all the components contribute to the total mass of the batch, $\Sigma m_i = m$, it follows that the aggregate density of the mixture ρ is the sum of all concentrations,

$$\rho = \sum C_i \tag{3}$$

The aggregate density ρ is the density encountered in the preceding chapters in connection with the mixture's ability to act as conveyor belt for energy.

The size of the fluid batch can be described, of course, in terms of its extensive properties mass (m) and volume (V). An alternative description, preferred with good reason by chemical engineers, involves the concept of *mole*: a two-mole batch of a certain substance is bigger than an one-mole batch of the same substance. By definition, a mole is the amount of substance in a batch (in a thermodynamic system) that contains as many elementary entities (e.g., molecules) as there are in 0.012 kg carbon-12 [2]. That special number of entities is *Avogadro's number*, 6.022×10^{23}.

The mole is not a mass unit, simply because the mass of one mole is not the same for all substances (five basketballs together weigh a lot more than five tennis balls taken together; note that in this example the group of five entities plays the role of one mole). The *molar mass M* of a mixture or a component (molecular species) in a mixture is the mass of one mole of that mixture or component. The units of molar mass M are g/mol or kg/kmol, where kmol represents 1000 moles. The total *number of moles n* found in a certain batch, is obtained by dividing the total mass of the batch by the mass of one mole

$$n = \frac{m}{M} \tag{4}$$

By the same token, the number of moles of a certain component (n_i) found in a mixture is equal to the mass of that component (m_i) divided by its molar mass (M_i),

$$n_i = \frac{m_i}{M_i} \tag{5}$$

A dimensionless way to describe the composition of a mixture is by use of the *mass fraction* of each constituent,

$$\Phi_i = \frac{m_i}{m} \tag{6}$$

where, clearly, $\sum \Phi_i = 1$. Note that the concept of mass fraction is the same as that of quality encountered in the study of liquid and vapor mixtures. (When the quality of wet steam is 0.9, the mass fraction of saturated steam in the two-phase mixture is 0.9.) Another dimensionless alternative to describing composition is by comparing the number of moles of each component (n_i) with the total number of moles found in the mixture (n),

$$x_i = \frac{n_i}{n}; \qquad \sum x_i = 1 \tag{7}$$

The ratio x_i is the *mole fraction* of component i. To summarize, we have seen three alternative ways to discuss composition—a dimensional concept (concentration) and two dimensionless ratios (mass fraction and mole fraction). The conversion formulas relating these three properties are

$$C_i = \rho \Phi_i = \rho \frac{M_i}{M} x_i \tag{8}$$

where the equivalent molar mass of the mixture (M) is related to the molar masses of all the constituents by

$$M = \sum M_i x_i \tag{9}$$

If the mixture can be modeled as an *ideal gas*, then its equation of state is

$$PV = mRT \quad \text{or} \quad PV = n\bar{R}T \tag{10}$$

where the mixture's ideal gas constant (R) and the universal gas constant

$(\overline{R} = 8.314 \text{ J/(mol K)})$ are related via

$$R = \frac{\overline{R}}{M} \tag{11}$$

The *partial pressure* P_i of constituent i is the pressure one would measure if constituent i alone were to fill the mixture volume (V) at the same temperature as mixture (T).

$$P_i V = m_i R T \quad \text{or} \quad P_i V = n_i \overline{R} T \tag{12}$$

Summing the above equations over i, we obtain *Dalton's Law*

$$P = \sum_i P_i \tag{13}$$

which states that the pressure of a mixture of gases at a specified volume and temperature is equal to the sum of the partial pressures of constituents. Of use in the calculation of concentrations is the relationship between partial pressure and mole fraction

$$\frac{P_i}{P} = x_i \tag{14}$$

which is obtained by dividing eqs. (12) and (10). The concentration of a certain gas (C_i) can then be related to the partial pressure of that gas via eq. (8).

The nomenclature reviewed in this section applies to a mixture in *equilibrium*, that is, to a fluid batch whose composition, pressure, and temperature do not vary from point to point. Beginning with the next section, we focus on something that departs fundamentally from the equilibrium mixture description: we focus on the flow of a mixture whose composition, pressure, and temperature may vary from one point to another. We will view this *nonequilibrium* mixture as a patchwork of small equilibrium batches of the type described in this section: the equilibrium state of each of these batches varies slightly as you shift from one batch to the next.

MASS CONSERVATION

The centerpiece in the mathematical analysis of convective mass transfer is the principle of mass conservation in (or *continuity* through) the control volume sketched in Fig. 9.1. In Chapter 1 we invoked the same principle in the case of a fluid of density ρ whose composition was not questioned: that fluid might very well have been a mixture of two or more fluids. In this section we apply the principle of mass conservation to each component or constituent in the mixture. Relative to Fig. 9.1 we argue that the net flow of constituent i into

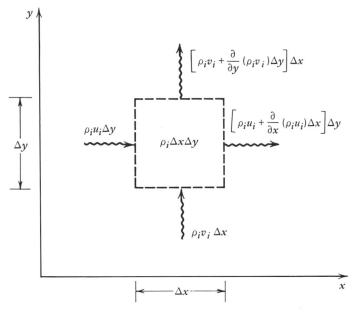

Figure 9.1 The conservation of the mass of component i in the flow of a mixture.

the control volume is equal to the rate of accumulation of constituent i inside the control volume.

$$\frac{\partial \rho_i}{\partial t} \Delta x \Delta y = \rho_i u_i \Delta y - \left[\rho_i u_i + \frac{\partial}{\partial x} (\rho_i u_i) \Delta x \right] \Delta y$$

$$+ \rho_i v_i \Delta x - \left[\rho_i v_i + \frac{\partial}{\partial y} (\rho_i v_i) \Delta y \right] \Delta x + \dot{m}_i''' \Delta x \Delta y \quad (15)$$

In the above balance, ρ_i is the number of kilograms of constituent i per cubic meter found locally in the mixture at point (x, y). The velocity components (u_i, v_i) account for the motion of constituent i relative to the control volume. The use of the (u_i, v_i) notation at this stage should not be taken as a suggestion that a motion with such velocity components can actually be seen or measured. Nevertheless, these velocity components do have a physical meaning, for example, the group $\rho_i u_i$ represents the net mass flux of constituent i (measured in kg s^{-1} m^{-2}) in the x direction. Concluding eq. (15) is the term containing $\dot{m}_i'''$, which is the volumetric rate of constituent i generation (the units of $\dot{m}_i'''$ are kg s^{-1} m^{-3}). This last term must be taken into account in reactive flows that generate constituent i locally as a product of reaction. If constituent i is consumed by the reaction, the generation rate $\dot{m}_i'''$ is negative.

The mass conservation statement (15) reduces to

$$\frac{\partial \rho_i}{\partial t} + \frac{\partial}{\partial x}(\rho_i u_i) + \frac{\partial}{\partial y}(\rho_i v_i) = \dot{m}_i''' \tag{16}$$

which, in the absence of constituent generation ($\dot{m}_i''' = 0$), has the same form as the mass conservation statement for mixture flow [eq. (3), Chapter 1]. Indeed, the mass conservation equation for mixture flow can be derived as in Fig. 9.1 by viewing the mixture flow as the superposition of all the flows involving one constituent at a time. Superposing the constituent flows, that is, summing eq. (16) over i and letting $\dot{m}_i''' = 0$, we obtain

$$\frac{\partial \rho}{\partial t} + \frac{\partial}{\partial x}\sum \rho_i u_i + \frac{\partial}{\partial y}\sum \rho_i v_i = 0 \tag{17}$$

This mass conservation statement can only be the same as the statement derived in Chapter 1,

$$\frac{\partial \rho}{\partial t} + \frac{\partial}{\partial x}(\rho u) + \frac{\partial}{\partial y}(\rho v) = 0 \tag{18}$$

A term-by-term comparison of eqs. (17) and (18) brings us to a very important concept in mass convection—the concept of *mass-averaged velocity* components (u, v)

$$u = \frac{1}{\rho}\sum \rho_i u_i, \qquad v = \frac{1}{\rho}\sum \rho_i v_i \tag{19}$$

What in the heat transfer part of this convection course was called the *velocity component* emerges in mass transfer as a weighted average of all the constituent velocities.

In general, the mass-averaged velocity differs from the velocity of each constituent. In order to see this, take a glass with some water in it and inject a layer of grenadine syrup in the lower portion of the glass. What happens is shown in Fig. 9.2: in time, the syrup *diffuses* upward, its place being taken by clear water diffusing downward. In any horizontal cross-section, the vertical velocity of each constituent is finite (one positive and the other negative) while the mass-averaged velocity—the velocity of the mixture as a whole—is zero (to the fluid mechanicist of Chapter 1, the water in the glass is stagnant). Clearly, each constituent moves relative to the mixture as a whole.

The velocity difference $(u_i - u)$ may be called the *diffusion velocity* of constituent i in the x direction [4]. The product $\rho_i(u_i - u)$ is the per unit area flowrate of constituent i in the x direction relative to the bulk motion of the mixture; a shorter name for this quantity is *diffusion flux*, $j_{x,i}$. Combining the

O hrs   7 hrs   15 hrs   25 hrs   38 hrs

Figure 9.2 Vertical diffusion of grenadine syrup in water.

diffusion flux definitions

$$j_{x,i} = \rho_i(u_i - u)$$

$$j_{y,i} = \rho_i(v_i - v) \tag{20}$$

with the mass continuity equation for constituent i [eq. (16)] yields

$$\frac{\partial \rho_i}{\partial t} + \frac{\partial}{\partial x}(\rho_i u) + \frac{\partial}{\partial y}(\rho_i v) = -\frac{\partial j_{x,i}}{\partial x} - \frac{\partial j_{y,i}}{\partial y} + \dot{m}_i''' \tag{21}$$

Reverting now to the more popular concentration notation [$C_i = \rho_i$, eqs. (1) and (2)] and assuming that the mixture flow may be treated as one with ρ = constant, the conservation of constituent i requires

$$\frac{\partial C_i}{\partial t} + u\frac{\partial C_i}{\partial x} + v\frac{\partial C_i}{\partial y} = -\frac{\partial j_{x,i}}{\partial x} - \frac{\partial j_{y,i}}{\partial y} + \dot{m}_i''' \tag{22}$$

The three-dimensional counterpart of this conclusion is

$$\frac{DC_i}{Dt} = -\nabla \cdot \mathbf{j}_i + \dot{m}_i''' \tag{23}$$

The diffusion flux vector $\mathbf{j}_i$ is driven by the concentration gradient ∇C_i, in the same manner that the conduction heat flux is driven by the local temperature gradient. This idea was put forth by the German physiologist Adolph Fick in 1855 [5, 6]: it has the merit of having triggered the modern analytical development of the field of mass transfer in the same way that Fourier's ideas on conduction made heat transfer a modern science. In a *two-component mixture*, Fick's law of mass diffusion is

$$\mathbf{j}_1 = -D_{12}\nabla C_1 \tag{24}$$

where $D_{12} = D_{21} = D$ is the mass diffusivity of component 1 into component 2 and *vice-versa*. The diffusivity D, whose units are $m^2\ s^{-1}$, is a transport property whose numerical value depends in general on the mixture pressure, temperature, and composition. In view of the thermodynamics of irreversible processes, it is worth keeping in mind that the diffusion flux is caused solely by the concentration gradient [as in eq. (24)] strictly in a fluid with uniform pressure and temperature [7, 8]; nevertheless, eq. (24) is a useful approximation even in combined mass and heat transfer problems. Substituting eq. (24) in the mass conservation statement (23) and dropping the subscript i yields

$$\frac{DC}{Dt} = D\nabla^2 C + \dot{m}''' \tag{25}$$

Throughout the remainder of this chapter, the concentration C refers to the component whose migration by mixture flow is of interest.

Analytically, the mass transfer problem consists of solving eq. (25) for the concentration field $C(x, y, z)$ and determining the mass fluxes associated with the concentration field from Fick's law (24). From the outset, it is worth noting the similarities between the mass convection problem and the energy convection problem formulated in Chapter 1. The latter consists of determining the temperature field $T(x, y, z)$ from the energy equation

$$\frac{DT}{Dt} = \alpha\nabla^2 T + \frac{q'''}{\rho c_P} \tag{26}$$

and the heat fluxes from Fourier's law of thermal diffusion, $\mathbf{q}'' = -k\nabla T$ [note that eq. (26) holds under special circumstances outlined in Chapter 1]. Equations (25) and (26) show that the concentration C occupies the place of temperature, while the mass diffusivity D replaces the thermal diffusivity α. The correspondence between mass transfer and heat transfer in convection will be exploited in the following sections, in order to streamline the presentation and to avoid repetition.

The mass conservation or *concentration equation* (25) has the following forms in three dimensions (Fig. 1.1):

Cartesian (x, y, z)

$$\frac{\partial C}{\partial t} + u\frac{\partial C}{\partial x} + v\frac{\partial C}{\partial y} + w\frac{\partial C}{dz} = D\left(\frac{\partial^2 C}{\partial x^2} + \frac{\partial^2 C}{\partial y^2} + \frac{\partial^2 C}{\partial z^2}\right) + \dot{m}''' \tag{27a}$$

Cylindrical (r, θ, z)

$$\frac{\partial C}{\partial t} + v_r \frac{\partial C}{\partial r} + \frac{v_\theta}{r} \frac{\partial C}{\partial \theta} + v_z \frac{\partial C}{\partial z} = D \left[\frac{1}{r} \frac{\partial}{\partial r} \left(r \frac{\partial C}{\partial r} \right) + \frac{1}{r^2} \frac{\partial^2 C}{\partial \theta^2} + \frac{\partial^2 C}{\partial z^2} \right] + \dot{m}'''$$

(27b)

Spherical (r, ϕ, θ)

$$\frac{\partial T}{\partial t} + v_r \frac{\partial C}{\partial r} + \frac{v_\phi}{r} \frac{\partial C}{\partial \phi} + \frac{v_\theta}{r \sin \phi} \frac{\partial C}{\partial \theta}$$

$$= D \left[\frac{1}{r^2} \frac{\partial}{\partial r} \left(r^2 \frac{\partial C}{\partial r} \right) + \frac{1}{r^2 \sin \phi} \frac{\partial}{\partial \phi} \left(\sin \phi \frac{\partial C}{\partial \phi} \right) + \frac{1}{r^2 \sin^2 \phi} \frac{\partial^2 C}{\partial \theta^2} \right] + \dot{m}'''$$

(27c)

Given the proportionality between concentration, mass fraction, and mole fraction as a means of quantizing composition [eq. (8)], the concentration equations formulated above can easily be replaced with mass fraction (Φ) equations

$$\frac{D\Phi}{Dt} = D\nabla^2 \Phi + \frac{\dot{m}'''}{\rho}$$

(28)

$$\mathbf{j} = -\rho D \nabla \Phi$$

(29)

or mole fraction (x) equations

$$\frac{Dx}{Dt} = D\nabla^2 x + \frac{M}{M_1} \frac{\dot{m}'''}{\rho}$$

(30)

$$\mathbf{j} = -\rho \frac{M_1}{M} D\nabla x$$

(31)

where M_1 is the molar mass of the constituent of interest (whose mole fraction is x). To be consistent, the material that follows refers only to one formulation, the concentration-type equations (24), (25), and (27).

LAMINAR FORCED CONVECTION

The analogy between mass transfer and heat transfer becomes even more apparent if we focus on the laminar boundary layer flow sketched in Fig. 9.3. A uniform stream U_∞ flows parallel to a solid surface coated or made out of a

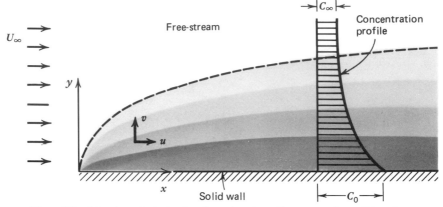

Figure 9.3 Mass transfer to laminar boundary layer flow over a flat, solid surface.

substance that is soluble in the stream. An example of such a flow configuration is the forced convection drying of a porous solid wall saturated with water. In this case, the air stream U_∞ is, in general, humid, its water vapor content far away from the wall described by the free-stream concentration C_∞. Following Prandtl and the boundary layer methodology presented in Chapter 2, we can expect a concentration boundary layer in the vicinity of the wall, that is, a concentration distribution that smoothes out the discrepancy between the relatively dry free-stream C_∞ and the relatively wet flow *lamina* immediately adjacent to the wall C_0. The concentration gradient between the wall and the free-stream sucks the soluble substance away from the wall: to predict the rate of such mass transfer is the engineering objective of this chapter.

It is important to keep in mind that the wall concentration C_0 is the concentration of a fluid batch adjacent to the wall, not the concentration of soluble substance inside the wall. The concentration C_0 is the concentration on the $y = 0^+$ side of the wall (Fig. 9.3). Returning to the wall drying example of the preceding paragraph, the concentration of H_2O can differ vastly from $y = 0^-$ to $y = 0^+$. The concentration inside the wall ($y = 0^-$) depends on the porosity of the wall and the degree to which the pores are filled with water. The concentration on the fluid side of the wall–stream interface is dictated by the equilibrium vapor pressure of water at the free-stream pressure and temperature (assuming $dP_\infty/dx = 0$ and no wall–stream temperature difference) [9].

The mass flux from the wall into the stream is (by Fick's law)

$$j_0 = -D\left(\frac{\partial C}{\partial y}\right)_{y=0} \tag{32}$$

The concentration field $C(x, y)$ is obtained by solving the *boundary layer* concentration equation

$$u\frac{\partial C}{\partial x} + v\frac{\partial C}{\partial y} = D\frac{\partial^2 C}{\partial y^2} \tag{33}$$

subject to the following boundary conditions

$$C = C_0 \quad \text{at } y = 0, \qquad C \to C_\infty \quad \text{as } y \to \infty \tag{34}$$

where both C_0 and C_∞ are constants. The mixture flow field (u, v) is known from Blasius' solution for laminar boundary layer flow (Chapter 2). The mass transfer problem [(33) and (34)] is obviously identical to that solved by Pohlhausen for heat transfer [see eqs. (93)–(95) of Chapter 2]; hence, the wall concentration gradient can be written immediately (noting the $T \to C$, $\alpha \to D$ transformation):

$$\left(\frac{\partial C}{\partial y}\right)_{y=0} = (C_\infty - C_0)\left(\frac{U_\infty}{\nu x}\right)^{1/2}\left\{\int_0^\infty \exp\left[-\frac{\nu/D}{2}\int_0^\gamma f(\beta)\,d\beta\right]d\gamma\right\}^{-1}$$

$$\tag{35}$$

In this solution we see the emergence of a new dimensionless group, the *Schmidt number*

$$\text{Sc} = \frac{\nu}{D} \tag{36}$$

in place of the Prandtl number ν/α of the original Pohlhausen solution. Recalling the two asymptotes of the nested integral of eq. (35), we conclude that

$$\frac{(\partial C/\partial y)_{y=0}}{(C_\infty - C_0)\left(\dfrac{U_\infty}{\nu x}\right)^{1/2}} = \begin{cases} 0.332\,\text{Sc}^{1/3}, & \text{for Sc} > 1 \\ 0.564\,\text{Sc}^{1/2}, & \text{for Sc} < 1 \end{cases} \tag{37}$$

Mimicking the Nusselt number nondimensionalization of temperature gradient (or wall heat flux) employed in convective heat transfer, the above conclusion can be put in dimensionless form as a so-called *local Sherwood number*,

$$\text{Sh} = \left(\frac{\partial C}{\partial y}\right)_{y=0}\frac{x}{C_\infty - C_0} = \frac{j_0}{C_0 - C_\infty}\frac{x}{D} \tag{38}$$

namely,

$$\text{Sh} = 0.332\,\text{Sc}^{1/3}\,\text{Re}_x^{1/2}, \qquad \text{Sc} > 1$$

$$\text{Sh} = 0.564\,\text{Sc}^{1/2}\,\text{Re}_x^{1/2}, \qquad \text{Sc} < 1 \tag{39}$$

Carrying the analogy between mass transfer and heat transfer one step further, the ratio $j_0/(C_0 - C_\infty)$ may be called the *mass transfer coefficient* h_m, so that the Sherwood number can also be defined as

$$\text{Sh} = \frac{h_m x}{D} \tag{40}$$

The symmetry between the mass transfer scaling laws (39) and their heat transfer correspondents [eqs. (102) and (104) of Chapter 2] is worth thinking about. It implies that if the bulk (mixture) flow configuration is the same in both problems, and if the wall boundary condition is the same (e.g., $T_0 =$ constant versus $C_0 =$ constant), the mass transfer result is obtained directly from the heat transfer result through the transformation

$$\text{Nu} \rightarrow \text{Sh}, \qquad \text{Pr} \rightarrow \text{Sc}, \qquad \text{Re}_x \rightarrow \text{Re}_x \tag{41}$$

Like their heat transfer counterparts, the mass transfer results for laminar boundary layer flow [eqs. (39)] are based on the assumption that the normal velocity component v vanishes at the wall. This assumption is not always appropriate in laminar boundary layer mass transfer, in view of the flow of matter (j_0) through the wall–stream interface (think, e.g., of the evaporation of a liquid film or the sublimation of the wall material itself in a gaseous laminar boundary layer). The solution to the general mass transfer problem with finite normal velocity at the wall–stream interface was reported by Hartnett and Eckert [10]. Relative to heat transfer, this general solution describes the effect of blowing or suction through the solid wall bathed by laminar boundary layer flow.

Mass transfer results for forced convection in laminar duct flow can be obtained by applying the transformation (41) to the heat transfer results of Chapter 3. Two examples of Sherwood numbers for fully developed laminar duct flow are given

$$\text{Sh} = 3.66, \quad \text{round tube}, \quad C_0 = \text{constant}$$

$$\text{Sh} = 7.54, \quad \text{parallel-plate channel}, \quad C_0 = \text{constant} \tag{42}$$

Values corresponding to other duct cross-sectional shapes may be found in Table 3.2. Once again, these Sherwood numbers are valid as long as the wall mass flux is small enough so that the mixture velocity normal to the duct wall may be regarded as zero at the wall.

LAMINAR NATURAL CONVECTION

Consider the vertical wall with heat and mass transfer sketched in Fig. 9.4, where T_0, T_∞, C_0, and C_∞ are all known constants. The vertical boundary layer flow is due to the buoyancy effect associated with the density difference

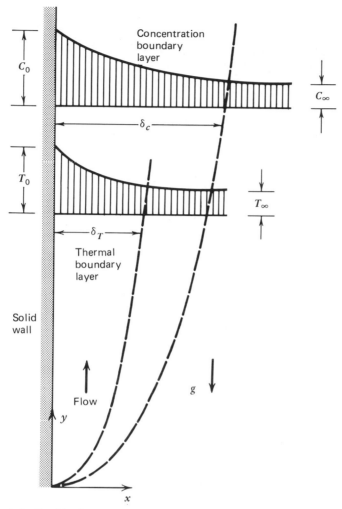

Figure 9.4 Combined mass and heat transfer to a buoyant vertical boundary layer.

between boundary layer fluid and unaffected (reservoir) fluid. As was shown in Chapter 4, the boundary layer momentum equation for this flow is

$$u\frac{\partial v}{\partial x} + v\frac{\partial v}{\partial y} = \nu\frac{\partial^2 v}{\partial x^2} + \frac{1}{\rho}(\rho_\infty - \rho)g \qquad (43)$$

In the case of heat transfer, we saw that the density difference $(\rho_\infty - \rho)$ is approximately proportional to the temperature difference $(T - T_\infty)$, in accordance with the Boussinesq approximation. In the presence of mass transfer, the driving density difference $(\rho_\infty - \rho)$ may also be due to the concentration profile (Fig. 9.4): a vertical buoyant layer will most certainly form if the wall releases a substance (constituent) less dense than the reservoir fluid mixture. Keeping in mind that the thermodynamic state of the fluid mixture depends on pressure, temperature, and composition, in the limit of small density variations at constant pressure, we can write

$$\rho \cong \rho_\infty + \left(\frac{\partial \rho}{\partial T}\right)_P (T - T_\infty) + \left(\frac{\partial \rho}{\partial C}\right)_P (C - C_\infty) + \cdots \tag{44}$$

Recalling the definition of thermal expansion coefficient

$$\beta = -\frac{1}{\rho}\left(\frac{\partial \rho}{\partial T}\right)_P \tag{45}$$

we introduce the *concentration expansion coefficient*

$$\beta_c = -\frac{1}{\rho}\left(\frac{\partial \rho}{\partial C}\right)_P \tag{46}$$

to obtain the equivalent of the Boussinesq approximation for the combined heat and mass transfer problem.[†] Based on this approximation, the boundary layer momentum equation (43) becomes

$$\underbrace{u\frac{\partial v}{\partial x} + v\frac{\partial v}{\partial y}}_{\text{Inertia}} = \underbrace{\nu\frac{\partial^2 v}{\partial x^2}}_{\text{Friction}} + \underbrace{g\beta(T - T_\infty)}_{\substack{\text{Body force} \\ \text{due to} \\ \text{nonuniform} \\ \text{temperature}}} + \underbrace{g\beta_c(C - C_\infty)}_{\substack{\text{Body force} \\ \text{due to} \\ \text{nonuniform} \\ \text{concentration}}} \tag{47}$$

The flow is thus linearly coupled to the temperature and concentration fields obtained by solving the boundary layer energy and concentration equations

$$u\frac{\partial T}{\partial x} + v\frac{\partial T}{\partial y} = \alpha\frac{\partial^2 T}{\partial x^2} \tag{48}$$

$$u\frac{\partial C}{\partial x} + v\frac{\partial C}{\partial y} = D\frac{\partial^2 C}{\partial x^2} \tag{49}$$

Mass-Transfer-Driven Flow

In order to determine the mass transfer rate between wall and fluid reservoir, we focus on two limiting situations. First, consider the limit of "no heat

---

[†]Coefficients β and β_c can be positive or negative, hence, in the scale analysis of this chapter $\beta(T_0 - T_\infty)$ means the absolute value of $\beta(T_0 - T_\infty)$.

transfer," that is, the case of a boundary layer driven solely by the concentration gradient. The problem reduces to solving eqs. (47) and (49), subject to the velocity and concentration boundary conditions sketched in Fig. 9.4. This problem is analytically identical to the heat transfer problem solved in Chapter 4: the local Sherwood number is obtained by subjecting the heat transfer results to the transformation Nu → Sh, $\alpha \to D$, Pr → Sc, $\beta(T_0 - T_\infty)$ → $\beta_c(C_0 - C_\infty)$. The mass transfer results are

$$\text{Sh} = 0.503\,\text{Ra}_{m,y}^{1/4}, \quad \text{for Sc} > 1$$

$$\text{Sh} = 0.6\left(\text{Ra}_{m,y}\text{Sc}\right)^{1/4}, \quad \text{for Sc} < 1 \tag{50}$$

where $\text{Ra}_{m,y}$ is the local Rayleigh number of a vertical boundary layer whose buoyancy is caused by mass transfer [7]

$$\text{Ra}_{m,y} = \frac{g\beta_c(C_0 - C_\infty)y^3}{\nu D} \tag{51}$$

Heat-Transfer-Driven Flow

The second limit of interest is the mass transfer to a vertical boundary layer driven by the wall–reservoir temperature difference. The length and velocity scales of such a layer have been summarized in Table 4.1. Below, we rely on pure scale analysis to derive the mass transfer rate or the Sherwood number.

Let δ_c be the boundary layer thickness scale of the concentration profile. In the flow region of thickness δ_c and height H, the concentration equation (49) requires

$$\frac{v}{H} \sim \frac{D}{\delta_c^2} \tag{52}$$

Note that v is the vertical velocity scale *in the region of thickness δ_c*: naturally, v will depend on the relative size of δ_c and the other (two) length scales of the ΔT-driven boundary layer flow. Four possibilities exist, as is shown in Fig. 9.5. We examine in some detail the first possibility (Pr > 1, $\delta_c < \delta_T$) and for the remaining three we list only the conclusions (see Table 9.1).

Table 4.1 shows that in a heat-transfer-driven boundary layer the vertical velocity reaches the order of magnitude $\alpha/H\,\text{Ra}_H^{1/2}$ at a distance of order $H\,\text{Ra}_H^{-1/4}$ away from the wall (if Pr > 1). Turning our attention to the first sketch of Fig. 9.5, we conclude that the velocity scale in the δ_c-thin layer is

$$v \sim \frac{\delta_c}{\delta_T}\frac{\alpha}{H}\text{Ra}_H^{1/2}, \quad \delta_c < \delta_T \tag{53}$$

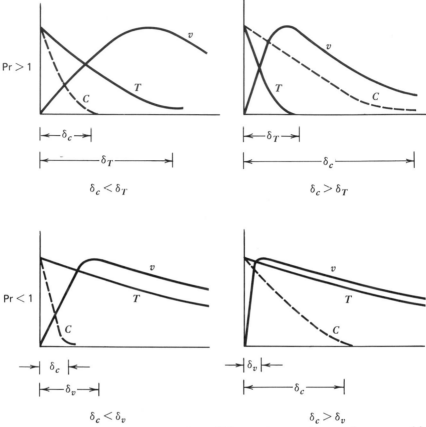

Figure 9.5 The relative size of boundary layer thicknesses in natural convection mass and heat transfer.

where $\delta_T \sim H \, \mathrm{Ra}_H^{-1/4}$ is the thermal boundary layer thickness. Combining eqs. (52) and (53), we obtain the δ_c scale,

$$\delta_c \sim H\left(\frac{D}{\alpha}\right)^{1/3} \mathrm{Ra}_H^{-1/4} \tag{54}$$

The order of magnitude of the overall Sherwood number is always inversely proportional to δ_c

$$\mathrm{Sh}_{0-H} = \frac{(j_0)_{0-H}}{C_0 - C_\infty} \frac{H}{D} \sim \frac{H}{\delta_c} \tag{55}$$

hence,

$$\mathrm{Sh}_{0-H} \sim \left(\frac{\alpha}{D}\right)^{1/3} \mathrm{Ra}_H^{1/4} \tag{56}$$

Table 9.1   Mass Transfer Rate Scales for a Vertical Boundary Layer Driven by Heat Transfer

Fluid	Overall Sherwood Number Scale, or Concentration Layer Slenderness Ratio H/δ_c	
$\Pr > 1$, $\delta_c < \delta_T$ (or $\mathrm{Le} > 1$)	$\mathrm{Le}^{1/3}\mathrm{Ra}_H^{1/4}$	
$\Pr > 1$, $\delta_c > \delta_T$ (or $\mathrm{Le} < 1$)	$\mathrm{Le}^{1/2}\mathrm{Ra}_H^{1/4}$	$(\mathrm{Sc} > 1)$
	$\mathrm{Le}\,\Pr^{1/2}\mathrm{Ra}_H^{1/4}$	$(\mathrm{Sc} < 1)$
$\Pr < 1$, $\delta_c < \delta_v$ (or $\mathrm{Sc} > 1$)	$\mathrm{Le}^{1/3}\Pr^{1/12}\mathrm{Ra}_H^{1/4}$	
$\Pr < 1$, $\delta_c > \delta_v$ (or $\mathrm{Sc} < 1$)	$\mathrm{Le}^{1/2}\Pr^{1/4}\mathrm{Ra}_H^{1/4}$	$(\mathrm{Le} > 1)$
	$\mathrm{Le}\,\Pr^{1/4}\mathrm{Ra}_H^{1/4}$	$(\mathrm{Le} < 1)$

Less approximate analyses could, of course, be carried out in order to refine this mass transfer result. However, based on the extensive comparison of scaling results with integral and similarity results (Chapters 2–5), it is reasonable to expect eq. (56) to be correct within 25 percent.

It is known that dimensional analysis turns up all the *dimensionless groups* that could be formed by algebraically combining the dimensional parameters of the problem [14]. Scale analysis, on the other hand, leads to only those dimensionless groups that have a physical meaning relative to the problem at hand. In this way, eqs. (55) and (56) teach us that the slenderness ratio of the concentration boundary layer is governed by a new dimensionless group,

$$\left(\frac{\alpha}{D}\right)^{1/3}\mathrm{Ra}_H^{1/4} \sim \frac{\text{height of boundary layer}}{\text{thickness of concentration layer}}$$

The dimensionless ratio α/D has been identified already by dimensional analysis and is called the *Lewis number*,

$$\mathrm{Le} = \frac{\alpha}{D} \quad \text{or} \quad \mathrm{Le} = \frac{\mathrm{Sc}}{\Pr} \tag{57}$$

Finally, we note that the $\delta_c < \delta_T$ assumption made in the beginning of this analysis means

$$H\,\mathrm{Le}^{-1/3}\mathrm{Ra}_H^{-1/4} < H\,\mathrm{Ra}_H^{-1/4} \tag{58}$$

or

$$\mathrm{Le} > 1 \tag{59}$$

In conclusion, the first case of Fig. 9.5 corresponds to fluid mixtures with both Prandtl number and Lewis number greater than unity. The remaining three possibilities are sketched in Fig. 9.5 and their corresponding mass transfer scaling laws are listed in Table 9.1. The student is urged to derive these results based on scale analysis (see Problem 3 and the *Solutions Manual*). Note that in the case of low Prandtl number fluids, the concentration thickness δ_c is compared with the viscous layer thickness δ_v (Chapter 4).

To summarize, we analyzed the combined heat and mass transfer problem in natural convection by first calculating mass transfer in a mass-transfer-driven flow [eqs. (50)] and later by deriving the mass transfer scales in boundary layers driven by heat transfer (Table 9.1). It remains to decide whether a given layer is actually driven by mass transfer or heat transfer, in other words, is the Sherwood number given by eqs. (50) or Table 9.1? One way to decide is to think of the wall surface, whose concentration is C_0, as being coated with two mass-insulating blankets, one $(\delta_c)_{MT}$-thick when the layer is driven by mass transfer, and the other with a thickness $(\delta_c)_{HT}$ when heat transfer is the driving mechanism (one blanket extends from $x = 0$ to $x = (\delta_c)_{MT}$ and the other from $x = 0$ to $x = (\delta_c)_{HT}$). The mass flux released by the wall will choose the shortest path to the mass sink (the reservoir); hence, the scales of Table 9.1 hold when their corresponding δ_c's are smaller than the concentration profile thickness in a mass-transfer-driven layer,

$$(\delta_c)_{HT} < (\delta_c)_{MT} \tag{60}$$

For example, for fluids with $Pr > 1$ and $Le > 1$ (the top line in Table 9.1), criterion (60) implies

$$\frac{\beta(T_0 - T_\infty)}{\beta_c(C_0 - C_\infty)} > Le^{-1/3} \tag{61}$$

as a condition for heat-transfer-driven natural convection mass transfer. Note that this criterion is not the same as the direct comparison of the scales of the last two terms in the boundary layer momentum equation (47) (this observation is discussed further in the last paragraph of this chapter).

In the above scale analysis, the competition between the temperature and concentration layers of Fig. 9.4 was illustrated by calculating the mass transfer rate from the wall to the fluid reservoir. The transition from concentration-driven to temperature-driven natural convection has also an interesting effect on the heat transfer rate calculation. This new aspect of the combined heat and mass transfer problem is examined in Problem 11 at the end of this chapter.

Similarity and integral solutions for vertical boundary layer natural convection driven by the combined effect of mass and heat transfer have been reported by Gebhart and Pera [11] and Somers [12], among others. A review of the theoretical and experimental progress in this research area has been published by Ostrach [13].

TURBULENT FLOW

The analogy between mass transfer and heat transfer also exists in the case of turbulent convection, again, under conditions of low wall mass transfer rates. The time-fluctuating concentration field can be smoothed out through the process of time-averaging of Chapter 7, which begins with the decomposition

$$C = \overline{C} + C' \tag{62}$$

where $\overline{C}(x, y, z)$ is the time-averaged concentration field. Substituting eq. (62) into the concentration equation (27a) with $\dot{m}''' = 0$, and observing the rules of time-averaging algebra, yields

$$\overline{u}\frac{\partial \overline{C}}{\partial x} + \overline{v}\frac{\partial \overline{C}}{\partial y} + \overline{w}\frac{\partial \overline{C}}{\partial z} = D\nabla^2\overline{C} - \frac{\partial}{\partial x}\left(\overline{u'C'}\right) - \frac{\partial}{\partial y}\left(\overline{v'C'}\right) - \frac{\partial}{\partial z}\left(\overline{w'C'}\right) \tag{63}$$

The time-averaged products of type $(\overline{v'C'})$ are the additonal unknowns that are responsible for the closure problem in turbulent convective mass transfer.

In a two-dimensional slender flow region such as the forced boundary layer sketched in Fig. 9.3, the time-averaged concentration equation (63) assumes the simpler form

$$\overline{u}\frac{\partial \overline{C}}{\partial x} + \overline{v}\frac{\partial \overline{C}}{\partial y} = D\frac{\partial^2 \overline{C}}{\partial y^2} - \frac{\partial}{\partial y}\left(\overline{v'C'}\right) \tag{64}$$

To bring into view the eddy mass transfer across the boundary layer, we write eq. (65) as

$$\overline{u}\frac{\partial \overline{C}}{\partial x} + \overline{v}\frac{\partial \overline{C}}{\partial y} = -\frac{\partial j_{app}}{\partial y} \tag{65}$$

where the apparent mass flux $j_{app}(x, y)$ is due to both molecular and eddy transport,

$$j_{app} = -D\frac{\partial \overline{C}}{\partial y} + \overline{v'C'} \tag{66}$$

Based on an argument similar to the one presented in Fig. 7.3 for heat transfer, the eddy mass flux term $\overline{v'C'}$ may be seen as being connected to the mean concentration gradient

$$-\overline{v'C'} = \varepsilon_m \frac{\partial \overline{C}}{\partial y}, \quad \text{eddy mass flux} \tag{67}$$

This notation is in essence a definition for the *mass eddy diffusivity* $\varepsilon_m(x, y)$, an empirical function (not to be confused with the momentum eddy diffusivity ε_M of Chapter 7). Summarizing the new notation, we can write the boundary layer concentration equation (64) as

$$\bar{u}\frac{\partial \bar{C}}{\partial x} + \bar{v}\frac{\partial \bar{C}}{\partial y} = \frac{\partial}{\partial y}\left[(D + \varepsilon_m)\frac{\partial \bar{C}}{\partial y}\right] \qquad (68)$$

The similarity between eq. (68) and the boundary layer energy equation (26) of Chapter 7 makes many of the turbulent heat transfer results convertible into mass transfer results. Thus, in the case of mass transfer in turbulent boundary layer flow from a wall surface of concentration C_0 to a free stream of concentration C_∞, we can combine the $\tau_{app} = \tau_0$ analysis of the momentum boundary layer with a similar assumption for the concentration layer

$$j_{app} = j_0, \quad \text{independent of } y \qquad (69)$$

to obtain the equivalent of eq. (77) of Chapter 7

$$\frac{h_m}{U_\infty} = \frac{\frac{1}{2}C_{f,x}}{0.9 + \left(\frac{1}{2}C_{f,x}\right)^{1/2}(13.2\,\text{Sc} - 10.25)} \qquad (70)$$

where $C_{f,x}$ is the local skin-friction coefficient (Fig. 7.6). Experimental measurements of the mass-transfer coefficient h_m are correlated very well by formulas like eq. (70) or the Colburn relation [14]

$$\frac{h_m}{U_\infty}\text{Sc}^{2/3} = \frac{1}{2}C_{f,x} \qquad (71)$$

In conclusion, experiment-aided formulas for the mass-transfer coefficient in turbulent boundary layer flow can be obtained by subjecting the heat-transfer correlations to the transformation

$$\frac{h}{\rho c_p U_\infty} \rightarrow \frac{h_m}{U_\infty}$$

$$\text{Pr} \rightarrow \text{Sc} \qquad (72)$$

The group h_m/U_∞ emerges as a Stanton-type number for turbulent mass transfer [note that the dimensions of h_m are m s^{-1}, eq. (40)].

For fully developed turbulent mass transfer inside a duct, we can derive in a straightforward manner the mass transfer equivalent of eq. (125) of Chapter 7. Simpler experimental correlations are the formulas obtained by changing the notation in the Colburn and Dittus–Boelter relations [eqs. (126) and (127),

Chapter 7] [14, 15],

$$\frac{h_m}{U}\,\mathrm{Sc}^{2/3} \cong \frac{f}{2} \tag{73}$$

$$\frac{h_m D_h}{D} \cong 0.024\,\mathrm{Sc}^{0.4}\,\mathrm{Re}_{D_h}^{0.8} \tag{74}$$

In these formulas D_h is the hydraulic diameter of the duct (not to be confused with the mass diffusivity D) and U is the cross-section-averaged velocity.

The turbulent mixing of a certain flow region surrounded by a region of different concentration is a phenomenon of great interest in environmental engineering. The mass transfer by free-stream turbulence can be analyzed based on the methodology constructed in Chapter 8. The first idea in this methodology is that free-stream mixing regions such as shear layers, jets, and plumes are *slender*, so that the boundary layer approximation applies. In addition, recognizing the overwhelming contribution of large eddies to turbulent transport across the stream, it is assumed that the eddy diffusivity is independent of the coordinate normal to the stream. Finally, based on empirical observations or the $\lambda_B/D = $ constant property of inviscid flow (Chapters 6 and 8; also Ref. 1), it is recognized that all these mixing regions grow linearly in the flow direction (Fig. 8.7).

For a two-dimensional flow proceeding in the x direction, the boundary layer-type concentration equation (68) reduces to

$$\bar{u}\frac{\partial \overline{C}}{\partial x} + \bar{v}\frac{\partial \overline{C}}{\partial y} = \varepsilon_m \frac{\partial^2 \overline{C}}{\partial y^2} \tag{75}$$

since in the free-stream region ε_m can be regarded as much greater than D. The identical form of eq. (75) for mass transfer and eq. (14) of Chapter 8 for heat transfer suggests that the time-averaged temperature fields derived in Chapter 8 can be converted to concentration fields by effecting the transformation

$$\overline{T} \rightarrow \overline{C}$$

$$\varepsilon_H \rightarrow \varepsilon_m \tag{76}$$

For example, the time-averaged concentration in a two-dimensional shear layer may be condensed in the expression

$$\frac{\overline{C} - C_\infty}{C_0 - C_\infty} \cong \frac{1}{2}\left[1 + \mathrm{erf}\left(\sigma\frac{y}{x}\right)\right] \tag{77}$$

where C_0, C_∞ are the constant concentrations found on either side of the shear layer, and where σ is an empirical constant accounting for the angle of the mixing region ($\sigma = 13.5$). Note that expression (77) rests also on the assumption that the *turbulent Schmidt number* is approximately equal to 1,

$$Sc_t = \frac{\varepsilon_M}{\varepsilon_m} \cong 1 \tag{78}$$

As a second example, the turbulent mixing zone created by the discharge of a round jet of concentration C_0 into a reservoir of concentration C_∞ has a concentration field described by [eq. (43), Chapter 8]

$$\frac{\bar{C} - C_\infty}{\bar{C}_c - C_\infty} \cong \exp\left[-\left(\frac{r}{0.127x} \right)^2 \right] \tag{79}$$

The centerline concentration $\bar{C}_c$ is related to the nozzle concentration C_0 by invoking the conservation of the constituent whose concentration $\bar{C}$ we are discussing

$$u_0(C_0 - C_\infty) \frac{r_0^2}{2} = \int_0^\infty \bar{u}(\bar{C} - C_\infty) r \, dr \tag{79'}$$

where u_0 is the cross-section-averaged velocity through the nozzle and $\bar{u}$ is the time-averaged velocity distribution in the conical mixing region [eq. (40) of Chapter 8].

As a third example, the concentration downstream from a point mass source bathed by a turbulent stream with uniform time-averaged velocity $\bar{U}_\infty$ and uniform eddy diffusivity ε_m (grid-generated turbulence, Chapter 8) is distributed as [6]

$$\bar{C} - C_\infty = \frac{\dot{m}}{4\pi\varepsilon_m x} \exp\left(-\frac{r^2 \bar{U}_\infty}{4\varepsilon_m x} \right) \tag{80}$$

In this expression, $\dot{m}$ is the strength of the point mass source measured in kilograms (of the constituent whose concentration is C) per second. The concentration field downstream from a line source of strength $\dot{m}'[\text{kg m}^{-1}\text{ s}^{-1}]$ is analogous to the temperature field behind a line heat source in a uniform turbulent stream [eqs. (68) and (71), Chapter 8],

$$\bar{C} - C_\infty = \frac{\dot{m}'}{\left(4\pi\bar{U}_\infty\varepsilon_m x \right)^{1/2}} \exp\left(-\frac{y^2 \bar{U}_\infty}{4\varepsilon_m x} \right) \tag{81}$$

An excellent treatment of turbulent mixing and environmental convection is offered in textbook form by Fisher, List, Koh, Imberger, and Brooks [6].

THE EFFECT OF CHEMICAL REACTION

In the laminar and turbulent mass transfer examples discussed so far, the mixture was not reacting chemically. Of interest to chemical engineers, in particular, is the case when a chemical reaction is present and the constituent generation term $\dot{m}'''$ in eq. (25) is finite. In some reactions the species of interest is a product of reaction ($\dot{m}''' > 0$), while in others the species is being consumed ($\dot{m}''' < 0$). In the case of *homogeneous reactions*, the volumetric mass rate of production of a species can be expressed as [16]

$$\dot{m}''' = k_n''' C^n \tag{82}$$

where n is the order of the reaction and k_n''' is the rate constant. In the case of a first-order reaction $n = 1$, the units of k_1''' are s^{-1}. In homogeneous reactions, the production or consumption of the species of interest takes place in the fluid, that is, wherever the species exists. In *heterogeneous reactions*, on the other hand, the reaction takes place on the surface of a catalyst; the rate of species production in this case may be expressed as

$$\dot{m}'' = k_n'' C_0^n \tag{83}$$

where C_0 is the concentration at the surface, n is the order of the reaction, and k_n'' is the reaction rate.

If a reaction can either generate or consume a species, it will certainly influence the distribution of that species in the flow of the mixture. Perhaps the best way to illustrate the effect of chemical reaction on mass convection is to focus on a very basic problem for which we already know the solution for the case when the chemical reaction is absent. We will then solve the problem allowing for the presence of a chemical reaction and, comparing the two solutions, we will develop a feeling for the effect of chemical reaction.

As test problem, consider the laminar boundary layer flow sketched in Fig. 9.3: mass is being swept away from a wall surface of concentration C_0 by a uniform stream containing none of the species released by the wall, $C_\infty = 0$. The concentration field for this flow configuration is known (see the Pohlhausen solution leading to eq. (35) in this chapter or an integral solution to the same problem, Chapter 2). With homogeneous reaction present, the problem assumes a slightly different statement

$$u\frac{\partial C}{\partial x} + v\frac{\partial C}{\partial y} = D\frac{\partial^2 C}{\partial y^2} - k_n''' C^n$$

$$C(x,0) = C_0, \quad C(x,\infty) = 0 \tag{84}$$

where the flow (u, v) is known from Chapter 2. Note the minus sign of the last term in eq. (84), which implies that in this example the chemical reaction consumes the species of interest. This homogeneous reaction problem was solved by Chambré and Young [17], and its counterpart with catalytic surface reaction by Chambré and Acrivos [18]. Below we outline and extend somewhat the integral solution reported by Bird, Stewart, and Lightfoot [19].

The integral version of the concentration equation (84) is

$$\frac{d}{dx}\int_0^\infty uC\,dy = -D\left(\frac{\partial C}{\partial y}\right)_{y=0} - k_n'''\int_0^\infty C^n dy \tag{85}$$

Assuming the simplest shape for the velocity and concentration profiles

$$u = U_\infty \frac{y}{\delta(x)}, 0 \le y \le \delta$$

$$C = C_0\left[1 - \frac{y}{\delta_c(x)}\right], 0 \le y \le \delta_c \tag{86}$$

eq. (85) reduces to

$$\frac{1}{Sc} = \frac{4}{3}x\frac{d(\Delta^3)}{dx} + \Delta^3 + \frac{12k_n'''C_0^{n-1}}{(n+1)U_\infty}x\Delta^2 \tag{87}$$

where

$$\Delta = \frac{\delta_c}{\delta} \le 1 \tag{88}$$

Note that δ_c is the unknown thickness of the concentration profile, while the velocity boundary layer thickness δ is listed in Table 2.1,

$$\delta = 3.46x\,\text{Re}_x^{-1/2} \tag{89}$$

In the absence of chemical reaction ($k_n''' = 0$), the solution to eq. (87) is $\Delta = Sc^{-1/3}$, as is known already from eq. (60) of Chapter 2. With chemical reaction, the thickness ratio Δ emerges as a function of two parameters

$$\Delta = \Delta(Sc, X) \tag{90}$$

where X is a dimensionless longitudinal variable

$$X = \frac{12k_n'''C_0^{n-1}}{(n+1)U_\infty}x \tag{91}$$

such that $X = 0$ represents the case of no chemical reaction.

The local Sherwood number or the local wall mass flux can be expressed as

$$Sh = \frac{j_0}{C_0}\frac{x}{D} = \frac{x}{\delta_c} = \frac{0.289}{\Delta}\text{Re}_x^{1/2} \tag{92}$$

In the absence of chemical reaction the local Sherwood number reduces to

$$Sh = 0.289 \, Sc^{1/3} Re_x^{1/2}, \qquad Sc > 1 \tag{93}$$

which is only 13 percent below the similarity solution (39). Now, the effect of chemical reaction on mass transfer can be presented as the ratio

$$\phi_r = \frac{Sh(\text{with chemical reaction})}{Sh(\text{without chemical reaction})} \tag{94}$$

which, dividing eq. (92) by eq. (93), means

$$\phi_r = \frac{1}{Sc^{1/3}\Delta} \quad \text{or} \quad \phi_r = \frac{\Delta(Sc, 0)}{\Delta(Sc, X)} \tag{95}$$

Function ϕ_r has been plotted in Fig. 9.6, after integrating eq. (87) numerically from the initial condition $\Delta = Sc^{-1/3}$ at $X = 0$. It is evident that, depending on the values of Sc and X, the effect of chemical reaction on the mass transfer rate can be significant. For a given fluid (Sc = constant), the mass transfer rate with chemical reaction is greater that the mass transfer rate in the absence of chemical reaction, and the discrepancy between the two rates increases in the downstream direction. This effect is due to the decrease in δ_c caused by the chemical reaction (i.e., by the consumption of the species of concentration C). The reverse effect would be observed if the homogeneous reaction would be generating the species in the boundary layer.

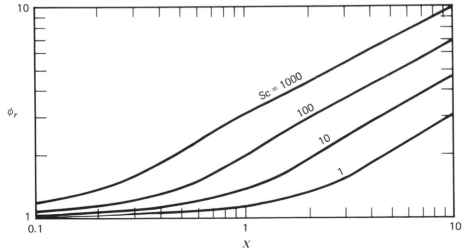

Figure 9.6 The effect of species-depleting chemical reaction on forced convection mass transfer to laminar boundary layer flow.

In conclusion, the local Sherwood number in $Sc > 1$ laminar boundary layer flow with homogeneous reaction that consumes the species can be calculated as

$$Sh = 0.289\phi_r Sc^{1/3} Re_x^{1/2} \tag{96}$$

where ϕ_r is given by Fig. 9.6. It can be shown [19] that sufficiently far downstream $(X > 10)$, the correction factor ϕ_r approaches asymptotically $Sc^{1/6}X^{1/2}$, so that local Sherwood number becomes

$$Sh = 0.289 \, Sc^{1/2} \left[\frac{12 k_n''' C_0^{n-1} x^2}{(n+1)\nu} \right]^{1/2} \tag{97}$$

This limit shows that sufficiently far downstream the mass transfer rate j_0 (or δ_c) becomes independent of longitudinal position.

FLOW THROUGH A POROUS MEDIUM

Like heat transfer, the phenomena of mass diffusion and mass convection can take place in the fluid mixture that saturates a porous solid matrix (think of the migration of moisture through a double wall filled with fiberglass wool). The basic principles of heat and fluid flow through porous media are outlined in Chapter 10; in this section we extend this discussion to some of the most basic problems in mass convection through porous media. Overall, this section points once more to the analogy between convective mass transfer and convective heat transfer, therefore, the reader is urged to read Chapter 10 first.

The *concentration* equation or the statement of mass conservation for one constituent in flow through a porous medium can be derived based on the one-dimensional, single-pore flow model sketched in Fig. 10.4. Constituent i enters the pore with a velocity $u_{i,p}$, where subscript p is a reminder that $u_{i,p}$ is what would be measured inside the pore. The statement that *inflow equals outflow plus accumulation inside the control volume* is

$$A_p \Delta x \frac{\partial \rho_{i,p}}{\partial t} = \rho_{i,p} u_{i,p} A_p - \left[\rho_{i,p} u_{i,p} A_p + A_p \Delta x \frac{\partial}{\partial x} (\rho_{i,p} u_{i,p}) \right] \tag{98}$$

where, as until now, $\rho_{i,p}$ is the number of kilograms of constituent i per cubic meter of fluid mixture (which is located inside the pore). According to eq. (2) of Chapter 10, the product $u_{i,p} A_p$ can be replaced by $u_i A$, where u_i is the volume-averaged velocity of constituent i. Equation (98) reduces to

$$\frac{A_p}{A} \frac{\partial \rho_{i,p}}{\partial t} = -\frac{\partial}{\partial x} (\rho_{i,p} u_i) \tag{99}$$

in which A_p/A is the porosity of the medium, $\phi = A_p/A$. The concentration inside the pore ($\rho_{i,p}$) is related to the concentration in the porous medium as a whole (ρ_i) via

$$\rho_i = \phi\rho_{i,p}, \qquad \left[\frac{\text{kilograms of } i}{\text{m}^3 \text{ of porous medium}}\right] \tag{100}$$

Combining eqs. (99) and (100) and assuming that the porous medium is homogeneous ($\phi = $ constant), we obtain

$$\phi\frac{\partial \rho_i}{\partial t} = -\frac{\partial}{\partial x}(\rho_i u_i) \tag{101}$$

Except for the porosity factor appearing on the left-hand side, eq. (101) is the one-dimensional equivalent of eq. (16) derived in the beginning of this chapter. Therefore, the analytical steps and the definitions that followed eq. (16) could also be repeated here; in the end, the general constituent conservation equation reads

$$\phi\frac{\partial C}{\partial t} + u\frac{\partial C}{\partial x} + v\frac{\partial C}{\partial y} + w\frac{\partial C}{\partial z} = D\nabla^2 C + \dot{m}''' \tag{102}$$

This equation accounts for the convection of mass in an imaginary laminar-like flow defined by the volume-averaging process explained in detail in the beginning of Chapter 10. In this flow, u, v, and w are the volume-averaged velocity components, C is the number of kilograms of constituent per unit volume of fluid-saturated porous medium, D is the mass diffusivity of the porous medium (with the fluid mixture in it), and $\dot{m}'''$ is the rate of constituent production per unit volume of fluid-saturated porous medium.

The concentration equation (102) is nearly identical to the energy equation [see eq. (34) of Chapter 10], therefore, we expect to be able to translate the heat transfer results of Chapter 10 into mass transfer results. Below, we illustrate this analogy based on two examples.

Forced Boundary Layers

Consider the uniform Darcy flow $u = U_\infty$, $v = 0$ parallel to a surface of constant concentration C_0 (Fig. 9.7). Sufficiently far from the wall, the homogeneous porous medium has the concentration C_∞. We are interested in the wall mass flux j_0 or the local Sherwood number: both quantities depend on the concentration field in the immediate vicinity of the wall. Treating the wall region as slender ($\delta_c \ll x$, Fig. 9.7), the steady state concentration equation for a nonreacting mixture reduces to

$$u\frac{\partial C}{\partial x} + v\frac{\partial C}{\partial y} = D\frac{\partial^2 C}{\partial y^2} \tag{103}$$

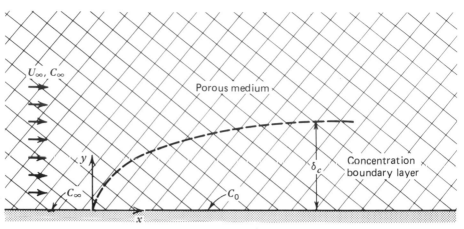

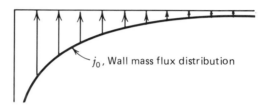

Figure 9.7   Forced convection mass transfer to a uniform flow through a fluid-saturated porous medium.

This equation and the concentration boundary conditions $C(x, 0) = C_0, C(x, \infty) = C_\infty$ are identical to their heat transfer counterparts [see Fig. 10.5 and eq. (42) of Chapter 10]. The concentration field is therefore obtained by subjecting the temperature field [given by eqs. (49) and (52) of Chapter 10] to the transformation $T \to C, \alpha \to D$:

$$\frac{C - C_0}{C_\infty - C_0} = \operatorname{erf}\left[\frac{y}{2x}\left(\frac{U_\infty x}{D}\right)^{1/2}\right] \qquad (104)$$

Note that the new dimensionless group $U_\infty x/D$ replaces the Peclet number $U_\infty x/\alpha$ revealed by the heat transfer solution. The local Sherwood number deduced from eq. (104) is

$$\mathrm{Sh} = \frac{j_0}{C_0 - C_\infty}\frac{x}{D} = 0.564\left(\frac{U_\infty x}{D}\right)^{1/2} \qquad (105)$$

The wall mass flux j_0 varies as $x^{-1/2}$, as indicated in the lower portion of Fig. 9.7. We note also that the mass transfer rate formula (105) is analogous to the heat transfer rate formula (54) of Chapter 10.

Natural Boundary Layers

As a second example of convection mass transfer in a porous medium, consider the buoyant boundary layer formed in the vicinity of a vertical surface of concentration C_0 = constant and temperature T_0 = constant, in a homogeneous porous medium (C_∞, T_∞). As shown in Fig. 9.8, in general, the concentration boundary layer thickness is not the same as the thermal boundary layer thickness, $\delta_c \neq \delta_T$. The flow owes its buoyancy to the density difference $[\rho(x, y) - \rho_\infty]$ which in turn may be caused by the temperature difference $(T - T_\infty)$, or the concentration difference $(C - C_\infty)$, or a combination of $(T - T_\infty)$ and $(C - C_\infty)$ [see eq. (44), earlier in this chapter]. It can be shown that, subject to the Boussinesq approximation (44), the boundary layer

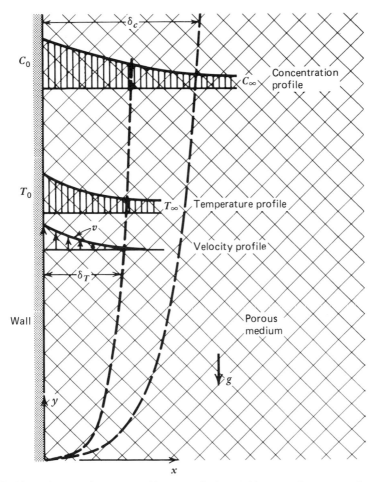

Figure 9.8 Natural convection mass and heat transfer in a fluid-saturated porous medium near a vertical wall.

momentum equation for the system of Fig. 9.8 is

$$\frac{\partial v}{\partial x} = \frac{gK}{\nu}\left(\beta\frac{\partial T}{\partial x} + \beta_c\frac{\partial C}{\partial x}\right) \tag{106}$$

where K [m^2] is the permeability of the medium, and where β and β_c are the expansion coefficients defined in eqs. (45) and (46). The momentum equation (106) can be integrated from the reservoir ($v = 0, T = T_\infty, C = C_\infty$) to a point near the wall to yield

$$v = \frac{gK}{\nu}[\beta(T - T_\infty) + \beta_c(C - C_\infty)] \tag{107}$$

This form shows that both the heat transfer and mass transfer from the wall can drive a buoyant boundary layer parallel to the wall.

The boundary layer concentration equation in the coordinates of Fig. 9.8 reads

$$u\frac{\partial C}{\partial x} + v\frac{\partial C}{\partial y} = D\frac{\partial^2 C}{\partial x^2} \tag{108}$$

again under conditions of steady state and nonreacting mixture. In order to determine the mass flux from the wall to the boundary layer flow, we focus on two limiting situations. First, consider the no heat transfer limit where the flow is *driven by mass transfer*. Thus, deleting the term $\beta(T - T_\infty)$ from eq. (107) we obtain two governing equations [eqs. (107) and (108)] that are analogous to the momentum and energy equations governing the heat transfer problem [eq. (63) and (64) of Chapter 10]. The Sherwood number formula can be deduced based on this analogy by applying the transformation ($T \rightarrow C, \alpha \rightarrow D, \beta \rightarrow \beta_c$) to the Nusselt number formula [eq. (76) of Chapter 10],

$$\text{Sh} = \frac{j_0}{C_0 - C_\infty}\frac{y}{D} = 0.444\,\text{Ra}_{D,y}^{1/2} \tag{109}$$

where $\text{Ra}_{D,y}$ is a new dimensionless group, the Darcy-modified Rayleigh number for a buoyant concentration boundary layer

$$\text{Ra}_{D,y} = \frac{g\beta_c(C_0 - C_\infty)Ky}{\nu D} \tag{110}$$

Note also that $\text{Ra}_{D,y}^{1/2}$ is the scale of y/δ_c, therefore, the physical meaning of $\text{Ra}_{D,y}$ is

$$\text{Ra}_{D,y} \sim \left(\frac{\text{height of concentration boundary layer}}{\text{thickness of concentration boundary layer}}\right)^2 \tag{111}$$

The second limit of interest is the *heat-transfer-driven* boundary layer, whose vertical velocity scale is (Chapters 10, 11)

$$v \sim \frac{g\beta K}{\nu}(T_0 - T_\infty) \tag{112}$$

In the concentration boundary layer region (dimensions $x \sim \delta_c$ and $y \sim H$), the concentration equation (108) requires

$$\frac{v}{H} \sim \frac{D}{\delta_c^2} \tag{113}$$

Combining eqs. (112) and (113), we conclude that the slenderness ratio of a ΔT-driven concentration boundary layer is

$$\frac{\delta_c}{H} \sim \left[\frac{gKH\beta(T_0 - T_\infty)}{\nu D}\right]^{-1/2} \tag{114}$$

Since the H-averaged Sherwood number scale is of order H/δ_c, we conclude that

$$\mathrm{Sh}_{0-H} \sim \left[\frac{gKH\beta(T_0 - T_\infty)}{\nu D}\right]^{1/2} \tag{115}$$

An important observation is that scales (114) and (115) are valid only if $\delta_c < \delta_T$ (which means Le > 1). In porous media where Le < 1, it can be shown that the concentration boundary layer thickness δ_c scales as δ_T/Le [20], where δ_T is the thermal boundary layer thickness ($\delta_T \sim H \mathrm{Ra}_H^{-1/2}$, in the notation employed in Chapter 10).

In conclusion, we have two alternatives for estimating the mass transfer rate to a buoyant boundary layer—eq. (109) for a mass-transfer-driven boundary layer and eq. (115) for a heat transfer-driven flow. The mass transfer rate is given by eq. (115) when

$$(\delta_c)_{\mathrm{HT}} < (\delta_c)_{\mathrm{MT}} \tag{116}$$

in other words, when

$$\beta(T_0 - T_\infty) > \beta_c(C_0 - C_\infty) \tag{117}$$

The transition criterion from one limit to another amounts to comparing directly the scales of the two body force terms in the momentum equation (107). This conclusion distinguishes the phenomenon of natural convection mass transfer in porous media from the natural convection mass transfer in a pure fluid mixture (discussed earlier in this chapter). Equation (61) showed that in fluids the transition from ΔC-driven to ΔT-driven mass transfer does not necessarily take place when the two body force terms are of the same order of magnitude (in fluids, the transition depends on an *additional* parameter, the Lewis number, for example). This basic difference between porous media and fluids has its roots in the fact that the velocity boundary layer in porous media has only one transversal length scale (δ_T, Fig. 9.8), whereas in fluid natural convection the velocity boundary layer has two scales (Table 4.1).

SYMBOLS

A	cross-sectional area
c_P	specific heat at constant pressure
C	concentration
C_0	concentration next to the wall
C_∞	concentration far from the wall
$C_{f,x}$	skin friction coefficient
D	mass diffusivity
D_h	hydraulic diameter
f	friction factor
g	gravitational acceleration
h_m	mass transfer coefficient [eq. (40)]
H	wall height
j	diffusion flux [eq. (20)]
j_0	wall mass flux
j_{app}	apparent mass flux [eq. (66)]
k_n'', k_n'''	reaction rates [eqs. (83, 82)]
K	permeability [eq. (106)]
Le	Lewis number (α/D or Sc/Pr)
m	mass
$\dot{m}$	strength of point mass source (kg s^{-1})
$\dot{m}'$	strength of line mass source (kg s^{-1} m^{-1})
$\dot{m}''$	mass generation rate per unit area (kg s^{-1} m^{-2})
$\dot{m}'''$	volumetric rate of mass generation (kg s^{-1} m^{-3})
M	molar mass
n	number of moles
n	order of reaction [eqs. (82) and (83)]
P	pressure
P_i	partial pressure
Pr	Prandtl number
r, θ, z	cylindrical coordinates (Fig. 1.1)
r, ϕ, θ	spherical coordinates (Fig. 1.1)
R	ideal gas constant
$\overline{R}$	universal gas constant

$\mathrm{Ra}_{D,y}$	Darcy-modified Rayleigh number for concentration-driven natural convection through a porous medium [eq. (110)]
Ra_H	Rayleigh number (Chapter 4)
$\mathrm{Ra}_{m,y}$	Rayleigh number for concentration-driven natural convection [eq. (51)]
Re_x	Reynolds number ($U_\infty x/\nu$)
Sc	Schmidt number (ν/D)
Sc_t	turbulent Schmidt number ($\varepsilon_M/\varepsilon_m$)
Sh	Sherwood number [eq. (38)]
T	temperature
u, v	mass-averaged velocity components [eq. (19)]
U	duct-averaged velocity
V	volume
x	mole fraction [eq. (7)]
X	longitudinal coordinate [eq. (91)]
x, y, z	cartesian coordinates
α	thermal diffusivity
β	thermal expansion coefficient [eq. (45)]
β_c	concentration expansion coefficient [eq. (46)]
δ	velocity boundary layer thickness [eq. (89)]
δ_c	concentration boundary layer thickness (Fig. 9.4)
δ_T	thermal boundary layer thickness (Fig. 9.4)
δ_v	viscous layer thickness (Pr < 1 fluids)
Δ	function [eq. (88)]
ε_H	thermal eddy diffusivity (Chapter 7)
ε_m	mass eddy diffusivity [eq. (67)]
ε_M	momentum eddy diffusivity (Chapter 7)
ν	kinematic viscosity
ρ	density
σ	empirical constant [eq. (77)]
ϕ	porosity [eq. (100)]
ϕ_r	ratio, showing the effect of chemical reaction on mass transfer [eq. (94)]
Φ	mass fraction [eq. (6)]
$\overline{(\ \ )}$	time-averaged part
$(\ \ )'$	fluctuating part
$(\ \ )_c$	property measured along the stream centerline
$(\ \ )_0$	property measured at the wall
$(\ \ )_\infty$	property measured far from the wall
$(\ \ )_i$	property of constituent i
$(\ \ )_p$	property measured inside the pore
$(\ \ )_{0-H}$	averaged from $y = 0$ to $y = H$
$(\ \ )_{\mathrm{HT}}$	flow driven by heat transfer
$(\ \ )_{\mathrm{MT}}$	flow driven by mass transfer

REFERENCES

1. A. Bejan, *Entropy Generation through Heat and Fluid Flow*, Wiley, New York, 1982; Preface.
2. R. W. Haywood, *Equilibrium Thermodynamics*, Wiley, New York, 1980, Chapter 16.
3. W. C. Reynolds and H. C. Perkins, *Engineering Thermodynamics*, 2nd ed., McGraw-Hill, New York, 1977, Chapters 10, 11.
4. R. B. Bird, W. E. Stewart, and E. N. Lightford, *Transport Phenomena*, Wiley, New York, 1960, Chapter 16.
5. A. Fick, On liquid diffusion, *Philos. Mag.*, Vol. 4, No. 10, 1855, pp. 30–39.
6. H. B. Fischer, E. J. List, R. C. Y. Koh, J. Imberger, and N. H. Brooks, *Mixing in Inland and Coastal Waters*, Academic Press, New York, 1979, Chapter 2.
7. E. R. G. Eckert and R. M. Drake, Jr., *Analysis of Heat and Mass Transfer*, McGraw-Hill, New York, 1972, p. 717.
8. I. Prigogine, *Thermodynamics of Irreversible Processes*, 3rd ed., Wiley, New York, 1967.
9. D. K. Edwards, V. E. Denny, and A. F. Mills, *Transfer Processes*, 2nd ed., Hemisphere, Washington, D.C., 1979, Chapter 3.
10. J. P. Hartnett and E. R. G. Eckert, Mass-transfer cooling in a laminar boundary layer with constant fluid properties, *Trans. ASME*, Vol. 79, 1957, pp. 247–254.
11. B. Gebhart and L. Pera, The nature of vertical natural convection flows resulting from the combined buoyancy effects of thermal and mass diffusion, *Int. J. Heat Mass Transfer*, Vol. 14, 1971, pp. 2025–2050.
12. E. V. Somers, Theoretical considerations of combined thermal and mass transfer from a vertical flat plate, *J. Appl. Mech.*, Vol. 23, 1956, pp. 295–301.
13. S. Ostrach, Natural convection with combined driving forces, *Physico-Chemical Hydrodynamics*, Vol. 1, 1980, pp. 233–247.
14. W. M. Rohsenow and H. Y. Choi, *Heat, Mass and Momentum Transfer*, Prentice-Hall, Englewood Cliffs, NJ, 1961, Chapter 17.
15. E. R. Gilliland and T. K. Sherwood, Diffusion of vapors into air streams, *Ind. Eng. Chem.*, Vol. 26, 1934, pp. 516–523.
16. R. B. Bird, W. E. Stewart, and E. N. Lightfoot, *op. cit.*, p. 520.
17. P. L. Chambré and J. D. Young, On the diffusion of a chemically reacting species in a laminar boundary layer flow, *Phys. Fluids*, Vol. 1, 1958, pp. 48–54.
18. P. L. Chambré and A. Acrivos, On chemical surface reactions in laminar boundary layer flows, *J. Appl. Phys.*, Vol. 27, 1956, pp. 1322–1328.
19. R. B. Bird, W. E. Stewart, and E. N. Lightfoot, *op. cit.*, pp. 605–608.
20. A. Bejan, The scales of combined mass and heat transfer in natural convection, Report CUMER-84, Mech. Eng. Dept., University of Colorado, Boulder, 1984.

PROBLEMS

1. Show that concentration (C_i), mass fraction (Φ_i), and mole fraction (x_i) are proportional to one another and that these proportionalities are given by eqs. (8) and (9).

2. Consider the configuration of Fig. 9.3 in which air passes in laminar boundary layer flow past a porous wall saturated with water. The wall is so porous that it contains 25 percent water on a volume basis. The air stream is at

atmospheric pressure and temperature ($T_\infty = 310$ K). Calculate the streamside water concentration C_0 and compare the result with the water concentration immediately below the wall–stream interface (Fig. 9.3).

3. Determine the overall Sherwood number scales for the last three configurations sketched in Fig. 9.5. In each case, determine the condition necessary for heat-transfer-driven natural convection mass transfer. [Note that for Pr > 1 and Le > 1, this condition is expressed by eq. (61).]

4. Consider the mass transfer with chemical reaction in laminar boundary layer flow in the limit Sc → 0. Construct an integral analysis based on eq. (84) and the assumption that the concentration layer is much thicker than the velocity boundary layer; hence $u = U_\infty$ and $v = 0$. Determine the local Sherwood number and explain how this parameter is influenced by the presence of chemical reaction.

5. Determine the mass flux of a gas i that diffuses into a liquid film that falls along a vertical solid wall. Assume that the liquid film flow is laminar, and that it has reached terminal velocity. Let $v_{max} = $ constant and $\delta = $ constant be the film velocity at the liquid–gas interface and the film thickness, respectively. Take a two-dimensional cartesian system of coordinates with the y axis pointing downward and $x = 0$ on the liquid–gas interface. The concentration of i at the interface (on the liquid side) is $C_0 = $ constant, and the solid surface is impermeable to i, $\partial C / \partial x = 0$ at $x = \delta$. State the complete mathematical problem that would allow you to determine the interface mass flux $j_0 = -D(\partial C / \partial x)_{x=0}$ as a function of altitude. Solve this problem in the limit of "small contact length y", that is, in the limit where the concentration boundary layer thickness (or penetration distance) δ_c is much smaller than the film thickness δ.

6. Use scale analysis or the order-of-magnitude conclusions of the preceding problem to estimate the rate of gas absorption from a gas bubble into the liquid pool through which the gas bubble rises. The gas concentration on the liquid side of the gas–liquid interface is C_0 and the mass diffusivity is D. The bubble has a diameter D_b and rises through the liquid pool with the terminal velocity v_{max}. State clearly all the assumptions on which your scale analysis is based.

7. A film of liquid solvent flows downward along a vertical wall, and in the process, dissolves the layer of paint with which the wall is coated. The liquid film falls at a terminal velocity (in fully developed laminar flow); hence, its velocity distribution and thickness δ are independent of vertical position. Attach a two-dimensional cartesian system to the wall, such that the y axis points downward and the plane $x = 0$ coincides with the wall surface. The liquid film extends from $x = 0$ to $x = \delta$. The concentration of paint in the film is $C_0 = $ constant at $x = 0$. Determine the local mass flux $-D(\partial C / \partial x)_{x=0}$, assuming that the concentration layer is much thinner than the layer of liquid solvent. Is the coat of paint being eroded evenly?

8. Illustrate the effect of a first-order chemical reaction on the mass transfer rate in forced convection through a porous medium ($C_\infty = 0$) adjacent to a surface of concentration C_0 (see Fig. 9.7). Show that the concentration equation in this case reduces to

$$U_\infty \frac{\partial C}{\partial x} = D \frac{\partial^2 C}{\partial y^2} \pm k_l''' C$$

where the "$+$" sign corresponds to a species-generating reaction and the "$-$" sign accounts for a species-depleting reaction. Based on integral analysis, determine the local Sherwood number and show how this result is influenced by the reaction rate k_1 and by the sign of k_1.

9. A vertical wall of height H heats a liquid pool in such a way that the local heat flux through the wall surface q'' is independent of position along the wall. This arrangement gives birth to a natural convection boundary layer flow along the wall. The liquid has the property to dissolve the wall material; the concentration of wall material is C_0 at the wall surface and $C_\infty = 0$ sufficiently far from the surface. Assuming the mass diffusivity of wall material known D, determine the order of magnitude of the mass transfer rate from the wall to the boundary layer flow. Base this analysis on the assumption that the boundary layer is driven by heat transfer.

10. Solve the preceding problem in the context of a vertical wall heating a porous medium saturated with fluid. What condition must be satisfied if the boundary layer flow is to be driven by heat transfer and not by mass transfer?

11. Consider the concentration and temperature boundary layers sketched in Fig. 9.4. If the flow is driven by mass transfer (i.e., not by heat transfer), then what is the wall–reservoir heat transfer rate? Answer the same question for the case where the reservoir is a porous medium saturated with fluid (see Fig. 9.8).

10

Principles of Convection Through Porous Media

In the next two chapters, we focus on a rapidly growing branch of fluid mechanics and heat transfer, namely, the flow of energy-carrying fluids through porous media. Porous materials such as sand and crushed rock underground are saturated with water which, under the influence of local pressure gradients, migrates and transports energy through the material. The transport properties of fluid-saturated porous materials are very important in the petroleum and geothermal industries. Further examples of convection through porous media may be found in manmade systems such as fiber and granular insulations, winding structures for high-power density electric machines, and the cores of nuclear reactors.

In the field of fluid mechanics, the dynamics of fluid flow through a porous medium is a relatively old topic, in view of the historical relevance of this topic to the proper management of the underground water table and the engineering of irrigation systems [1–3]. Indeed, the conceptual centerpiece in this branch of fluid mechanics—the Darcy flow model—originated in the nineteenth century in connection with the engineering of public fountains [4]. The convective heat transfer potential of flows through porous media is a relatively new topic, as the technologies of porous insulation, gas-cooled electric machinery, and nuclear reactors grew out of the late twentieth century concern with the cost of exergy (or fuel). In Chapter 11 we focus in greater detail on the energy-engineering aspects of convection through porous media; in the present chapter we develop the basic principles of fluid mechanics and heat transfer through a fluid-saturated porous structure, in the same manner in which we developed the foundations of convective heat transfer in Chapter 1.

MASS CONSERVATION

The analytical treatment of convective heat transfer through a porous medium begins with the observation that the actual heat and fluid flow picture is complicated: Figure 10.1 shows two examples of manmade porous materials used in the home-building industry and, clearly, the flow geometry differs unpredictably from one region of the material to another. If we try to imagine the flow of fluid and energy through such a complicated labyrinth, it is very tempting to disregard the local complication and unpredictability of the

Figure 10.1 Examples of granular porous materials used in the construction industry; top: crushed limestone (~ 1 cm size); bottom: Liapor® spheres (~ 0.5 cm diameter). (Courtesy of Franz Keller, Käser and Beck Betonwerk-Baustoffe, 8542 Eckersmühlen, West Germany.)

phenomenon and, instead, to concentrate on the overall capability of this system to transport fluid and energy. This temptation is, of course, fueled by the engineering interest in the overall performance of finite-size layers or spaces filled with porous materials. But, at the same time, the decision to assume away (to smooth out) the complicated features of the actual phenomenon is in fact an admission of defeat in the face of Nature. This decision is analogous to the idea of time-averaging a turbulent flow field (Chapter 7) in order to do away with the flow complications called *eddies*. These decisions do not simplify the respective flows—as smooth (laminar-like) porous media and turbulent flows do not exist—instead, they simplify and often trivialize the discussion devoted to "explaining" such phenomena.

For these reasons, the analysis of convection through porous media is largely an empirical exercise. The method consists of applying the conservation principles of Chapter 1 to a "gray" medium visualized by holding the photographs of Fig. 10.1 sufficiently far away so that the grain (the pebbles and the air channels) become indistinguishable. A two-dimensional version of such a medium is sketched in Fig. 10.2, which shows that a small enough control volume $\Delta x \Delta y$ retains the irregular features of the grainy structure photographed in Fig. 10.1. Although the flow and heat transfer through the gray medium may be regarded as two-dimensional, the picture in the small control volume $\Delta x \Delta y$ differs from one $z = $ constant plane to another. Locally, in regions of size comparable with the channel size and solid grain size, the flow is always three-dimensional. This is another similarity between porous medium flow and turbulent flow: in the latter, the time-averaged flow may be two-dimensional when, in fact, the real (eddy) flow is always three-dimensional.

From an overall two-dimensional flow we can isolate inside a control surface a chunk ($\Delta x \Delta y W$) with W so much larger than either Δx or Δy so that, for the purpose of mass-flow accounting, the important flowrates are in the x and y directions only (the flow cross-sections in the x, y directions are $W\Delta y$ and $W\Delta x$, and both are much bigger than the flow cross-section in the z direction, $\Delta x \Delta y$). Consider first the mass flowrate entering the $\Delta x \Delta y W$ chunk of porous material from the left through the $x = $ constant plane

$$\dot{m}_x = \rho \int_y^{y+\Delta y} \int_0^W u_p \, dz \, dy \tag{1}$$

where $u_p(y, z)$ is the uneven x velocity distribution over the void patches of the $x = $ constant plane. Imagining a control surface $W\Delta y$ sufficiently larger than the pore and solid grain cross-sections, we define the area-averaged velocity in the x direction.

$$u = \frac{1}{W\Delta y} \int_y^{y+\Delta y} \int_0^W u_p(y, z) \, dz \, dy \tag{2}$$

In other words,

$$\dot{m}_x = \rho u (W\Delta y) \tag{3}$$

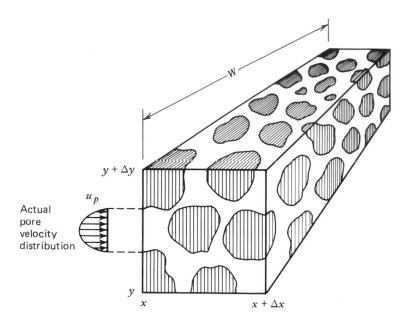

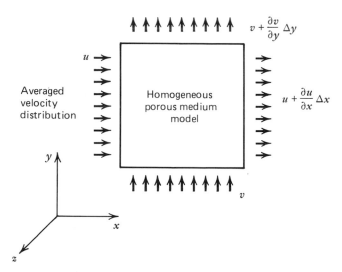

Figure 10.2 The averaging of the pore velocity distribution, as a basis for the homogeneous porous medium model.

The area-averaged velocity in the y direction is defined in the same way

$$v = \frac{1}{W\Delta x} \int_x^{x+\Delta x} \int_0^W v_p(x, z)\, dz\, dx \tag{4}$$

so that the mass flowrate in the y direction can be expressed as in Chapter 1,

$$\dot{m}_y = \rho v (W\Delta x). \tag{5}$$

Note that in deriving eqs. (3) and (5) we treated the density ρ as constant in the $\Delta x\,\Delta y$ element of the two-dimensional flow: this does not mean that ρ is constant throughout the x-y field.

The reward for smoothing out the complications of the channel flow and introducing the area-averaged velocities u, v is that, with expressions like (3) and (5), the averaged flow looks like any other homogeneous fluid flow. Therefore, applying the mass conservation principle [eq. (1) of Chapter 1] to the $\Delta x\,\Delta y\,W$ element yields

$$\frac{\partial \rho}{\partial t} + \frac{\partial(\rho u)}{\partial x} + \frac{\partial(\rho v)}{\partial y} = 0 \tag{6}$$

In general, for a three-dimensional averaged flow the mass conservation statement reads

$$\frac{D\rho}{Dt} + \rho \nabla \cdot \mathbf{v} = 0 \tag{7}$$

where $\mathbf{v}$ is the volume-averaged velocity vector (u, v, w). Note that eq. (7) is the same as eq. (6) of Chapter 1: this coincidence is not accidental since, historically, the concept of area-averaged velocity was introduced precisely in order to be able to apply the pure-fluids mathematical apparatus (Chapter 1) to flows through porous media.

THE DARCY FLOW MODEL

In the fluid mechanics of porous media, the place of momentum equations or force balances is occupied by the numerous experimental observations summarized mathematically as the *Darcy law*. These observations were first reported by Darcy [4] who, based on measurement alone, discovered that the area-averaged fluid velocity through a column of porous material is directly proportional to the pressure gradient established along the column. Subsequent experiments proved that the area-averaged velocity is, in addition, inversely proportional to the viscosity (μ) of the fluid seeping through the porous material. Therefore, in relation to the one-dimensional forced flow sketched in

Fig. 10.3, the Darcy observations amount to writing

$$u = \frac{K}{\mu}\left(-\frac{dP}{dx} \right) \tag{8}$$

where K is an empirical constant called *permeability*. From eq. (8), which is in essence the definition of permeability much in the same way as the Fourier Law of heat conduction is the definition of thermal conductivity, we learn that the dimensions of K must be

$$[K] = \frac{[\mu][u]}{[-dP/dx]} = (\text{length})^2 \tag{9}$$

There is an obvious similarity between eq. (8) and the formula for average velocity in Hagen–Poiseuille flow [eqs. (23) in Chapter 3]: this similarity suggests that the Darcy flow is the macroscopic manifestation of a highly viscous flow through the pores of the permeable structure, and that $K^{1/2}$ is a length scale representative of the effective pore diameter. In fact, by postulating a certain small-scale network of channels of known geometry and assuming

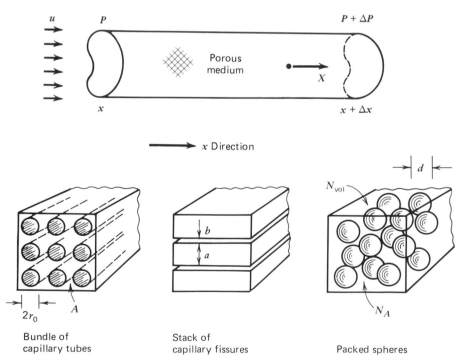

Figure 10.3 The Darcy flow experiment and three possible models for calculating the porous medium permeability K.

Hagen–Poiseuille flow through each channel, it is possible to derive eq. (8) where K will emerge as a function of the network geometry. Analyses of this kind are proposed in Problems 2–4. Using $K^{1/2}$ as a length scale to define the Reynolds number

$$\text{Re} = \frac{uK^{1/2}}{v} \tag{10}$$

and the friction factor

$$f = \frac{\left(-\dfrac{dP}{dx}\right)K^{1/2}}{\rho u^2} \tag{11}$$

the Darcy law (8) can be rewritten as

$$f_{\text{Darcy}} = \frac{1}{\text{Re}} \tag{12}$$

Experimental measurements [5] have shown that eqs. (8) and (12) are valid as long as $O(\text{Re}) < 1$; if the Reynolds number based on $K^{1/2}$ exceeds $O(1)$, inertial effects flatten the $f(\text{Re})$ curve in a manner reminiscent of the friction factor curve in turbulent flow over a rough surface (Fig. 7.9).

$$f = \frac{1}{\text{Re}} + C \tag{13}$$

where C is an empirical constant approximately equal to 0.55 [5]. The more general friction factor expression (13) follows from Forschheimer's [6] modification of the Darcy law (see also Cheng [7])

$$-\frac{dP}{dx} = \frac{\mu}{K}u + b\rho u^2 \tag{14}$$

where b is another empirical constant.

In the present treatment of convection through porous media, we rely exclusively on the simple Darcy flow model (8, 12), in other words, we consistently make the assumption that the flow is slow enough or the pores are small enough so that

$$O\left(\frac{uK^{1/2}}{v}\right) < 1 \tag{15}$$

The study of porous medium flow with inertial effects [eqs. (13) and (14)] can be pursued in similar manner, by starting with eq. (14) instead of eq. (8) (e.g. [8]).

In the presence of a body force per unit volume ρg_x, the Darcy law (8) becomes

$$u = \frac{K}{\mu}\left(-\frac{\partial P}{\partial x} + \rho g_x\right) \tag{16}$$

acknowledging the fact that the flow through the porous column of Fig. 10.3 stops when the externally controlled pressure gradient dP/dx matches the hydrostatic gradient ρg_x. In vectorial notation, the three-dimensional generalization of eq. (16) is

$$\mathbf{v} = \frac{K}{\mu}(-\nabla P + \rho\mathbf{g}) \tag{17}$$

where $\mathbf{v}$ is the velocity vector (u, v, w) and $\mathbf{g}$ the body acceleration vector (g_x, g_y, g_z).

In many problems involving only the seepage flow of water through soil, ρ and μ may be regarded as constant. With the y axis oriented upward against the gravitational acceleration g, the body acceleration vector is $(0, -g, 0)$, in other words, eq. (17) becomes

$$\mathbf{v} = -\frac{K}{\mu}\nabla\phi \tag{18}$$

where $\phi(x, y, z)$ is the new function

$$\phi = P + \rho g y \tag{19}$$

Note that under the same conditions ($\rho = $ constant), the mass conservation statement (7) reduces to

$$\nabla \cdot \mathbf{v} = 0 \tag{20}$$

Combining eqs. (18) and (20), we find that seepage flows are governed by the Laplace-type equation

$$\nabla^2\phi = 0 \tag{21}$$

which in the absence of free surfaces can be solved in the (x, y, z) space according to, for example, the classical methods of steady state conduction heat transfer [9]. For many solutions of eq. (21) for seepage flows, and for a special transformation designed to handle the free surface ($P = $ constant) boundary condition, the reader is directed to Muskat's treatise [1] (see also Yih [3]).

THE FIRST LAW OF THERMODYNAMICS

Perhaps the simplest way to derive the main components of the energy equation for a porous medium is by considering the one-dimensional heat and fluid flow model of Fig. 10.4. The figure shows the elementary building block suggested earlier by models such as the capillary tube bundle and the capillary fissures of Fig. 10.3. The void space contained in the volume element $A\,\Delta x$ is $A_p\,\Delta x$; the volume element is defined such that the ratio $(A_p\,\Delta x)/(A\,\Delta x)$ matches the porosity ratio of the porous medium from which the elementary volume has been isolated. Therefore, in the system of Fig. 10.4, the *porosity* is defined as the ratio

$$\phi = \frac{A_p\,\Delta x}{A\,\Delta x} \tag{22}$$

To derive the energy equation for a porous medium regarded as a homogeneous medium, we start with the energy equations for the solid and fluid parts and average these equations over the elementary volume $A\,\Delta x$. For the solid part we have

$$\rho_s c_s \frac{\partial T}{\partial t} = k_s \frac{\partial^2 T}{\partial x^2} + q_s''' \tag{23}$$

where $(\rho, c, k)_s$ are the properties of the solid matrix and q_s''' is the rate of internal heat generation per unit volume of solid material. Assuming that the temperature T does not vary within the solid volume, the integral of eq. (23)

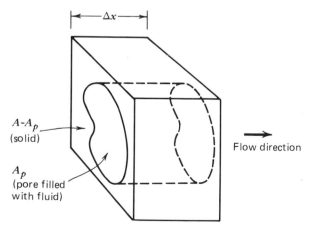

Figure 10.4 One-dimensional element for deriving the First Law of Thermodynamics for a homogeneous porous medium.

over the space occupied by the solid yields

$$\Delta x \left(A - A_p \right) \rho_s c_s \frac{\partial T}{\partial t} = \Delta x \left(A - A_p \right) k_s \frac{\partial^2 T}{\partial x^2} + \Delta x \left(A - A_p \right) q_s''' \quad (24)$$

The energy conservation equation at any point in the space occupied by fluid is [eq. (39) from Chapter 1]

$$\rho_f c_{P_f} \left(\frac{\partial T}{\partial t} + u_p \frac{\partial T}{\partial x} \right) = k_f \frac{\partial^2 T}{\partial x^2} + \mu \Phi \quad (25)$$

where $(\rho, c_P, k)_f$ are fluid properties.[†] It is assumed that the compressibility term $\beta T DP/Dt$ is negligible in eq. (39) of Chapter 1. In eqs. (23) and (25), it is also assumed that $(c, k)_s$ and $(c_P, k)_f$ are known constants. Note further that temperature T is the temperature of both parts, solid and fluid, in other words, the fluid and the porous structure are assumed to be in *local thermal equilibrium*. This assumption, although adequate for small-pore media such as geothermal reservoirs and fibrous insulation, must be relaxed in the study of nuclear reactor cores and electrical windings where the temperature difference between solid and fluid (coolant) is a very important safety parameter.

Integrating eq. (25) over the pore volume $A_p \Delta x$ yields

$$\Delta x A_p \rho_f c_{P_f} \frac{\partial T}{\partial t} + \Delta x A \rho_f c_{P_f} u \frac{\partial T}{\partial x} = \Delta x A_p k_f \frac{\partial^2 T}{\partial x^2} + \Delta x \mu \iint_{A_p} \Phi \, dA_p \quad (26)$$

It is worth noting that in the second term on the left-hand side of the above equation we made use of the definition of average velocity [eq. (2)]: in the present case, $Au = \iint_{A_p} u_p dA_p$. The last term on the right-hand side represents the internal heating associated with viscous dissipation or entropy generation. The dissipation term in eq. (26) equals the mechanical power needed to extrude the viscous fluid through the pore; this power requirement is equal to the mass flowrate times the externally maintained pressure drop, divided by the fluid density; hence

$$\Delta x \mu \iint_{A_p} \Phi \, dA_p = Au \left(-\frac{\partial P}{\partial x} + \rho_f g_x \right) \Delta x \quad (27)$$

It is easy to prove this identity in the case of known Hagen–Poiseuille flows through pores with simple cross-sections (Problem 6); however, eq. (27) holds for any unspecified pore geometry.

[†] The subscript $(\ )_f$ is used to distinguish the fluid properties only in the energy equation; thus, ρ_f is the same as the fluid density ρ used in the mass conservation equation (7) and in the Darcy law (17).

Volumetric averaging of the energy conservation statement is achieved by adding eqs. (24) and (26) side-by-side and dividing by the volume element $A \Delta x$ of the porous structure regarded as a homogeneous medium:

$$
\left[\phi \rho_f c_{P_f} + (1 - \phi) \rho_s c_s \right] \frac{\partial T}{\partial t} + \rho_f c_{P_f} u \frac{\partial T}{\partial x}
$$

$$
= \left[\phi k_f + (1 - \phi) k_s \right] \frac{\partial^2 T}{\partial x^2} + (1 - \phi) q_s''' + u \left(-\frac{\partial P}{\partial x} + \rho_f g_x \right) \quad (28)
$$

The thermal conductivity of the porous medium k emerges as a combination of the conductivities of the two constituents

$$
k = \phi k_f + (1 - \phi) k_s \quad (29)
$$

This simple expression, however, is the result of the one-dimensional model of Fig. 10.4 which amounts to a parallel conduction model. In general, k must be measured experimentally, as the thermal conductivity of the porous matrix filled with fluid.

The thermal inertia of the medium depends on the inertias of the solid and the fluid: this complication is accounted for by introducing the capacity ratio [7]

$$
\sigma = \frac{\phi \rho_f c_{P_f} + (1 - \phi) \rho_s c_s}{\rho_f c_{P_f}} \quad (30)
$$

Finally, the internal heat generation rate per unit volume of porous medium q''' decreases as the porosity increases,

$$
q''' = (1 - \phi) q_s''' \quad (31)
$$

With the new notation defined in eqs. (29)–(31), the energy equation for the *homogeneous* porous medium reads

$$
\rho_f c_{P_f} \left(\sigma \frac{\partial T}{\partial t} + u \frac{\partial T}{\partial x} \right) = k \frac{\partial^2 T}{\partial x^2} + q''' + \frac{\mu}{K} u^2 \quad (32)
$$

A similar derivation conceived for a three-dimensional flow model yields

$$
\rho_f c_{P_f} \left(\sigma \frac{\partial T}{\partial t} + \mathbf{v} \cdot \nabla T \right) = k \nabla^2 T + q''' + \frac{\mu}{K} (\mathbf{v})^2 \quad (33)
$$

where $\mathbf{v}$ is the average velocity vector (u, v, w). In situations without internal heat generation q''' and negligible viscous dissipation effect $(\mu/K)(\mathbf{v})^2$, the First Law of Thermodynamics reduces to

$$
\sigma \frac{\partial T}{\partial t} + u \frac{\partial T}{\partial x} + v \frac{\partial T}{\partial y} + w \frac{\partial T}{\partial z} = \alpha \left(\frac{\partial^2 T}{\partial x^2} + \frac{\partial^2 T}{\partial y^2} + \frac{\partial^2 T}{\partial z^2} \right) \quad (34)
$$

The thermal diffusivity of the homogeneous porous medium α is defined as the ratio

$$\alpha = \frac{k}{\rho_f c_{P_f}} \tag{35}$$

Note that k is an aggregate property of the fluid-saturated porous medium, whereas $(\rho_f c_{P_f})$ is a property of the fluid only.

In the present treatment, we rely on eqs. (7), (17), and (34) as governing equations for convection through a homogeneous porous medium. The assumptions made in deriving these equations are:

1. The medium is homogeneous, in other words, the solid material and the fluid permeating through the pores are distributed evenly throughout the porous medium.

2. The medium is isotropic, meaning that transport properties such as K and k do not depend on the direction of the experiment from which they are measured. If the medium is anisotropic, then the Darcy law (17) assumes the form

$$u = \frac{K_x}{\mu}\left(-\frac{\partial P}{\partial x} + \rho g_x\right)$$

$$v = \frac{K_y}{\mu}\left(-\frac{\partial P}{\partial y} + \rho g_y\right) \tag{36}$$

$$w = \frac{K_z}{\mu}\left(-\frac{\partial P}{\partial z} + \rho g_z\right)$$

and the energy equation (34) is replaced by

$$\sigma\frac{\partial T}{\partial t} + u\frac{\partial T}{\partial x} + v\frac{\partial T}{\partial y} + w\frac{\partial T}{\partial z} = \alpha_x\frac{\partial^2 T}{\partial x^2} + \alpha_y\frac{\partial^2 T}{\partial y^2} + \alpha_z\frac{\partial^2 T}{\partial z^2} \tag{37}$$

where $(\alpha_x, \alpha_y, \alpha_z) = (k_x, k_y, k_z)/(\rho_f c_{P_f})$.

3. At any point in the porous medium, the solid matrix is in thermal equilibrium with the fluid filling the pores.

4. The local Reynolds number based on averaged velocity and $K^{1/2}$ does not exceed $O(1)$, meaning that the Darcy law applies in its original form (17).

THE SECOND LAW OF THERMODYNAMICS

Convection processes through fluid-saturated porous media are inherently irreversible, partly due to the transfer of heat in the direction of finite temperature gradients and partly due to the highly viscous flow through the pores. The Second Law of Thermodynamics [eq. (47) in Chapter 1] may be applied to the one-dimensional flow model discussed in the preceding section (Fig. 10.4) to yield the entropy generation rate per unit volume of homogeneous porous medium

$$S_{gen}''' = \frac{k}{T^2}\left(\frac{\partial T}{\partial x}\right)^2 + \frac{\mu u^2}{KT} \geq 0 \tag{38}$$

assuming that $q''' = 0$. The analysis leading to eq. (38) is proposed as an exercise using Ref. 10 as a guide (Problem 7). For a general three-dimensional convection process, the local rate of entropy generation becomes

$$S_{gen}''' = \underbrace{\frac{k}{T^2}(\nabla T)^2}_{\geq 0} + \underbrace{\frac{\mu}{KT}(\mathbf{v})^2}_{\geq 0} \geq 0 \tag{39}$$

where, it must be stressed, T represents *absolute* temperature. Note that the viscous irreversibility term in eq. (39) may not be negligible even in cases when the viscous dissipation term can be neglected in the energy equation (33).

FORCED BOUNDARY LAYERS

The basic problem in heat convection through porous media consists of predicting the heat transfer rate between a differentially heated, solid impermeable surface and a fluid-saturated porous medium. It is important to realize that, although in nature many porous media interact thermally with one another, porous media ultimately make contact with humans and the human realm across solid surfaces. For this reason, it is appropriate to begin the study of convection through porous media by focusing on the simplest possible *wall heat transfer* problem, namely, the interaction between a solid–wall and the parallel flow permeating through the porous material confined by the wall.

Constant Wall Temperature

Relative to the geometry and two-dimensional coordinate system of Fig. 10.5, the steady state governing equations are

$$\frac{\partial u}{\partial x} + \frac{\partial v}{\partial y} = 0 \tag{40}$$

$$u = -\frac{K}{\mu}\frac{\partial P}{\partial x}; \qquad v = -\frac{K}{\mu}\frac{\partial P}{\partial y} \tag{41}$$

$$u\frac{\partial T}{\partial x} + v\frac{\partial T}{\partial y} = \alpha\frac{\partial^2 T}{\partial y^2} \tag{42}$$

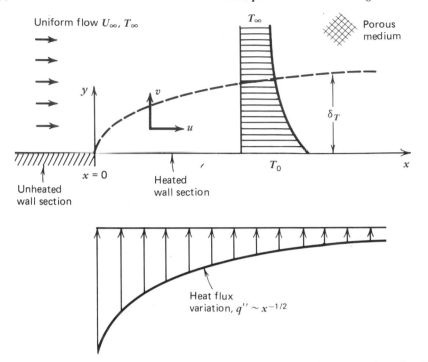

Figure 10.5 Thermal boundary layer development in a porous medium near a heated wall.

where it has been assumed that ρ is constant, that the boundary layer is slender, and that the gravity effect is negligible. Consider now the uniform parallel flow

$$u = U_\infty, \qquad v = 0, \qquad P(x) = -\frac{\mu}{K} U_\infty x + \text{constant} \qquad (43)$$

that satisfies the fluid mechanics part of the problem [eqs. (40) and (41)]. If the temperature of the fluid-saturated medium is T_∞ and if the wall temperature downstream of some point $x = 0$ is T_0, then what is the heat transfer rate between the $x > 0$ wall and the porous medium? We answer this question based on scale analysis. Let δ_T be the thickness of the slender layer of length x that effects the temperature transition from T_0 to T_∞; we refer to δ_T as the thermal boundary layer thickness, keeping in mind the analogy between the present problem and the Blasius–Pohlhausen problem of Chapter 2. However, unlike in Chapter 2, this time we do not encounter a velocity boundary layer thickness, because in $U_\infty K^{1/2}/\nu < O(1)$ porous media the wall friction effect is not felt beyond a few pore lengths $K^{1/2}$ in the y direction [11].

Writing $\Delta T = T_0 - T_\infty$, the energy equation (42) reveals the following balance between enthalpy flow in the x direction and thermal diffusion in the y

direction

$$U_\infty \frac{\Delta T}{x} \sim \alpha \frac{\Delta T}{\delta_T^2} \tag{44}$$

where it has been assumed that the thermal boundary layer region is *slender*, $\delta_T \ll x$. The relevant heat transfer implications of eq. (44) are

$$\frac{\delta_T}{x} \sim Pe_x^{-1/2} \tag{45}$$

$$Nu_x = h\frac{x}{k} \sim \frac{x}{\delta_T} \sim Pe_x^{1/2} \tag{46}$$

with the local Peclet number defined as

$$Pe_x = \frac{U_\infty x}{\alpha} \tag{47}$$

In conclusion, the thermal boundary layer thickness δ_T increases as $x^{1/2}$ downstream from the point where wall heating begins and the local heat-transfer coefficient (or the local heat flux q'') decreases as $x^{-1/2}$. Since results (45) and (46) are based on the slender thermal boundary layer assumptions, they are valid only when $Pe_x^{1/2} > O(1)$, that is, sufficiently far downstream from $x = 0$.

The *similarity* solution to the heat transfer problem defined by eqs. (40)–(42) and the boundary conditions of Fig. 10.5 is developed by introducing the similarity variable suggested by the scaling law (45),

$$\eta = \frac{y}{x} Pe_x^{1/2} \tag{48}$$

The similarity temperature profile is

$$\frac{T - T_0}{T_\infty - T_0} = \theta(\eta) \tag{49}$$

With this notation, the energy equation and its boundary conditions become

$$\theta'' + \tfrac{1}{2}\eta\theta' = 0 \tag{50}$$

$$\theta(0) = 0, \qquad \theta(\infty) = 1 \tag{51}$$

Solving eqs. (50) and (51) by separation of variables, we find

$$\theta = \text{erf}\left(\frac{\eta}{2}\right) \tag{52}$$

hence,

$$\left(\frac{d\theta}{d\eta}\right)_{\eta=0} = \pi^{-1/2} = 0.564. \tag{53}$$

Cheng [12] found the same result by integrating (50) and (51) numerically. According to this similarity solution, the local Nusselt number is

$$\mathrm{Nu}_x = \frac{q''}{T_0 - T_\infty}\frac{x}{k} = \left(\frac{d\theta}{d\eta}\right)_{\eta=0}\mathrm{Pe}_x^{1/2} = 0.564\,\mathrm{Pe}_x^{1/2} \tag{54}$$

which agrees within a factor of order $O(1)$ with the scale law (46). Averaging the heat-transfer coefficient over the heat wall length L we obtain

$$\mathrm{Nu}_{0-L} = h_{0-L}\frac{L}{k} = 1.13\,\mathrm{Pe}_L^{1/2} \tag{55}$$

Constant Wall Heat Flux

If the impermeable wall is subjected to the uniform heat flux condition $q'' = $ constant, then it is easy to show that the temperature difference $T_0(x) - T_\infty$ varies as $x^{1/2}$ downstream from $x = 0$. As is shown in Problem 10, the similarity solution in this case requires numerical integration; the final result is

$$T(x,y) - T_\infty = \frac{q''/k}{\left(-d\theta_{q''}/d\eta\right)_{\eta=0}}\left(\frac{\alpha x}{U_\infty}\right)^{1/2}\theta_{q''}(\eta) \tag{56}$$

where the $\theta_{q''}$ similarity profile is shown in Fig. 10.6. The local heat-transfer coefficient based on local temperature difference $T_0(x) - T_\infty$ and heat flux (q'') is

$$\mathrm{Nu}_x = \frac{q''}{T_0(x) - T_\infty}\frac{x}{k} = 0.886\,\mathrm{Pe}_x^{1/2} \tag{57}$$

since, according to the solution to Problem 10, $(-d\theta_{q''}/d\eta)_{\eta=0} = 0.886$. The average Nusselt number based on the wall–fluid temperature difference averaged between $x = 0$ and $x = L$ is

$$\mathrm{Nu}_{0-L} = \frac{q''L/k}{(T_0)_{0-L} - T_\infty} = 1.329\,\mathrm{Pe}_L^{1/2} \tag{58}$$

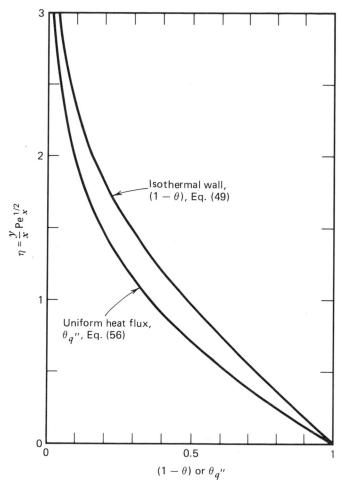

Figure 10.6 Similarity temperature profiles for forced convection boundary layer flow through a homogeneous porous medium.

Other Conditions

In the preceding two examples, the flow was uniform and parallel to the impermeable wall. A more general class of problems emerges if we consider the uniform flow (U_∞, T_∞) incident to a wedge-shaped impermeable obstacle of included angle $m\pi$. As in the classical Falkner–Skan flows [13] discussed at the end of Chapter 2, potential flow theory reveals that the velocity along each side of the wedge varies as $u = Cx^n$, where $n = m/(2 - m)$. Cheng [12] showed that in such cases a similarity solution exists for the heat transfer problem if the wall temperature varies as $T_0(x) = T_\infty + Ax^n$, with x measured downstream from the tip. The isothermal wall problem considered first in this section is the $n = 0$ special case of the class of wedge flows documented by Cheng [12]. The ($q'' = $ constant, $U_\infty = $ constant) problem also considered in this section and in Problem 10 does not appear to have been solved previously.

NATURAL BOUNDARY LAYERS

In this section we review briefly the porous medium analog of the boundary layer flows of Chapter 4, namely, the heat transfer between a vertical heated surface and a fluid-saturated semiinfinite porous reservoir (Fig. 10.7). This, the simplest boundary layer model for natural convection in porous media, was

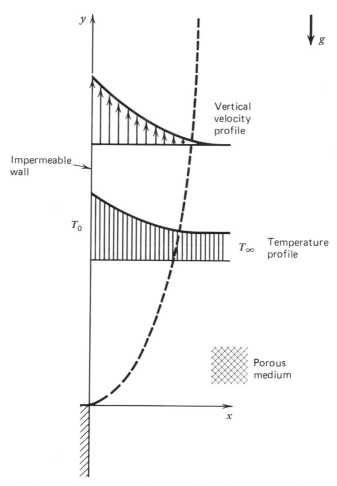

Figure 10.7   Natural convection boundary layer flow through a porous medium near a heated vertical impermeable wall.

studied first by Cheng and Minkowycz [14]. Taking the gravitational acceleration g in the negative y direction, the Darcy flow satisfies

$$u = -\frac{K}{\mu}\frac{\partial P}{\partial x}, \qquad v = -\frac{K}{\mu}\left(\frac{\partial P}{\partial y} + \rho g\right) \qquad (59)$$

or, by eliminating P

$$\frac{\partial u}{\partial y} - \frac{\partial v}{\partial x} = \frac{Kg}{\mu}\frac{\partial \rho}{\partial x} \qquad (60)$$

Defining the streamfunction ψ

$$u = \frac{\partial \psi}{\partial y}, \qquad v = -\frac{\partial \psi}{\partial x} \qquad (61)$$

so that the mass continuity equation (40) is satisfied identically, the Darcy law
(60) becomes

$$\frac{\partial^2 \psi}{\partial x^2} + \frac{\partial^2 \psi}{\partial y^2} = \frac{Kg}{\mu}\frac{\partial \rho}{\partial x} \qquad (62)$$

If the boundary layer of Fig. 10.7 is slender, then the governing balances
[force (62) and energy (34)] reduce to

$$\frac{\partial^2 \psi}{\partial x^2} = -\frac{Kg\beta}{\nu}\frac{\partial T}{\partial x} \qquad (63)$$

$$\frac{\partial \psi}{\partial y}\frac{\partial T}{\partial x} - \frac{\partial \psi}{\partial x}\frac{\partial T}{\partial y} = \alpha\frac{\partial^2 T}{\partial x^2} \qquad (64)$$

In writing eq. (63) we used already the Boussinesq approximation of Chapter
4, $\rho = \rho_0[1 - \beta(T - T_0)]$, to effect the *coupling* between the flow field $\psi(x, y)$
and the temperature field $T(x, y)$.

Constant Wall Temperature

We consider first the model sketched in Fig. 10.7 where the temperature of the
vertical impermeable wall is uniform, T_0; in this case, the approximate boundary
conditions are

$$T = T_0, \qquad \psi = 0, \quad \text{at } x = 0$$
$$(65)$$
$$T \to T_\infty, \qquad \partial\psi/\partial x \to 0, \quad \text{as } x \to \infty$$

The scale analysis of problem (63)–(65) reveals the order of magnitude of the heat transfer rate between the wall and the semiinfinite porous reservoir. From eqs. (63) and (64) we have

$$\frac{\psi}{\delta_T^2} \sim \frac{Kg\beta}{\nu}\frac{\Delta T}{\delta_T} \tag{66}$$

$$\frac{\psi\Delta T}{y\delta_T} \sim \alpha\frac{\Delta T}{\delta_T^2} \tag{67}$$

where δ_T is the boundary layer thickness (the x scale) and $\Delta T = T_0 - T_\infty$. Combining expressions (66) and (67), we conclude that

$$\frac{\delta_T}{y} \sim \mathrm{Ra}_y^{-1/2}, \qquad \psi \sim \alpha\mathrm{Ra}_y^{1/2} \tag{68}$$

$$\mathrm{Nu}_y = h\frac{y}{k} \sim \frac{y}{\delta_T} \sim \mathrm{Ra}_y^{1/2} \tag{69}$$

where the heat-transfer coefficient scales as $h \sim q''/\Delta T \sim k/\delta_T$, and Ra_y is the Darcy-modified Rayleigh number

$$\mathrm{Ra}_y = \frac{Kg\beta y\Delta T}{\alpha\nu} \tag{70}$$

Natural boundary layers in porous media have a single length scale δ_T [eq. (68)]: this feature distinguishes them from their counterparts in pure fluids (Chapter 4) that are characterized by two length scales (see Fig. 4.3 and Table 4.1).

The similarity formulation of the isothermal wall problem starts with recognizing from eq. (68) the similarity variable

$$\eta = \frac{x}{y}\mathrm{Ra}_y^{1/2} \tag{71}$$

Introducing the similarity profiles

$$\frac{\psi}{\alpha\mathrm{Ra}_y^{1/2}} = f(\eta), \qquad \frac{T - T_\infty}{T_0 - T_\infty} = \theta(\eta) \tag{72}$$

the problem statement (63)–(65) becomes

$$f'' = -\theta' \tag{73}$$

$$f\theta' = 2\theta'' \tag{74}$$

$$\theta(0) = 1, \qquad f(0) = 0$$
$$\theta(\infty) \to 0, \qquad f'(\infty) \to 0 \tag{75}$$

Numerical integration yields [14]

$$\text{Nu}_y = \frac{q''}{T_0 - T_\infty}\frac{y}{k} = 0.444\,\text{Ra}_y^{1/2} \tag{76}$$

or, averaged over a wall of height H

$$\text{Nu}_{0-H} = h_{0-H}\frac{H}{k} = 0.888\,\text{Ra}_H^{1/2} \tag{77}$$

These heat transfer results are compatible with the scaling law (69).

Constant Wall Heat Flux

If the vertical impermeable wall is characterized by a uniform heat flux q'', then the local temperature difference $T_0(y) - T_\infty$ and the boundary layer thickness δ_T must vary such that

$$q'' \sim k\frac{T_0(y) - T_\infty}{\delta_T} = \text{constant} \tag{78}$$

Combining this with the earlier scaling results [eqs. (66)–(68)], we conclude that

$$\frac{\delta_T}{y} \sim \text{Ra}_{*\,y}^{-1/3} \tag{79}$$

where $\text{Ra}_{*\,y}$ is the Darcy-modified Rayleigh number based on heat flux,

$$\text{Ra}_{*\,y} = \frac{Kg\beta y^2 q''}{\alpha\nu k} \tag{80}$$

The local heat transfer rate must therefore scale as

$$\text{Nu}_y = \frac{q''}{T_0(y) - T_\infty}\frac{y}{k} \sim \text{Ra}_{*\,y}^{1/3} \tag{81}$$

The numerical solution to the similarity for simulation of this problem [14] confirms this result (see Problem 14)

$$\text{Nu}_y = \frac{q''}{T_0(y) - T_\infty}\frac{y}{k} = 0.772\,\text{Ra}_{*\,y}^{1/3} \tag{82}$$

$$\text{Nu}_{0-H} = \frac{q''}{(T_0)_{0-H} - T_\infty}\frac{H}{k} = 1.03\,\text{Ra}_{*\,H}^{1/3} \tag{83}$$

The Effect of Wall Inclination

The results presented above for a vertical wall apply, subject to a slight modification, to the more general case where the wall is inclined relative to the vertical direction. In such a case the gravitational acceleration acts in both x and y directions; hence,

$$u = -\frac{K}{\mu}\left(\frac{\partial P}{\partial x} - \rho g_x\right), \qquad v = -\frac{K}{\mu}\left(\frac{\partial P}{\partial y} - \rho g_y\right) \tag{84}$$

where g_x, g_y are the respective components of gravitational acceleration. Introducing the Boussinesq approximation and eliminating P between eqs. (84) yields

$$\frac{\partial^2 \psi}{\partial x^2} + \frac{\partial^2 \psi}{\partial y^2} = \frac{K\beta}{\nu}\left(g_y \frac{\partial T}{\partial x} - g_x \frac{\partial T}{\partial y}\right) \tag{85}$$

which is a more general version of eq. (63).

In the boundary layer regime, we have $x \sim \delta_T$ and $y \gg \delta_T$, therefore, the boundary layer approximation of eq. (85) is

$$\frac{\partial^2 \psi}{\partial x^2} = \frac{K g_y \beta}{\nu} \frac{\partial T}{\partial x} \tag{86}$$

This approximation is valid as long as g_x is not order-of-magnitude greater than g_y. Note that the above equation is the same as eq. (63) employed earlier. In conclusion, if the impermeable wall makes an angle γ with the vertical direction, then the results developed for vertical walls are applicable as long as (g) is replaced by $(g \cos \gamma)$ in the Rayleigh number calculations.

Conjugate Boundary Layers

Natural boundary layers are rarely driven by surfaces with *known* temperature or heat flux, as in Fig. 10.7; more often they are the consequence of the heat transfer interaction between two fluid systems separated by a vertical (or inclined) impermeable wall. The wall may face fluid-saturated porous media on both sides (Fig. 10.8a) or on only one side (Fig. 10.8b). In either case, if a temperature difference exists between the two fluid systems, conjugate boundary layers form on both sides of the wall.

The problem of conjugate boundary layers on both sides of a solid wall inserted in a fluid-saturated porous medium was treated in Ref. 15. The problem was solved analytically based on the Oseen-linearization method described in Chapter 5: it was found that the coefficient in the (Nusselt number) ~ (Rayleigh number)$^{1/2}$ proportionality decreases steadily as the wall thickness parameter ω increases (Fig. 10.9). Parameter ω is defined as

$$\omega = \frac{W}{H} \frac{k}{k_w} \mathrm{Ra}_H^{1/2} \tag{87}$$

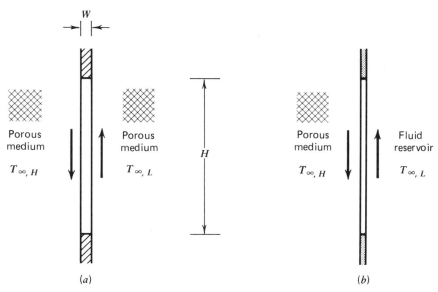

Figure 10.8 Conjugate boundary layers on the two sides of a vertical impermeable partition separating: (*a*) two porous media; (*b*) a porous medium and a fluid reservoir.

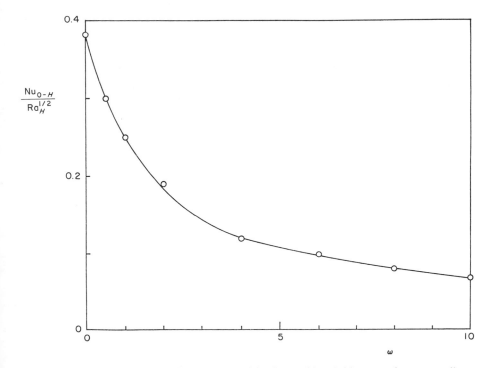

Figure 10.9 Heat transfer through a vertical partition inserted in a fluid-saturated porous medium [15].

where W and H are the dimensions of the wall cross-section and k_w is the thermal conductivity of the wall material. Parameter Ra_H is the Rayleigh number based on wall height (H) and the temperature difference between the two systems ($\Delta T = T_{\infty,H} - T_{\infty,L}$).

$$\mathrm{Ra}_H = \frac{Kg\beta H}{\alpha v}(T_{\infty,H} - T_{\infty,L}) \tag{88}$$

The Nusselt number Nu_{0-H} of Fig. 10.9 is based on wall-averaged heat flux and overall temperature difference.

$$\mathrm{Nu}_{0-H} = \frac{q''_{0-H}}{T_{\infty,H} - T_{\infty,L}} \frac{H}{k} \tag{89}$$

In the limit of negligible wall thermal resistance ($\omega = 0$), the heat transfer rate through the impermeable wall is [15]

$$\mathrm{Nu}_{0-H} = 0.383\,\mathrm{Ra}_H^{1/2} \tag{90}$$

If one side of the wall faces a fluid reservoir (as the surface of a double wall filled with fiberglass insulation in a house), then the conjugate boundary layer problem consists of the interaction of a porous medium layer with a wall jet of the type studied in Chapter 4. This problem was solved in Ref. 16; the heat transfer results are reproduced in Fig. 10.10. The analysis of Ref. 16 reveals the

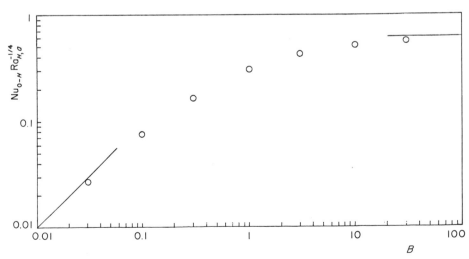

Figure 10.10 Heat transfer through the interface between a porous medium and a fluid reservoir [16].

important role played by a new dimensionless group

$$B = \frac{k \, \mathrm{Ra}_H^{1/2}}{k_a \mathrm{Ra}_{H,a}^{1/4}},$$ (91)

in determining whether the conjugate problem is dominated by porous medium convection or pure fluid convection. In the definition of the B number [eq. (91)], k_a, and $\mathrm{Ra}_{H,a}$ represent the fluid conductivity and the Rayleigh number on the pure fluid side (the "air" side; the right side of Fig. 10.8b). Both Rayleigh numbers, Ra_H and $\mathrm{Ra}_{H,a}$, and the H-averaged Nusselt number Nu_{0-H} are based on the overall temperature difference $T_{\infty,H} - T_{\infty,L}$ (Fig. 10.8b).

The B number effect on the fluid mechanics of this conjugate boundary layer problem is illustrated in Fig. 10.11, where the vertical velocities, temperatures, and positions away from the impermeable interface have been nondimensionalized in accordance with the appropriate scaling for each boundary layer. As the B number increases, the heat transfer problem is dominated by the flow on the pure fluid (a) side: this effect is due to the fact that B is the ratio of the porous-side thermal conductance divided by the open-side (a) thermal conductance.

Thermal Stratification

If the fluid-saturated porous medium of Fig. 10.7 is not truly infinite in the x and y directions, then the discharge of the boundary layer into the medium leads, in time, to the thermal stratification of the $x > \delta_T$ region. It is shown in the next chapter that the stable thermal stratification is a characteristic feature of confined porous media in the so-called core region situated sufficiently far from the surrounding boundary layers. According to Fig. 10.12, if the bottom (or *starting*) temperature difference $T_0 - T_{\infty,0}$ remains fixed, then as the positive temperature gradient $\gamma = dT_\infty/dy$ increases, the average temperature difference between the wall and the porous medium decreases. Therefore, we should expect a steady decrease in the total heat transfer rate as γ increases.

The effect of stable thermal stratification on natural boundary layers in porous media was studied in the context of two conjugate boundary layers on both sides of an impermeable partition inserted vertically into a porous medium [15] (Fig. 10.8a). We focus on this case later in this subsection. The much simpler problem of a vertical or slightly inclined impermeable wall facing a linearly stratified porous medium does not appear to have been considered so far in the literature. A brief, integral-type solution to this problem is presented below.

In accordance with the previous notation and the new temperature conditions of Fig. 10.12, the Darcy law (63) integrated once requires

$$T = \frac{\nu}{Kg\beta} v + \text{function}(y)$$ (92)

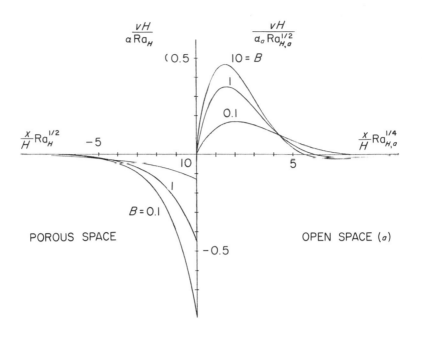

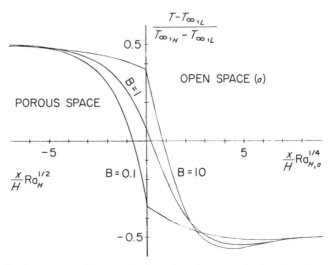

Figure 10.11 Conjugate natural convection boundary layers at the interface between a porous medium (left) and a fluid reservoir (right) [16]: top, velocity profiles; bottom, temperature profiles.

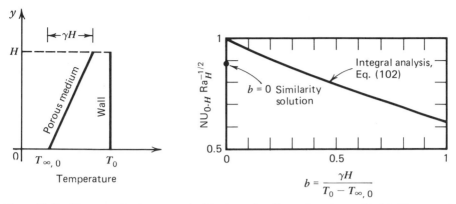

Figure 10.12 Heat transfer from a vertical isothermal wall to a linearly stratified fluid-saturated porous medium.

Therefore, unlike in the Karman–Pohlhausen integral procedure employed in Chapter 2, in the present case we have the freedom to choose only *one* profile shape (e.g., v) as the second profile follows immediately from eq. (92). Let the vertical velocity profile be

$$v = v_0 \exp\left(-\frac{x}{\delta_T}\right) \tag{93}$$

where both v_0 and δ_T are unknown functions of altitude. Then, using the Darcy law (92) and the temperature boundary conditions

$$T(0, y) = T_0, \qquad T(\infty, y) = T_{\infty,0} + \gamma y \tag{94}$$

the corresponding temperature profile is

$$T(x, y) = (T_0 - T_{\infty,0} - \gamma y)\exp\left(-\frac{x}{\delta_T}\right) + T_{\infty,0} + \gamma y \tag{95}$$

with the maximum (wall) vertical velocity

$$v_0 = \frac{Kg\beta}{\nu}(T_0 - T_{\infty,0} - \gamma y) \tag{96}$$

The integral form of the boundary layer energy equation is obtained by integrating eq. (64) across the boundary layer from $x = 0$ to $x \to \infty$. Thus, we obtain

$$(u)_{x\to\infty}(T)_{x\to\infty} + \frac{d}{dy}\int_0^\infty vT\,dx = -\alpha\left(\frac{\partial T}{\partial x}\right)_{x=0} \tag{97}$$

where $(T)_{x \to \infty} = T_{\infty,0} + \gamma y$ and, from the mass conservation equation

$$(u)_{x \to \infty} = -\frac{d}{dy} \int_0^\infty v \, dx \tag{98}$$

Substituting the assumed v and T profiles into the energy integral equation (97) yields

$$\frac{d\delta_*}{dy_*} = \frac{2}{\delta_*(1 - by_*)} \tag{99}$$

with the following dimensionless notation

$$b = \frac{\gamma H}{T_0 - T_{\infty,0}}, \quad \text{the stratification parameter}$$

$$y_* = \frac{y}{H} \tag{100}$$

$$\delta_* = \frac{\delta_T}{H} \left[\frac{g\beta H(T_0 - T_{\infty,0})}{\alpha v} \right]^{1/2}$$

Integrating eq. (99) from $\delta_*(0) = 0$, we obtain

$$\delta_*(y_*) = \left[-\frac{4}{b} \ln(1 - by_*) \right]^{1/2} \tag{101}$$

which reduces to the expected result $(\delta_* \sim y_*^{1/2})$ as b approaches zero.

The total heat transfer rate can then be calculated in the usual manner by averaging the heat flux over the wall height H; the result can be arranged as

$$\frac{\mathrm{Nu}_{0-H}}{\mathrm{Ra}_H^{1/2}} = \int_0^1 \frac{(1 - by_*) \, dy_*}{\left[-\frac{4}{b} \ln(1 - by_*) \right]^{1/2}} \tag{102}$$

where both Nu_{0-H} and Ra_H are based on the maximum (*starting*) temperature difference

$$\mathrm{Nu}_{0-H} = \frac{q''_{0-H} H}{k(T_0 - T_{\infty,0})}; \quad \mathrm{Ra}_H = \frac{Kg\beta H}{\alpha v}(T_0 - T_{\infty,0}) \tag{103}$$

Equation (102) is plotted in Fig. 10.12: as expected, the coefficient in the $\mathrm{Nu}_{0-H} \sim \mathrm{Ra}_H^{1/2}$ proportionality decreases monotonically as b increases.

The accuracy of the above integral solution can be assessed by comparing its $b = 0$ limit

$$\frac{\mathrm{Nu}_{0-H}}{\mathrm{Ra}_H^{1/2}} = 1, \quad (b = 0) \tag{104}$$

with the similarity average Nusselt number for an isothermal wall adjacent to an isothermal porous medium [eq. (77)]. Therefore, in the $b = 0$ limit, the discrepancy between the two solutions is 12.6 percent.

The effect of stable stratification on both sides of an impermeable wall was studied analytically in [15] based on the Oseen-linearization method. The heat transfer result is reproduced in Fig. 10.13 along with the postulated temperature conditions sufficiently far from the vertical wall sketched in Fig. 10.8a. The stratification parameter b', defined in Fig. 10.13, is the same for both porous sides. The average Nusselt number Nu_{0-H} and the Rayleigh number are this time based on the temperature difference ΔT between the two fluid-saturated porous media (note that ΔT is a constant independent of y).

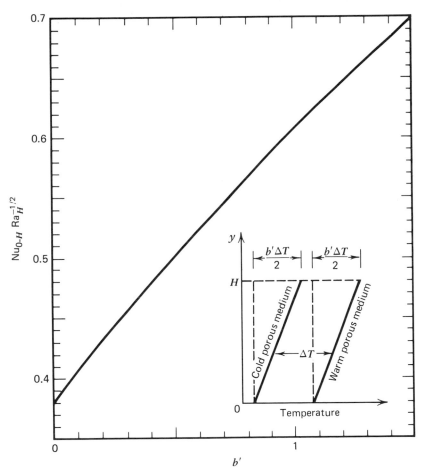

Figure 10.13 The effect of stable stratification on natural convection heat transfer through a vertical impermeable wall imbedded in a porous medium [15].

CONCENTRATED HEAT SOURCES

The heat transfer from small-size heat sources buried inside large-size conducting media is already a relatively large and important chapter in conduction heat transfer [17]. If the conducting medium is saturated with fluid, as, for example, the ground beneath us, then the heat released by concentrated sources migrates in accord with the principles of convection through porous media. There are obvious applications of this class of convection problems, from heat transfer calculations for the cooling of underground electric cables to the environmental impact of underground explosions and buried nuclear (heat-generating) waste.

There are many heat source and porous media configurations that can occur in real life: the treatment of all these possibilities is better suited for individual research projects, as is well beyond the classroom tutorial level adopted for the present chapter. Useful reviews of this growing chapter of porous media convection are already available [7, 18]. Below, we focus on what is perhaps the most basic configuration, namely, the point heat source buried in an infinite porous medium. We consider two distinct regimes of the phenomenon, the *low Rayleigh number* and the *high Rayleigh number*, separately and in the manner in which each were historically tackled for the first time (in Refs. 19 and 20, respectively).

The Low Rayleigh Number Regime

Consider the spherical system of coordinates $(r, \phi, \theta$, Fig. 1.1) attached to a point heat source of strength q [W]. The gravitational acceleration g acts parallel to the direction $\phi = 0$. Noting the θ-symmetry of the resulting temperature and flow field, the equations for mass conservation, Darcy flow, and transient energy conservation are, in order [19]

$$\frac{\partial}{\partial r}\left(r^2 v_r \sin\phi\right) + \frac{\partial}{\partial \phi}\left(r v_\phi \sin\phi\right) = 0 \tag{105}$$

$$v_r = -\frac{K}{\mu}\left(\frac{\partial P}{\partial r} + \rho g \cos\phi\right), \qquad v_\phi = -\frac{K}{\mu}\left(\frac{1}{r}\frac{\partial P}{\partial \phi} - \rho g \sin\phi\right) \tag{106}$$

$$\frac{1}{\alpha}\left(\sigma\frac{\partial T}{\partial t} + v_r\frac{\partial T}{\partial r} + \frac{v_\phi}{r}\frac{\partial T}{\partial \phi}\right) = \frac{1}{r^2}\frac{\partial}{\partial r}\left(r^2\frac{\partial T}{\partial r}\right) + \frac{1}{r^2\sin\phi}\frac{\partial}{\partial \phi}\left(\sin\phi\frac{\partial T}{\partial \phi}\right) \tag{107}$$

Introducing the streamfunction $\psi(r, \phi)$

$$v_r = \frac{1}{r^2\sin\phi}\frac{\partial \psi}{\partial \phi}, \qquad v_\phi = -\frac{1}{r\sin\phi}\frac{\partial \psi}{\partial r} \tag{108}$$

and eliminating the pressure between the two Darcy flow equations (106) yields

$$\frac{1}{R^2}\frac{\partial}{\partial\phi}\left(\frac{1}{\sin\phi}\frac{\partial\Psi}{\partial\phi}\right) + \frac{1}{\sin\phi}\frac{\partial^2\Psi}{\partial R^2} = \text{Ra}\left(\cos\phi\frac{\partial\theta}{\partial\phi} + R\sin\phi\frac{\partial\theta}{\partial R}\right) \quad (109)$$

$$\frac{\partial\theta}{\partial\tau} + \frac{1}{R^2\sin\phi}\left(\frac{\partial\Psi}{\partial\phi}\frac{\partial\theta}{\partial R} - \frac{\partial\Psi}{\partial R}\frac{\partial\theta}{\partial\phi}\right)$$

$$= \frac{1}{R^2}\frac{\partial}{\partial R}\left(R^2\frac{\partial\theta}{\partial R}\right) + \frac{1}{R^2\sin\phi}\frac{\partial\theta}{\partial\phi}\left(\sin\phi\frac{\partial\theta}{\partial\phi}\right) \quad (110)$$

with the following dimensionless notation

$$\tau = \frac{\alpha t}{\sigma K}, \qquad R = \frac{r}{K^{1/2}}$$

$$\theta = (T - T_\infty)\frac{kK^{1/2}}{q}, \qquad \Psi = \frac{\psi}{\alpha K^{1/2}} \quad (111)$$

$$\text{Ra} = \frac{Kg\beta q}{\alpha\nu k}, \quad \text{Rayleigh number based on the point source strength}$$

The initial condition for this transient convection problem is

$$v_r = v_\phi = 0 \quad \text{and} \quad T = T_\infty \quad \text{at } t = 0 \quad (112)$$

meaning *no flow* and *isothermal medium* before the point source is turned on. The appropriate boundary conditions are

$$v_r, v_\phi \to 0, \qquad T \to T_\infty \quad \text{as } r \to \infty$$

$$\partial v_r/\partial\phi, v_\phi, \partial T/\partial\phi = 0 \quad \text{at } \phi = 0, \pi \quad (113)$$

with the observation that v_r, v_ϕ, and T blow up as $1/r$ as $r \to 0$; this behavior is revealed, for example, by the energy balance over the spherical control volume containing the origin

$$\lim_{r\to 0}\left[-k(4\pi r)\frac{\partial T}{\partial r}\right] = q = \text{constant} \quad (114)$$

The nondimensional formulation of this problem is based on *pure conduction* scaling, that is, the balance between diffusion and thermal inertia in the energy equation. This scaling is valid for sufficiently small values of Ra: note that setting Ra $= 0$ in eq. (109) leads to the pure conduction problem (i.e., $\Psi = 0$). We expand the unknowns θ, Ψ as a power series of small Ra

$$(\theta, \Psi) = (\theta, \Psi)_0 + \text{Ra}(\theta, \Psi)_1 + \text{Ra}^2(\theta, \Psi)_2 + \cdots \quad (115)$$

We then substitute these expressions into the governing equations (109) and

(110) and collect the terms containing the same power of Ra. Setting each group equal to zero leads to a sequence of solvable problems; the results for the first two such problems are [19]

$$\theta_0 = \frac{1}{4\pi R} \text{erfc}\left(\frac{R}{2\tau^{1/2}}\right)$$

$$\Psi_0 = 0 \tag{116}$$

and

$$\Psi_1 = \frac{1}{8\pi}\tau^{1/2}\sin^2\phi\left(2\eta\,\text{erfc}\,\eta + \frac{1}{\eta}\text{erf}\,\eta - \frac{2}{\pi^{1/2}}e^{-\eta^2}\right)$$

$$\tag{117}$$

$$\theta_1 = \frac{\cos\phi}{64\pi^2\tau^{1/2}}\left(\frac{1}{\eta} - \frac{4}{3\pi^{1/2}} + \frac{6}{5\pi^{1/2}}\eta^2 - \frac{16}{45\pi}\eta^3 - \frac{152}{315\pi^{1/2}}\eta^4 + \cdots\right)$$

where η is the conduction similarity variable $\eta = R/(2\tau^{1/2})$. The power series for θ_1 was calculated up to the η^9 term in Ref. 19.

The above two-term approximate solution for the transient state [eqs. (116) and (117)] reveals the first convection correction to the conduction dominated phenomenon. Figure 10.14 shows that as soon as the heat source is turned on,

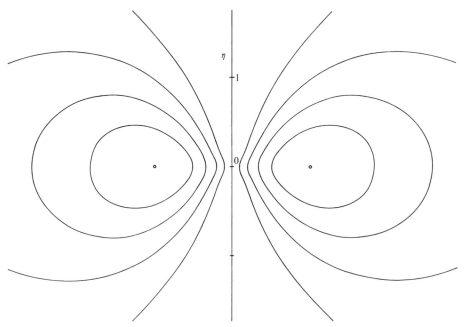

Figure 10.14 Transient natural convection flow pattern around a point heat source [19]; the lines correspond to equal increments of $\psi_1\tau^{-1/2}$.

a vortex ring forms around the source. The radius of the ring on Fig. 10.14 is $\eta = 0.881$, in other words, the actual radius grows in time as $1.762\,(\alpha t/\sigma)^{1/2}$.

A perturbation solution valid in a relatively wider Ra range is possible for the *steady state* [19]. Solving the preceding problem without the t dependence yields

$$\Psi = \frac{R}{8\pi}\left[\sin^2\!\phi\,\mathrm{Ra} + \frac{1}{24\pi}\sin\phi\sin 2\phi\,\mathrm{Ra}^2 - \frac{5}{18432\pi^2}(8\cos^4\!\phi - 3)\mathrm{Ra}^3 + \cdots\right]$$

$$\theta = \frac{1}{4\pi R}\left[1 + \frac{1}{8\pi}\cos\phi\,\mathrm{Ra} + \frac{5}{768\pi^2}\cos 2\phi\,\mathrm{Ra}^2\right.$$

$$\left. + \frac{1}{55296\pi^3}\cos\phi\,(47\cos^2\!\phi - 30)\mathrm{Ra}^3 + \cdots\right] \qquad (118)$$

This solution is valid for source strength Rayleigh numbers Ra of the order of 20 or less. The temperature field is illustrated in Fig. 10.15 by the isothermal

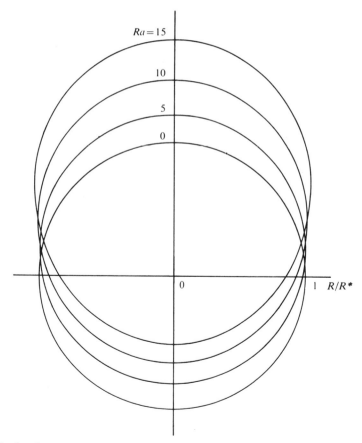

Figure 10.15 Steady temperature distribution around a point heat source [19]; the lines represent the $(4\pi R^*)\theta = 1$ isotherm, for increasing values of Ra.

surface $\theta = 1/(4\pi R^*)$, where $r^* = R^* K^{1/2}$ is a given radial distance. The figure shows that the warm region, originally spherical around the point source, shifts upward and becomes elongated like the flame of a candle as Ra increases. It is conceivable that if Ra is sufficiently large, the $\theta = $ constant region becomes slender enough to justify an analysis of the boundary layer type. We investigate this possibility in the subsection on high-Rayleigh number convection.

The low-Ra convection analysis of the point heat source presented above was applied recently to the study of a horizontal line source buried in a porous medium [21]. This study shows that in cross-section the transient flow looks very much like Fig. 10.14, in other words, the line source is sandwiched between two parallel line vortices whose centers move outward as $t^{1/2}$. Unlike the case of a point heat source, for which a steady state solution exists ([19], also eqs. (118) here), the horizontal line source problem does not admit a steady state flow field [21].

The High-Rayleigh Number Regime

When the heated region above the point heat source is slender enough, it is permissible to study its development based on boundary layer theory. This study of the point source was alluded to in one paragraph by Wooding [20] who, in the same paper, derived explicitly the corresponding form of the two-dimensional flow above a horizontal line source. In what follows we carry out in some detail the boundary layer analysis for the plume above a point heat source in a porous medium, as this analysis is not available in the literature.

Imagine a slender plume above the point heat source of Fig. 10.16, and attach a cylindrical system of coordinates (r, z) and (v_r, v_z) to the plume so that the z axis passes through the heat source and points against gravity. The governing equations for this θ-symmetric convection problem are

$$\frac{\partial v_r}{\partial r} + \frac{v_r}{r} + \frac{\partial v_z}{\partial z} = 0 \tag{119}$$

$$v_r = -\frac{K}{\mu}\frac{\partial P}{\partial r}, \qquad v_z = -\frac{K}{\mu}\left(\frac{\partial P}{\partial z} + \rho g\right) \tag{120}$$

$$v_r\frac{\partial T}{\partial r} + v_z\frac{\partial T}{\partial z} = \alpha\left[\frac{1}{r}\frac{\partial}{\partial r}\left(r\frac{\partial T}{\partial r}\right) + \frac{\partial^2 T}{\partial z^2}\right] \tag{121}$$

In the slender plume (i.e., boundary layer) regime, $r \sim \delta_T$ and $z \sim H \gg \delta_T$; therefore, after eliminating the pressure terms, eqs. (120) and (121) reduce to

$$\frac{\partial v_z}{\partial r} = \frac{Kg\beta}{\nu}\frac{\partial T}{\partial r} \tag{122}$$

$$v_r\frac{\partial T}{\partial r} + v_z\frac{\partial T}{\partial z} = \frac{\alpha}{r}\frac{\partial}{\partial r}\left(r\frac{\partial T}{\partial r}\right) \tag{123}$$

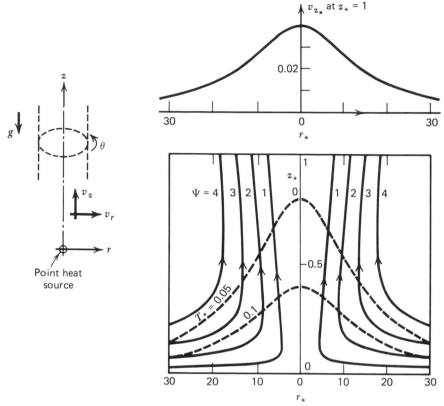

Figure 10.16 High Rayleigh number convection above a point heat source in a fluid-saturated porous medium.

The scale analysis of these two equations dictates

$$v_z \sim \frac{Kg\beta}{\nu} \Delta T \quad \text{and} \quad v_z \sim \frac{\alpha H}{\delta_T^2} \tag{124}$$

where ΔT is the plume-ambient temperature difference, $T - T_\infty$ = function (z). A third scaling law follows from the fact that energy released by the point source q [W] is convected upward through the plume flow; hence,

$$q \sim \rho v_z \delta_T^2 c_P \Delta T \tag{125}$$

Combining relations (124) and (125) yields the wanted plume scales

$$v_z \sim \frac{\alpha}{H} \text{Ra}, \qquad \delta_T \sim H\,\text{Ra}^{-1/2}, \qquad \Delta T \sim \frac{q}{kH} \tag{126}$$

where Ra is the source strength Rayleigh number already defined in eqs. (111).

The above scale analysis suggests the following dimensionless formulation of the problem:

Dimensionless variable

$$z_* = \frac{z}{H}, \qquad r_* = \frac{r}{H} Ra^{1/2}$$

$$v_{z_*} = \frac{v_z}{(\alpha/H)Ra}, \qquad v_{r_*} = \frac{v_r}{(\alpha/H)Ra^{1/2}} \qquad (127)$$

$$T_* = \frac{(T - T_\infty)}{q/kH}$$

Equations

$$\frac{\partial v_{r_*}}{\partial r_*} + \frac{v_{r_*}}{r_*} + \frac{\partial v_{z_*}}{\partial z_*} = 0 \qquad (128a)$$

$$\frac{\partial v_{z_*}}{\partial r_*} = \frac{\partial T_*}{\partial r_*} \qquad (128b)$$

$$v_{r_*}\frac{\partial T_*}{\partial r_*} + v_{z_*}\frac{\partial T_*}{\partial z_*} = \frac{\partial^2 T_*}{\partial z_*^2} \qquad (128c)$$

Boundary conditions

$$v_{r_*} = 0, \qquad \frac{\partial T_*}{\partial r_*} = 0 \quad \text{at} \quad r_* = 0$$

$$v_{z_*} \rightarrow 0, \qquad T_* \rightarrow 0 \quad \text{as} \quad r_* \rightarrow \infty \qquad (129)$$

Integrating eq. (128b) subject to the $r_* \rightarrow \infty$ boundary conditions (129), we find that

$$v_{z_*} = T_* \qquad (130)$$

Replacing T_* by v_{z_*} in eqs. (128c) and (129), we obtain a problem identical to the boundary layer treatment of a laminar round jet discharging into a constant-pressure reservoir. The round jet problem was solved by Schlichting [22]: applied to the present problem, his method consists of introducing the similarity variable η and streamfunction profile $F(\eta)$ such that

$$\eta = \frac{r_*}{z_*}, \qquad \psi = z_* F(\eta) \qquad (131)$$

where ψ is the streamfunction

$$v_{r_*} = -\frac{1}{r_*}\frac{\partial \psi}{\partial z_*}, \qquad v_{z_*} = \frac{1}{r_*}\frac{\partial \psi}{\partial r_*} \tag{132}$$

Substituting eqs. (131) and (132) into the energy equation (128c) yields, after some algebra,

$$\frac{d}{d\eta}\left(F'' - \frac{F'}{\eta} + \frac{FF'}{\eta}\right) = 0 \tag{133}$$

Integrating this result once and invoking the $\eta \to \infty$ condition we obtain the second-order equation

$$FF' = F' - \eta F'' \tag{134}$$

Schlichting showed that the solution satisfying both eq. (134) and boundary conditions (129) is

$$F = \frac{(C\eta)^2}{1 + (C\eta/2)^2} \tag{135}$$

where constant C is determined (in our case) from the energy conservation integral

$$q = \int_0^{2\pi}\int_0^{\infty} \rho c_p v_z (T - T_\infty) r\,dr\,d\theta \tag{136}$$

We find that $C = 1/4(\pi)^{1/2} = 0.141$, therefore, the final solution is

$$T_* = v_{z_*} = \frac{2C^2}{z_*}\frac{1}{1 + (C\eta/2)^2}$$

$$v_{r_*} = \frac{C}{z_*}\frac{C\eta - \frac{1}{4}(C\eta)^3}{\left[1 + (C\eta/2)^2\right]^2} \tag{137}$$

$$\psi = 4z_*\ln\left[1 + \left(\frac{C\eta}{2}\right)^2\right]$$

Figure 10.16 shows the traces of the $\psi, T_* = $ constant surfaces left in any $\theta = $ constant cut through the point source. Note that the $T_* = $ constant trace has the same shape as the vertical velocity profile v_{z_*}, in accordance with the first of eqs. (137). Note further that the flow and temperature field is presented in dimensionless form (in the r_*, z_* domain): whether or not the actual flow field is *slender* is determined by the slenderness condition $\delta_T/H < O(1)$ which, from eq. (126), translates into $Ra^{1/2} > O(1)$. Therefore, the low-Ra steady state solution of Ref. 19 and Fig. 10.15 coupled with the high-Ra solution presented as eqs. (137) and Fig. 10.16 cover fairly well the entire Ra range for convection around a point heat source in a saturated porous medium.

SYMBOLS

A	cross-sectional area
b	empirical constant [eq. (14)]
b	stratification parameter [eq. (100)]
b'	stratification parameter (Fig. 10.13)
B	dimensionless group [eq. (91)]
c	specific heat of incompressible substance
c_P	specific heat at constant pressure
f	friction factor
f, F	similarity streamfunction profiles [eqs. (72) and (131)]
g	gravitational acceleration
H	wall height
k	thermal conductivity
K	permeability
L	wall length
$\dot{m}$	mass flowrate
Nu_x	local Nusselt number [eq. (46)]
P	pressure
Pe_x	Peclet number ($U_\infty x/\alpha$)
q	strength of point heat source [W]
q''	heat flux [W/m^2]
q'''	volumetric heat generation rate [W/m^3]
r, ϕ, θ	spherical coordinates (Fig. 1.1)
R	dimensionless radial position [eq. (111)]
Ra	Rayleigh number based on q [eq. (111)]
Ra_y	Darcy-modified Rayleigh number [eq. (70)]
Ra_{*y}	Darcy-modified Rayleigh number based on q'' [eq. (80)]
Re	pore Reynolds number [eq. (10)]
S_{gen}'''	volumetric entropy generation rate
t	time
T	temperature
T_0	wall temperature
T_∞	temperature far from the wall
ΔT	temperature difference
u, v	volume-averaged velocity components
U_∞	uniform flow velocity
W	width

x, y cartesian coordinates
z vertical position (Fig. 10.16)
α thermal diffusivity of fluid-saturated porous medium [eq. (35)]
β coefficient of thermal expansion
γ vertical temperature gradient
δ_T thermal boundary layer thickness
η similarity variable
θ similarity temperature profile
μ viscosity
ν kinematic viscosity
ρ density
σ capacity ratio [eq. (30)]
τ dimensionless time [eq. (111)]
ϕ function [eq. (19)]
ϕ porosity [eq. (22)]
Φ viscous dissipation function [eq. (25)]
ψ streamfunction
Ψ dimensionless streamfunction [eq. (111)]
ω wall parameter [eq. (87)]
$(\ )_f$ property of fluid component
$(\ )_s$ property of solid component
$(\ )_p$ property measured inside the pore
$(\ )_{0-H}$ averaged from $y = 0$ to $y = H$
$(\ )_{0-L}$ averaged from $x = 0$ to $x = L$
$(\ )_H$ high temperature side
$(\ )_L$ low temperature side

REFERENCES

1. M. Muskat, *The Flow of Homogeneous Fluids through Porous Media*, McGraw-Hill, New York, 1937; 2nd printing by Edwards, Ann Arbor, Michigan, 1946.

2. J. Bear, *Dynamics of Fluids in Porous Media*, American Elsevier, New York, 1972.

3. C. S. Yih, *Fluid Mechanics*, McGraw-Hill, New York, 1969, pp. 379–383.

4. H. Darcy, *Les Fontaines Publiques de la Ville de Dijon*, Victor Dalmont, Paris, 1856.

5. J. C. Ward, Turbulent flow in porous media, *J. Hydraul. Div. ASCE*, Vol. 90, 1964, No. HY5, pp. 1–12.

6. P. H. Forschheimer, *Z. Ver. Dtsch. Ing.*, Vol. 45, 1901, pp. 1782–1788.

7. P. Cheng, Heat transfer in geothermal systems, *Adv. Heat Transfer*, Vol. 14, 1978, pp. 1–105.

8. A. Bejan and D. Poulikakos, The nonDarcy regime for vertical boundary layer natural convection in a porous medium, *Int. J. Heat Mass Transfer*, Vol. 27, 1984, pp. 717–722.

9. V. S. Arpaci, *Conduction Heat Transfer*, Addison-Wesley, Reading, MA, 1966.

10. A. Bejan, *Entropy Generation Through Heat and Fluid Flow*, Wiley, New York, 1982, Chapter 5.

11. G. Dagan, The generalization of Darcy's Law for nonuniform flows, *Water Resour. Res.*, Vol. 15, 1979, pp. 1–7.

12. P. Cheng, Combined free and forced convection flow about inclined surfaces in porous media, *Int. J. Heat Mass Transfer*, Vol. 20, 1977, pp. 807–814.

13. V. M. Falkner and S. W. Skan, Some approximate solutions of the boundary layer equations, *Philos. Mag.*, Vol. 12, 1931, p. 865.

14. P. Cheng and W. J. Minkowycz, Free convection about a vertical flat plate embedded in a saturated porous medium with application to heat transfer from a dike, *J. Geophys. Res.*, Vol. 82, 1977, pp. 2040–2044.

15. A. Bejan and R. Anderson, Heat transfer across a vertical impermeable partition imbedded in a porous medium, *Int. J. Heat Mass Transfer*, Vol. 24, 1981, pp. 1237–1245.

16. A. Bejan and R. Anderson, Natural convection at the interface between a vertical porous layer and an open space, *J. Heat Transfer*, Vol. 105, 1983, pp. 124–129.

17. H. S. Carlsaw and J. C. Jaeger, *Conduction of Heat in Solids*, Oxford University Press, Oxford, England, 1959, Chapter X.

18. C. E. Hickox and H. A. Watts, Steady thermal convection from a concentrated source in a porous medium, *J. Heat Transfer*, Vol. 102, 1980, pp. 248–253.

19. A. Bejan, Natural convection in an infinite porous medium with a concentrated heat source, *J. Fluid Mech.*, Vol. 89, 1978, pp. 97–107.

20. R. A. Wooding, Convection in a saturated porous medium at large Rayleigh or Péclet number, *J. Fluid Mech.*, Vol. 15, 1963, pp. 527–544.

21. D. A. Nield and S. P. White, Natural convection in an infinite porous medium produced by a line heat source, *Mathematical Models in Engineering Science*, A. McNabb, R. Wooding, and M. Rosser, eds., Department of Scientific and Industrial Research, New Zealand, 1982.

22. H. Schlichting, *Boundary Layer Theory*, 4th ed., McGraw-Hill, New York, 1960, p. 181.

PROBLEMS

1. Derive the two-dimensional mass conservation equation (6) by invoking the instantaneous conservation of mass in the $\Delta x \Delta y$ porous element of Fig. 10.2. (Hint: Start with eq. (1) of Chapter 1.)

2. The unidirectional flow through a porous medium can be modeled as the flow through a bundle of capillary tubes of diameter $2r_0$, as shown in Fig. 10.3. Assume that the density of such tubes per unit frontal area is N/A [tubes/m^2]. Assume further that the flow through each tube can be modeled Hagen–Poiseuille (Chapter 3). Demonstrate that, based on these assumptions, the Darcy law (8) can be derived analytically, and that the effective permeability K of the capillary tube bundle porous medium is

$$K = \frac{\pi r_0^4}{8} \frac{N}{A}$$

3. Another way of deriving the proportionality between mean velocity and pressure gradient, the Darcy Law (8), is to employ the capillary fissure model of Fig. 10.3. Assume the existence of parallel cracks (fissures) separated by a distance a; the thickness of each crack is $b = $ constant. Assuming that in each crack the flow can be modeled as Hagen–Poiseuille through a parallel-plate channel, derive eq. (8) and show that the effective permeability of the medium is

$$K = \frac{b^3}{(12)(a + b)}$$

4. Model the porous column of Fig. 10.3 as a swarm of spherical particles all of diameter d; the packing is such that the number of spheres per unit volume is N_{vol} [particles/m^3], and the number of spheres per unit frontal area is N_A [particles/m^2]. For simplicity, assume that the fluid velocity u_p is uniform through the pores (void spaces) left between the N_A particles on the frontal area. If $u_p d/\nu$ is of order $O(1)$ or less, then the drag force F_1 exerted by the flow on each particle is given by Stokes' equation.

$$\frac{F_1/(\pi d^2/4)}{\frac{1}{2}\rho u_p^2} = C_D = \frac{24}{u_p d/\nu}$$

where C_D is the drag coefficient. Summing up these forces over the N_{vol} particles, derive the Darcy law (8) and show that for this model the permeability is

$$K = \frac{1 - (\pi/4) N_A d^2}{3\pi N_{vol} d}$$

5. Derive the friction factor equation (13) from Forschheimer's modification of the Darcy law [eq. (14)].

6. Model the elementary pore flow of Fig. 10.4 as Hagen–Poiseuille flow through a cylinder of radius r_0 and length Δx. Recalling that the volumetric dissipation rate anywhere in the fluid is (Chapter 1)

$$\mu \Phi = \mu \left(\frac{\partial u}{\partial r} \right)^2$$

demonstrate that eq. (27) is correct. Repeat this proof by modeling the pore flow as Hagen–Poiseuille through a fissure (parallel-plate channel) of thickness b.

7. Derive the local entropy generation rate formula (39) for a homogeneous porous medium. Start by applying eq. (47) of Chapter 1 to the one-dimensional convection model of Fig. 10.4; write eq. (47) of Chapter 1 for the solid part and the fluid part separately. Integrate each S_{gen}''' expression over their respective volumes, $(A - A_p)\Delta x$ and $A_p \Delta x$, and then average the sum of the two entropy generation integrals over the total volume $A \Delta x$. To obtain eqs. (38) and (39), make use of the appropriate mass and energy conservation statements and the canonical relation between entropy, enthalpy, and internal energy (consult [10] for more details).

8. In the one-dimensional porous medium flow of Fig. 10.3 (the top drawing), the following scales are known: $x \sim L$, $\partial T / \partial x \sim \Delta T / L$, and $u \sim U$. In addition, the transport properties (k, K, μ) are known. Based on this information, it is found that the viscous dissipation term may be neglected in the energy equation (33); in other words

$$ k \frac{\Delta T}{L^2} > \frac{\mu U^2}{K} $$

Show that the above scaling conclusion does *not* imply that the viscous contribution to irreversibility can be neglected in the S_{gen}''' formula (39).

9. Develop the heat transfer results for uniform porous medium fluid flow along an isothermal wall [eqs. (54) and (55)]. Derive first the similarity form of the energy equation [eq. (50)] and integrate this equation, keeping in mind the error function notation reviewed in the Appendix.

10. Determine the local heat transfer coefficient for uniform porous medium fluid flow parallel to an impermeable wall with uniform heat flux [eq. (57)]. Recognizing that the scale analysis represented by eqs. (44)–(47) is general and that the similarity variable η for this problem is given by eq. (48), show that the similarity formulation of the boundary layer energy equation is

$$ \theta_{q''}'' + \frac{\eta}{2} \theta_{q''}' - \frac{1}{2} \theta_{q''} = 0 $$

where $\theta_{q''}$ assumes the form defined in eq. (56). Solve the similarity energy equation numerically subject to $\theta_{q''} = 1$ at $\eta = 0$ and $\theta_{q''} = \theta_{q''}' = 0$ as $\eta \to \infty$. Divide the η domain into equal intervals of size $\Delta \eta$. At each level $\eta_i = i\,\Delta\eta$, approximate the $\theta_{q''}$ function and its derivatives by finite differences (consult Chapter 12). Substituting these finite-difference approximations into the energy equation yields a recurrence formula for calculating $\theta_{q''}$ at location i based on the $\theta_{q''}$ values at the preceding two locations $(i - 1, i - 2)$. Determine the $\theta_{q''}(\eta)$ profile by marching from $\eta = 0$ to a sufficiently large η; this can be done by guessing the value of $(d\theta/d\eta)_{\eta=0}$ and "shooting" to satisfy the outer boundary conditions $\theta_{q''} = \theta_{q''}' = 0$ at as large a value of η as possible. Compare your result with the results summarized in the attached table. Note that the accuracy of the numerical integration depends on the value chosen for $\Delta\eta$.

Step Size $\Delta\eta$	$(-d\theta_{q''}/d\eta)_{\eta=0}$ or, from eq. (57), $\mathrm{Nu}_x \mathrm{Pe}_x^{-1/2}$	Value of η Where $\theta_{q''} = \theta'_{q''} \simeq 0$
0.1	0.898	5.5
0.05	0.892	5.25
0.01	0.887	4.94
0.005	0.886	4.68
0.001	0.8863	5.722

11. Consider a fluid-saturated porous medium with $u = Cx^n$ near a solid impermeable wall of temperature $T_0 = T_\infty + Ax^\lambda$. The temperature T_∞ is the porous medium temperature sufficiently far away from the wall, outside the thermal boundary layer. Referring to eqs. (40)–(42), prove that a similarity temperature profile exists if $n = \lambda$. Derive the similarity form of the boundary layer energy equation (42), and show that this form matches eq. (50) if $n = 0$ (i.e., if the wall is isothermal).

12. Develop an integral solution for the natural boundary layer along the vertical isothermal wall of Fig. 10.7. Assume the vertical velocity profile

$$v = v_0 \exp\left(-\frac{x}{\delta_T}\right)$$

where both v_0 and δ_T are unknown functions of altitude (y). Determine the temperature profile based on the above assumption and the boundary layer approximation of the Darcy law [eq. (63)]. Finally, integrate the energy equation (64) across the boundary layer and determine $v_0(y)$ and $\delta_T(y)$. Verify that these results agree with the results of the scale analysis [eqs. (68)]. Calculate the local Nusselt number Nu_y, and estimate the percent departure of your result from the similarity result of eq. (77).

13. Develop the dimensionless similarity formulation of the problem of natural boundary layer convection along a vertical wall with uniform wall heat flux q''. Begin your analysis with the boundary layer equations (63) and (64) and use the scale analysis (78)–(81) in order to define the appropriate similarity variables. Compare your similarity problem statement with the corresponding work published in Ref. 14.

14. For the local heat transfer coefficient in natural convection along a $q'' = $ constant wall, Cheng and Minkowycz [14] reported (see also Ref. 12)

$$\mathrm{Nu}_y = 0.6788\, \mathrm{Ra}_y^{1/2}$$

where Nu_y and Ra_y are defined as in eqs. (70) and (76). Keeping in mind, however, that $(T_0 - T_\infty)$ is this time a function of y and that the appropriate scales for uniform heat flux convection are given by eqs. (78)–(81), translate the above expression into the language of eq. (81); compare your result with the local Nusselt number given in eq. (82).

15. Consider the heat transfer between two isothermal porous media separated by an impermeable partition (Fig. 10.8a). Develop an approximate estimate for the overall heat transfer rate by modeling the partition as isothermal. Report your result in the dimensionless notation employed in eq. (90). Repeat this approximate calculation by modeling the partition as a surface with uniform heat flux. Comparing both results with eq. (90), which partition model is better—the constant temperature or the constant heat flux?

16. Consider the natural convection heat transfer between a porous medium and fluid reservoir separated by a vertical impermeable surface (Fig. 10.8b). Calculate the overall Nusselt number based on H and overall temperature difference $(T_{\infty,H} - T_{\infty,L})$ by first modeling the surface as isothermal. Show that the B number [eq. (91)] emerges from this calculation as the ratio of the thermal resistances for each side. Repeat the analysis by modeling the interface as a surface with uniform heat flux. Compare both results with the data of Fig. 10.10 and decide which of the two interface models is better.

17. Repeat the integral analysis (92)–(103) for an isothermal wall facing a linearly stratified porous medium (Fig. 10.12) by assuming the parabolic velocity profile

$$v = v_0 \left(1 - \frac{x}{\delta_T}\right)^2$$

in place of the exponential profile (93). Based on this exercise, determine how susceptible the heat transfer results of Fig. 10.12 are to the selection of profile shapes for the integral analysis. Comparing your result with the similarity solution known for $b = 0$, decide which choice of profile shape yields more accurate predictions (the exponential or the parabolic?).

18. Complete the steps missing from the development of the two-term series solution for low-Rayleigh number transient convection around a point source [eqs. (116) and (117)]. For the derivation of the pure conduction component [eqs. (116)], consult Ref. 17 as a guide; for the first-order convection correction [eqs. (117)], consult Ref. 19.

19. Derive the first two terms, $O(Ra^0)$ and $O(Ra^1)$, of the steady state solution for natural convection around a point heat source in a fluid saturated porous medium [eqs. (118)]. Comment on the Ra range in which this shorter version of eqs. (118) might be valid.

20. Consider the temperature and flow field around a horizontal line heat source of strength q' [W/m] buried in a fluid-saturated porous medium. Attach a cylindrical coordinate system to the line source. In this system write the mass conservation equation, the Darcy law, and the *transient* energy conservation equation. Noting that this convection problem is two-dimensional in (r, θ), derive the flow and temperature fields valid in the limit Ra → 0, that is, the equivalent of eqs. (116) and (117) for the point heat source [the Rayleigh number for this problem turns out to be Ra = $K^{3/2}g\beta q'/(\alpha\nu k)$]. Consult Ref. 21 for further hints.

21. Determine the boundary-layer-type flow and temperature field above a horizontal line heat source of strength q' [W/m]. Recognizing that this flow is two-dimensional, attach the coordinate system $(x, y), (u, v)$ perpendicularly to the line source so that the y axis points upward. Proceeding in the same manner as in the analysis of high-Ra convection above a point source [eqs. (119)–(137)], show that the vertical velocity distribution is the same as in a plane incompressible momentum jet ([22], p. 164). Consult Refs. 7 and 20 for the correct answer, as well as for two alternative ways of deriving the same result.

11

Natural Convection in Confined Porous Media

The interest in natural convection through confined porous media is fueled by diverse and important engineering applications, for example, geothermal energy conversion and porous insulation development. A segment of the research work in this area has been summarized in a review article by Cheng [1]. The fundamental heat transfer research undertaken so far aligns itself under two key topics of current interest:

1. Natural convection in a layer with vertical sides at different temperatures.

2. Natural convection in a porous layer heated from below.

Configuration 1 is most representative of porous insulation layers oriented vertically, as in building technology, industrial cold-storage installations, and cryogenic engineering. The study of configuration 2 is essential to understanding the functioning of geothermal systems and fibrous insulations of the type used in attics. In this chapter we focus on the fundamentals of the natural convection phenomenon in both configurations, as well as in a number of other circumstances of engineering interest.

TRANSIENT HEATING FROM THE SIDE

As in the study of natural convection in enclosed spaces filled with fluid (Chapter 5), we begin with a theoretical discussion of the relevant time and length scales describing the flow in a porous layer heated from the side. The system of interest is shown in Fig. 11.1: a two-dimensional rectangular space of height H and horizontal dimension L is filled with a fluid-saturated porous matrix of permeability K. In accordance with the homogeneous porous medium

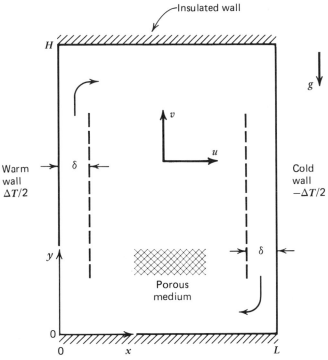

Figure 11.1 Schematic of a two-dimensional porous layer heated from the side.

model developed in the preceding chapter, the equations governing the conservation of mass, momentum, and energy are

$$\frac{\partial u}{\partial x} + \frac{\partial v}{\partial y} = 0 \tag{1}$$

$$u = -\frac{K}{\mu}\frac{\partial P}{\partial x} \tag{2}$$

$$v = -\frac{K}{\mu}\left(\frac{\partial P}{\partial y} + \rho g\right) \tag{3}$$

$$\sigma\frac{\partial T}{\partial t} + u\frac{\partial T}{\partial x} + v\frac{\partial T}{\partial y} = \alpha\left(\frac{\partial^2 T}{\partial x^2} + \frac{\partial^2 T}{\partial y^2}\right) \tag{4}$$

where u, v, μ, P, and T are the fluid velocity components, viscosity, pressure, and temperature. Parameter σ is the heat capacity ratio (Chapter 10)

$$\sigma = \frac{\phi(\rho c_P)_f + (1 - \phi)(\rho c_P)_s}{(\rho c_P)_f} \tag{5}$$

where ϕ is the porosity of the medium and $(\rho c_P)_f, (\rho c_P)_s$ are the heat capacities of the fluid and solid matrix, respectively. Parameter α is the effective thermal diffusivity of the medium

$$\alpha = \frac{k}{(\rho c_P)_f} \tag{6}$$

where k is the thermal conductivity of the porous matrix while saturated with fluid.

It is convenient to eliminate the pressure P between eqs. (2) and (3) and to introduce the Boussinesq approximation $\rho = \rho_0[1 - \beta(T - T_0)]$ in the body force term of eq. (3). We obtain a single momentum conservation statement

$$\frac{\partial u}{\partial y} - \frac{\partial v}{\partial x} = -\frac{Kg\beta}{\nu} \frac{\partial T}{\partial x} \tag{7}$$

where β and ν are the fluid thermal expansion coefficient and the kinematic viscosity (μ/ρ_0).

Consider now the following transient heating experiment [2]. Initially, the porous layer is isothermal $(T = 0)$ and the fluid permeating through it is motionless $(u = v = 0)$. At some point in time, the vertical wall temperatures are instantly changed to $+\Delta T/2$ and $-\Delta T/2$, respectively. The object of the following discussion is the evolution of the flow through the porous layer, subject to the boundary conditions

$$u = 0, \qquad T = \Delta T/2 \qquad \text{at } x = 0$$

$$u = 0, \qquad T = -\Delta T/2 \quad \text{at } x = L \tag{8}$$

$$v = 0, \qquad \frac{\partial T}{\partial y} = 0 \qquad \text{at } y = 0, H$$

These conditions account for the impermeability of the rectangular frame and for the fact that the two horizontal walls are insulated (Fig. 11.1).

Since the fluid is initially motionless, the vertical wall effect will first propagate into the porous space through pure conduction. The energy equation (4) dictates a balance between the thermal inertia and heat conduction into a layer of thickness $\delta(t)$

$$\sigma \frac{\Delta T}{t} \sim \alpha \frac{\Delta T}{\delta^2} \tag{9}$$

hence,

$$\delta \sim (\alpha t/\sigma)^{1/2} \tag{10}$$

The growth of the δ layer (Fig. 11.1) gives rise to a horizontal temperature gradient of order

$$\frac{\partial T}{\partial x} \sim \frac{\Delta T}{\delta} \tag{11}$$

This development makes the buoyancy term in the momentum equation (6) finite: the scales of the three terms appearing in eq. (7) are

$$\frac{u}{H}, \qquad \frac{v}{\delta}, \qquad \frac{Kg\beta}{\nu}\frac{\Delta T}{\delta} \tag{12}$$

Noting that mass conservation [eq. (1)] requires

$$\frac{u}{\delta} \sim \frac{v}{H} \tag{13}$$

we conclude that $(u/H)/(v/\delta) \sim (\delta/H)^2$. Therefore, if the vertical layer is slender $(\delta < H)$, the momentum equation dictates a balance between the second and third terms in expression (12); the result of this balance is the vertical velocity scale in the vicinity of each vertical wall

$$v \sim \frac{Kg\beta}{\nu}\Delta T \tag{14}$$

An interesting aspect of this scale is that it is time-independent. However, in view of the earlier result for $\delta(t)$ [eq. (10)], we learn that the flowrate driven by the heated wall $(v\delta)$ increases in time as $t^{1/2}$.

Once fluid motion is initiated, the energy equation (4) is ruled by three different scales

$$\sigma\frac{\Delta T}{t}, \qquad v\frac{\Delta T}{H}, \qquad \alpha\frac{\Delta T}{\delta^2} \tag{15}$$

$$\begin{array}{ccc} \text{Inertia} & \text{Convection} & \text{Conduction} \\ (t^{-1}), & (t^{1/2}), & (t^{-1}) \end{array}$$

Below each scale we see the evolution of each effect in time; since conduction from the wall will always be present, the convection effect eventually takes the place of inertia in the energy balance. The time t_f when the vertical layer becomes convective is given by

$$\sigma\frac{\Delta T}{t_f} \sim v\frac{\Delta T}{H} \tag{16}$$

hence,

$$t_f \sim \frac{\sigma H}{v} \tag{17}$$

Beyond this point the boundary layer thickness ceases to grow: its steady state scale is

$$\delta_f \sim H \, \mathrm{Ra}_H^{-1/2} \tag{18}$$

where Ra_H is the Darcy-modified Rayleigh number based on height,

$$\mathrm{Ra}_H = \frac{Kg\beta H \Delta T}{\alpha v} \tag{19}$$

The necessary condition for the existence of distinct vertical boundary layers in the steady state is $\delta_f < L$, in other words,

$$\frac{L}{H} \mathrm{Ra}_H^{1/2} > 1 \tag{20}$$

This condition is plotted on the $H/L - \mathrm{Ra}_H$ field of Fig. 11.2.

The flow scales reported as eqs. (14) and (18) can be used to predict the existence of distinct horizontal layers along the horizontal adiabatic walls (Fig. 11.1). The scaling argument for this prediction is analogous to the one leading to criterion [eq. (22) of Chapter 5] for distinct horizontal layers in cavities filled with fluid. The volumetric flowrate driven horizontally, in counterflow is of order $v\delta_f$; this stream carries enthalpy between the two vertical walls at a rate

$$Q_{\substack{\text{convection} \\ \text{left} \to \text{right} \\ \text{(Fig.11.1)}}} \sim v\delta_f (\rho c_P)_f \Delta T \tag{21}$$

The two branches of this horizontal counterflow experience heat transfer by thermal diffusion at a rate

$$Q_{\substack{\text{conduction} \\ \text{top} \to \text{bottom} \\ \text{(Fig.11.1)}}} \sim kL \frac{\Delta T}{H} \tag{22}$$

As in a poorly designed counterflow heat exchanger, one stream will travel the entire length of the porous layer (L) without a significant change in temperature when the vertical conduction rate [eq. (22)] is negligible relative to the horizontal convection rate [eq. (21)],

$$kL \frac{\Delta T}{H} < v\delta_f (\rho c_P)_f \Delta T \tag{23}$$

Therefore, the criterion for distinct horizontal layers is

$$\frac{H}{L}\mathrm{Ra}_H^{1/2} > 1 \tag{24}$$

Figure 11.2 shows the four natural convection regimes possible in a porous layer heated from the side. These regimes are analogous to regimes (I)–(IV) present in cavities heated from the side: in fact, the main features of the four circulation patterns presented in Fig. 5.4 apply unchanged to the four regimes identified in a porous layer (Fig. 11.2).

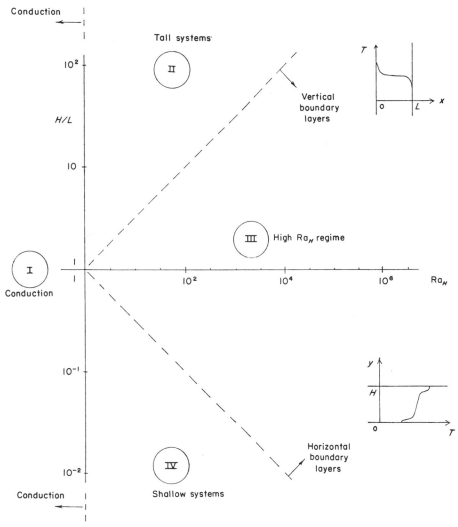

Figure 11.2 Chart showing the four heat transfer regimes for natural convection in a two-dimensional porous layer heated from the side.

Of interest in thermal insulation engineering is the net heat transfer rate Q across the overall ΔT. The scaling results of this section suggest the following heat transfer rate scales

$$
\begin{array}{lll}
\text{(I)} & \text{Pure conduction} & Q \sim kH\Delta T/L \\
\text{(II)} & \text{Tall layers} & Q \gtrsim kH\Delta T/L \\
\text{(III)} & \text{High-Ra}_H \text{ convection} & Q \sim kH\Delta T/\delta_f \\
\text{(IV)} & \text{Shallow layers} & Q \lesssim kH\Delta T/\delta_f
\end{array}
\tag{25}
$$

In the following two sections, we focus on regimes (III) and (IV) where the heat transfer rate is dominated by convection.

THE BOUNDARY LAYER REGIME

Perhaps, the most important convection regime in thermal insulation application is regime (III) (Fig. 11.2), where the resistance to heat transfer scales with the thickness of the vertical boundary layer δ_f. As is shown by Weber [3], an analytical solution to the flow and heat transfer problem in this regime can be developed based on the Oseen-linearization procedure presented in Chapter 5 for enclosures filled with fluid.

We focus on the region of thickness δ_f near the left (warm) wall and define the following dimensionless variables:

$$
x_* = \frac{x}{\delta_f}, \qquad y_* = \frac{y}{H}
$$

$$
\psi_* = \frac{\psi}{\alpha(\mathrm{Ra}_H)^{1/2}}, \qquad T_* = \frac{T - \frac{1}{2}(T_{\mathrm{warm}} + T_{\mathrm{cold}})}{T_{\mathrm{warm}} - T_{\mathrm{cold}}}
\tag{26}
$$

where ψ is the streamfunction ($u = -\partial\psi/\partial y$, $v = \partial\psi/\partial x$). When $\mathrm{Ra}_H > 1$, the dimensionless momentum and energy equations (7) and (4) reduce to

$$
\frac{\partial^2 \psi_*}{\partial x_*^2} = \frac{\partial T_*}{\partial x_*}
\tag{27}
$$

$$
\frac{\partial \psi_*}{\partial x_*}\frac{\partial T_*}{\partial y_*} - \frac{\partial \psi_*}{\partial y_*}\frac{\partial T_*}{\partial x_*} = \frac{\partial^2 T_*}{\partial x_*^2}
\tag{28}
$$

Treating $\partial\psi_*/\partial y_*$ and $\partial T_*/\partial y_*$ as unknown functions of y_* only, it is easy to show that

$$
\psi_* = \psi_\infty(1 - e^{-\lambda x_*})
\tag{29}
$$

$$
T_* = T_\infty + \left(\tfrac{1}{2} - T_\infty\right)e^{-\lambda x_*}
\tag{30}
$$

where T_∞, ψ_∞, and λ are all functions of y_*. The three unknowns are determined based on the following three conditions:

1. Momentum conservation [eq. (27)]

$$\lambda \psi_\infty = \tfrac{1}{2} - T_\infty \tag{31}$$

2. The $\int_0^\infty dx_*$ integral of the energy equation (28)

$$\frac{d}{dy_*}\left[\frac{1}{2\lambda}\left(\frac{1}{2} - T_\infty \right)^2 \right] + \psi_\infty \frac{dT_\infty}{dy_*} = \lambda\left(\frac{1}{2} - T_\infty \right) \tag{32}$$

3. The centrosymmetry condition, meaning that the streamlines and iso-therms must be symmetric relative to the geometric center of the rectangular porous layer ($L/2, H/2$). In the core, where $\psi_* \to \psi_\infty$ and $T_* \to T_\infty$, this condition amounts to recognizing ψ_∞ as an even function of z and T_∞ as an odd function of z (note that the vertical coordinate z is measured from midheight, $z = y_* - 1/2$). Introducing odd and even functions of z to account for the centrosymmetry of the core flow and using the preceding two condi-tions [eqs. (31) and (32)] yields [3]

$$\psi_* = C(1 - q^2)\left\{ 1 - \exp\left[-\frac{x_*}{2C(1+q)} \right] \right\} \tag{33}$$

$$T_* = \frac{1}{2}\left\{ q + (1-q)\exp\left[-\frac{x_*}{2C(1+q)} \right] \right\} \tag{34}$$

where the odd function $q(z)$ is given implicitly by

$$z = C^2(q - q^3/3) \tag{35}$$

Constant C represents the backbone of the boundary layer solution, in the same manner in which a similar constant controls Gill's solution for a rectangular cavity [see eqs. (37) and (38) of Chapter 5]. Weber [3] determined C by invoking the impermeable wall condition $\psi_* = 0$ at $z = \pm 1/2$; thus he obtained

$$C = \frac{\sqrt{3}}{2} \quad \text{or} \quad q = \pm 1 \quad \text{at } z = \pm\tfrac{1}{2} \tag{36}$$

Although not shown in [3], streamlines and isotherms can be plotted by combining eqs. (33)–(36): the result is shown here as Figs. 11.3 and 11.4. We see a thermal boundary layer flow discharging horizontally into a thermally stratified core. The flow field in the high-Ra_H regime is very similar to the corresponding flow in a cavity filled with fluid (Fig. 5.5). Combining the

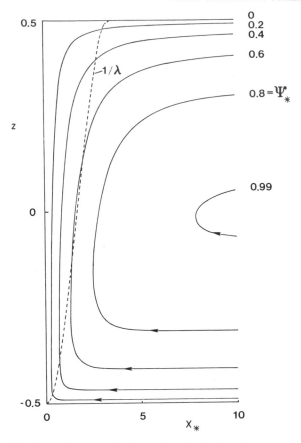

Figure 11.3 Streamlines near the heated wall in the boundary layer regime.

boundary layer patterns of Figs. 11.3 and 11.4 with their centrosymmetric copies we obtain a fairly accurate picture of the flow in the entire porous layer. In Fig. 11.5 we see the patterns corresponding to $H/L = 2.25$ and $\mathrm{Ra}_H = 337.5$. These patterns have been verified by a number of numerical solutions to the same problem (e.g., see Refs. 4 and 5).

The net heat transfer rate from T_{warm} to T_{cold} can be expressed as a conduction-based Nusselt number

$$\mathrm{Nu} = \frac{1}{kH\Delta T/L} \int_0^H - k\left(\frac{\partial T}{\partial x}\right)_{x=0,\,L} dy \tag{37}$$

The Weber solution (33)–(36) yields

$$\mathrm{Nu} = \frac{1}{\sqrt{3}} \frac{L}{H} \mathrm{Ra}_H^{1/2} = 0.577 \frac{L}{H} \mathrm{Ra}_H^{1/2} \tag{38}$$

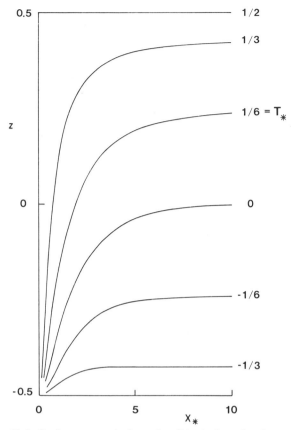

Figure 11.4 Isotherms near the heated wall in the boundary layer regime.

Noting that $L/H\, \mathrm{Ra}_H^{1/2} = L/\delta_f$, we learn from eq. (38) that the convective heat transfer greatly exceeds the conductive estimate when the boundary layer thickness is much smaller than the thickness of the porous system L. Thus, $\mathrm{Nu} > 1$ is another way of expressing the criterion for vertical boundary layers [eq. (20)].

A summary of experimental and numerical heat transfer results is presented in Fig. 11.6, based on Nu data from three sources [5–7]. The trend $\mathrm{Nu} \sim (L/H)\mathrm{Ra}_H^{1/2}$ predicted from scaling [eq. (25)] appears to be correct in the high-Ra_H limit. However, Weber's result consistently overpredicts the Nusselt number. It has been shown [8] that the discrepancy between theory and experiment is due to the way in which constant C was determined for the Weber solution [eq. (36)]. A more reasonable approach is to take *both* impermeable and adiabatic conditions at $z = \pm 1/2$ into account: this is accomplished approximately by invoking the zero energy flow condition $Q_y = 0$ [eq. (41) of Chapter 5] along the top and bottom walls. The value of constant C

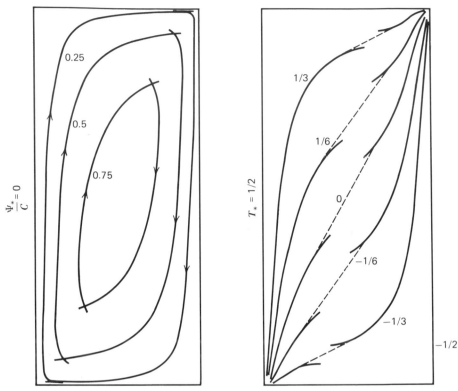

Figure 11.5 Streamlines and isotherms in a porous layer heated from the side ($H/L = 2.25$, $\mathrm{Ra}_H = 337.5$).

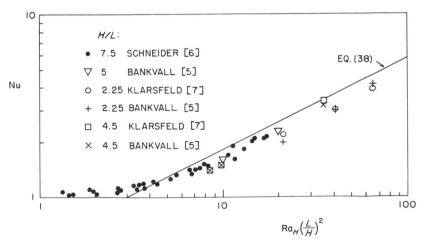

Figure 11.6 Compilation of experimental results for heat transfer through a porous layer heated from the side.

from this condition is [8]

$$C = \left(1 - q_e^2\right)^{-2/3} \mathrm{Ra}_H^{-1/6} \left(\frac{H}{L}\right)^{-1/3} \tag{39}$$

where $q_e = q(1/2)$, in other words, from eq. (35)

$$\frac{1}{2} = C^2 \left(q_e - \frac{q_e^3}{3}\right) \tag{40}$$

The resulting value of C is shown in Fig. 11.7: C approaches the Weber value $\sqrt{3}/2$ as the new group $(H/L)^{2/3}\mathrm{Ra}_H^{1/3}$ tends to infinity, that is, as the porous layer becomes tall.

The Nusselt number information of Fig. 11.7 is presented explicitly as an engineering chart in Fig. 11.8, as Nu versus $\mathrm{Ra}_H(L/H)^2$ for fixed values of H/L. The theoretical curves fall in the area covered by experimental and numerical results in Fig. 11.6. The accuracy of Fig. 11.8 as a tool for predicting the Nusselt number was demonstrated by Simpkins and Blythe [9]. Figure 11.9

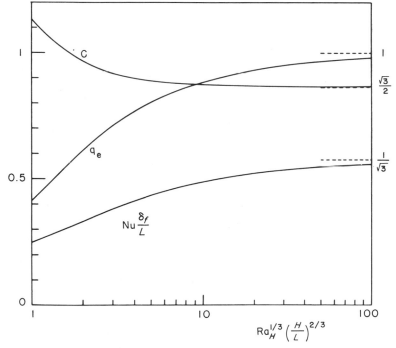

Figure 11.7 Summary of the boundary layer solution for natural convection heat transfer in a porous layer heated from the side [8].

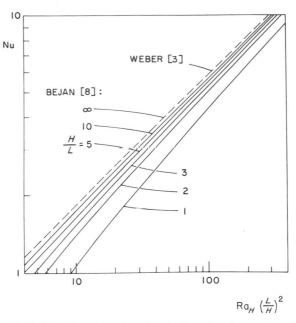

Figure 11.8 Nusselt number chart for the boundary layer regime [8].

(reproduced from Ref. 9) shows excellent agreement between the Nusselt number of Fig. 11.8 and the data reported by Bankvall [5] and Klarsfeld [7]. Also more accurate than eq. (38) is the integral boundary layer solution reported in Ref. 9: the success of this integral solution may be attributed to the fact that it also incorporates the zero energy flow condition proposed in the development of Fig. 11.8 [8].

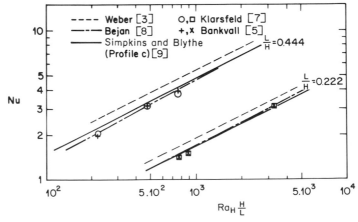

Figure 11.9 Comparison of theoretical and experimental results for the Nusselt number in the boundary layer regime [9].

SHALLOW LAYERS

The behavior of shallow porous layers ($H/L < 1$) is of interest not only in geothermal engineering but in thermal insulation design where one could imagine the use of horizontal impermeable baffles in order to inhibit the natural convection loop forming in a tall porous layer. Natural convection under regime (IV) differs from regime (III) in that horizontal boundary layers are absent (Fig. 11.2). (Note the presence of distinct horizontal layers in Fig. 11.3, regime (III): most of the fluid bypasses the core of the porous layer and flows along the top and bottom walls.) The main characteristics of regime (IV) are shown in Fig. 11.10 [10]: vertical end layers exist, however, a sizeable temperature drop is registered across the slender core itself. Note that Fig. 11.10, which is reproduced from Ref. 10, has the warm end of the layer on the right-hand side, unlike in Fig. 11.1.

The first published study of convection in shallow porous layers with different end-temperatures was conducted by Bejan and Tien [10]. This study showed that in the core region the circulation consists of a thermally stratified counterflow described by

$$u = -\frac{\alpha}{H}\mathrm{Ra}_H\frac{H}{L}K_1\left(y_* - \frac{1}{2}\right) \tag{41}$$

$$v = 0 \tag{42}$$

$$\frac{T - T_{\text{cold}}}{T_{\text{warm}} - T_{\text{cold}}} = K_1\frac{x}{L} + K_2 + \mathrm{Ra}_H\left(\frac{H}{L}\right)^2 K_1^2\left(\frac{y_*^2}{4} - \frac{y_*^3}{6}\right) \tag{43}$$

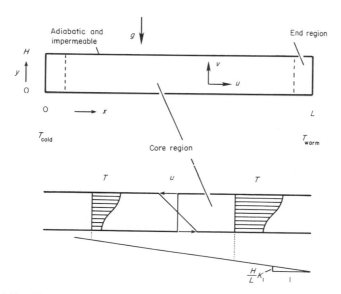

Figure 11.10   The structure of a horizontal porous layer with different end-temperatures [10].

where $y_* = y/H$. This flow and temperature distribution are shown schematically in Fig. 11.10. The same core flow was also reported by Walker and
Homsy [11]. Parameters K_1 and K_2 follow from the matching of the core flow
[eqs. (41)–(43)] to the flow in the two end regions.

The net heat transfer rate in the $T_{\text{warm}} \rightarrow T_{\text{cold}}$ direction is calculated by
integrating the conductive and enthalpy fluxes over any $x = $ constant cut
through the core region. The conduction-based Nusselt number becomes

$$\text{Nu} - \frac{Q}{kH\Delta T/L} = \frac{1}{kH\Delta T/L} \int_0^H \left[k\frac{\partial T}{\partial x} - (\rho c_P)_f uT \right] dy$$

$$= K_1 + \frac{1}{120} K_1^3 \left(\text{Ra}_H \frac{H}{L} \right)^2 \tag{44}$$

In infinitely shallow layers, the core accounts for the entire temperature drop
across the layer, hence, $K_1 = 1$ and [10, 11]

$$\text{Nu} = 1 + \frac{1}{120} \left(\text{Ra}_H \frac{H}{L} \right)^2 ; \qquad H/L \rightarrow 0 \tag{45}$$

In the most general case where H/L is finite, the core temperature gradient K_1
is a function of both H/L and Ra_H. Bejan and Tien [10] determined the
$K_1(H/L, \text{Ra}_H)$ function parametrically by matching the core solution to
integral solutions of the end regions. The result is

$$\frac{1}{120} \delta_e \text{Ra}_H^2 K_1^3 \left(\frac{H}{L} \right)^3 = 1 - K_1 \tag{46}$$

$$\frac{1}{2} K_1 \frac{H}{L} \delta_e \left(\frac{1}{\delta_e^2} - 1 \right) = 1 - K_1 \tag{47}$$

where δ_e is the ratio (end region thickness)$/H$. The Nusselt number is shown
plotted in Fig. 11.11 next to the numerical results published by Hickox and
Gartling [12]. Representative patterns of streamlines and isotherms in shallow
layers may be examined in [12].

An interesting aspect of the shallow layer Nu theory plotted in Fig. 11.11 is
that it collapses into a $\text{Nu} \sim (L/H)\text{Ra}_H^{1/2}$ proportionality in the $\text{Ra}_H \rightarrow \infty$
limit, as expected from scaling [eq. (25)]. Taking expressions (44), (46), and
(47) to the limit $\text{Ra}_H \rightarrow \infty$ (or $\delta_e \ll 1$, $K_1 \ll 1$), we find

$$\text{Nu} = 0.508 \frac{L}{H} \text{Ra}_H^{1/2}; \qquad \text{Ra}_H \rightarrow \infty \tag{48}$$

which is nearly identical to Weber's result [eq. (38)]. Therefore, the
$\text{Nu}(\text{Ra}_H, H/L)$ function represented by eqs. (44), (46), and (47) is adequate

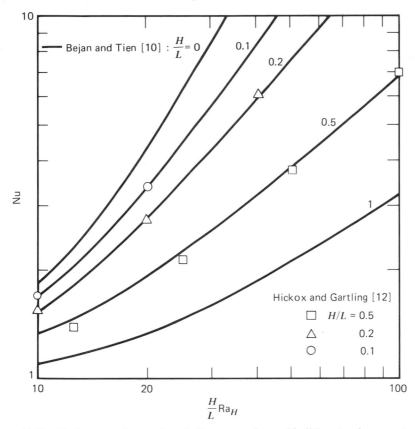

Figure 11.11 The heat transfer rate in a shallow porous layer with different end-temperatures.

for heat transfer calculations in both shallow and tall layers, at low and high Rayleigh numbers.

PARTIAL PENETRATION INTO ISOTHERMAL POROUS LAYERS

A separate class of buoyancy-driven flows through porous media stems from the interaction of finite-size, isothermal, porous layers with a localized source of heat. The basic configuration can be modeled as a two-dimensional layer of size $H \times L$, with three sides at one temperature. The fourth side is permeable and in communication with a fluid reservoir of different temperature. In this section we focus on two possible orientations of this configuration, a shallow layer with lateral heating (Fig. 11.12a) and a tall layer with heating from below or cooling from above (Fig. 11.12b). In both cases, natural convection penetrates the porous medium over a length dictated by the Rayleigh number alone and not by the geometric ratio H/L.

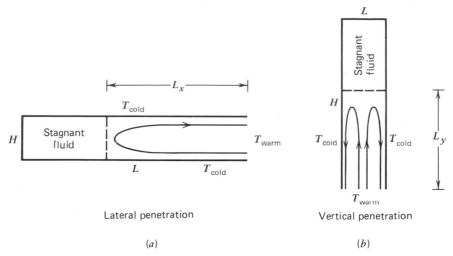

Lateral penetration   Vertical penetration

(a)   (b)

Figure 11.12 Lateral and vertical penetration of natural convection into an isothermal porous space heated from one end.

Lateral Penetration

To determine the scale of the lateral penetration length L_x (Fig. 11.12a), consider the governing equations (1), (4), and (7)

Mass:
$$\frac{u}{L_x} \sim \frac{v}{H} \tag{49}$$

Energy:
$$u\frac{\Delta T}{L_x} \sim \alpha\frac{\Delta T}{L_x^2}, \qquad \alpha\frac{\Delta T}{H^2} \tag{50}$$

Momentum:
$$\frac{u}{H}, \qquad \frac{v}{L_x} \sim \frac{Kg\beta}{\nu}\frac{\Delta T}{L_x} \tag{51}$$

Above, we have three equations for the three unknown scales u, v, and L_x. Assuming that the flow penetrates the layer such that $L_x > H$, it is easy to show that

$$L_x \sim H\,\mathrm{Ra}_H^{1/2} \tag{52}$$

The convective heat transport between the isothermal porous layer and the heat reservoir positioned laterally scales as

$$Q \sim (\rho c_P)_f Hu\,\Delta T$$

$$\sim k\,\Delta T\,\mathrm{Ra}_H^{1/2} \tag{53}$$

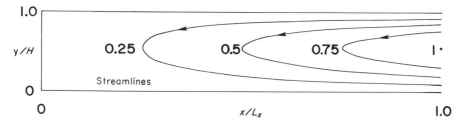

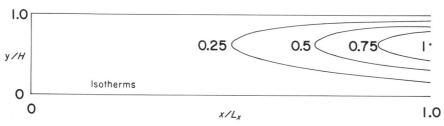

Figure 11.13 Streamlines and isotherms in the region of lateral penetration into a two-dimensional porous space [13].

This heat transfer result demonstrates that the actual length of the porous layer (L) does not influence the heat transfer rate; Q as well as L_x are set by the Rayleigh number Ra_H.

The actual flow and temperature patterns associated with the lateral penetration phenomenon can be determined analytically as a similarity solution [13]. Figure 11.13 shows the dimensionless streamfunction and temperature in the region of penetration only. The penetration length and heat transfer rate predicted by the similarity solution are [13]

$$L_x = 0.158H \, \mathrm{Ra}_H^{1/2} \tag{54}$$

$$\frac{Q}{k \, \Delta T} = 0.319 \, \mathrm{Ra}_H^{1/2} \tag{55}$$

Reference 13 documents the effect of anisotropy in the medium and the effect of temperature variation along the horizontal walls of the porous layer.

Vertical Penetration

Consider the two-dimensional layer of Fig. 11.12b, where the bottom wall is permeable and in communication with a different heat reservoir. We learn in the next section that in porous layers heated from below (or cooled from above) fluid motion is possible only above a critical Rayleigh number. In Fig. 11.12b, however, fluid motion will set in as soon as a ΔT is imposed between the bottom surface and vertical walls. Fluid motion will be present because no matter how small the ΔT, the porous medium experiences a finite-temperature gradient of order $\Delta T/L$ in the horizontal direction near the heated wall [see eq. (7)].

Let L_y be the distance of vertical penetration. From eqs. (1), (4), and (7), we have the following balances

Mass:
$$\frac{u}{L} \sim \frac{v}{L_y} \tag{56}$$

Energy:
$$u\frac{\Delta T}{L} \sim \alpha\frac{\Delta T}{L^2}, \qquad \alpha\frac{\Delta T}{L_y^2} \tag{57}$$

Momentum:
$$\frac{u}{L_y}, \qquad \frac{v}{L} \sim \frac{Kg\beta}{v}\frac{\Delta T}{L} \tag{58}$$

Assuming vertical penetration over a distance L_y greater than L, we conclude that

$$L_y \sim L\,\mathrm{Ra}_L \tag{59}$$

where Ra_L is the Rayleigh number based on L, $\mathrm{Ra}_L = Kg\beta L\Delta T/(\alpha v)$. The net heat transfer rate through the bottom wall of the system scales as

$$Q_y \sim (\rho c_P)_f Lv\Delta T$$

$$\sim k\,\Delta T\,\mathrm{Ra}_L \tag{60}$$

We learn that both L_y and Q_y are proportional to Ra_L, unlike the corresponding quantities in the case of lateral penetration, which are proportional to $\mathrm{Ra}_H^{1/2}$. Once again, the imposed temperature difference (ΔT) and the transversal dimension of the layer (L) determine the longitudinal extent (L_y) of the penetrative flow; the physical height of the porous layer (H) does not influence the phenomenon as long as it is greater than L_y.

The phenomenon of partial vertical penetration was studied in the cylindrical geometry in an attempt to model geothermal flows or the flow of air through the grain stored in a silo [14]. Representative streamlines of the flow in the penetration region L_y are reproduced in Fig. 11.14. The vertical penetration length and net heat transfer rate are [14]

$$\frac{L_y}{r_0} = 0.0847\,\mathrm{Ra}_{r_0} \tag{61}$$

$$\frac{Q_y/r_0}{k\,\Delta T} = 0.255\,\mathrm{Ra}_{r_0} \tag{62}$$

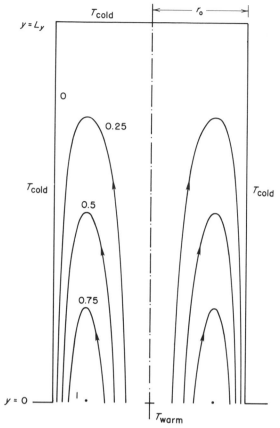

Figure 11.14 Streamlines in the region of vertical penetration into a cylindrical porous medium [14].

where $Ra_{r_0} = Kg\beta r_0 \Delta T/(\alpha \nu)$ is the Rayleigh number based on the dimension normal to the penetrative flow. Note that results (61) and (62) are predicted by the scaling analysis [expressions (59) and (60)] (for a two-dimensional layer, Q_y is the heat transfer rate per unit length in the direction normal to the x-y plane, Fig. 11.12).

POROUS LAYERS HEATED FROM BELOW

We now consider another class of flows that stand out in the field of natural convection through porous media: the Bénard-type flow, that is, the cellular convection that *may* take place through a porous layer heated from below and cooled from above. Figure 11.15 shows the simplest model in which to study the possibility of thermal convection in a system with zero horizontal temperature gradient [15]. We look at an infinite horizontal layer of thickness H with a

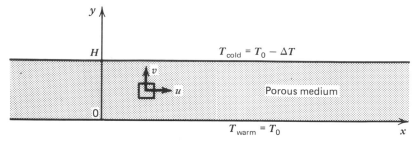

Figure 11.15   Two-dimensional porous layer heated from below.

warm bottom (T_0) and a cold top ($T_0 - \Delta T$). The steady state equations (1), (4), and (7) admit the obvious solution (no-flow and pure conduction)

$$u_b = v_b = 0 \tag{63}$$

$$T_b = T_0 - \Delta T \frac{y}{H} \tag{64}$$

The question is whether this no-flow solution will prevail forever, regardless of how high a ΔT we impose. We answer this question by running a stability experiment of the type described in Chapter 6 in connection with the laminar–turbulent flow transition. The stability experiment consists of dynamically disturbing the base solution [(63) and (64)] and observing under what conditions the imposed disturbance grows in amplitude. Thus, we substitute

$$T(x, y, t) = T_b(y) + \qquad T'(x, y, t)$$

$$u(x, y, t) = \qquad 0 + \qquad u'(x, y, t) \tag{65}$$

$$\underbrace{v(x, y, t)}_{\substack{\text{Transient} \\ \text{flow}}} = \underbrace{0 +}_{\substack{\text{Base} \\ \text{solution}}} \underbrace{v'(x, y, t)}_{\text{Disturbance}}$$

into the transient governing equations (1), (4), and (7). We obtain

$$\frac{\partial u'}{\partial x} + \frac{\partial v'}{\partial y} = 0 \tag{66}$$

$$\sigma \frac{\partial T'}{\partial t} + u' \frac{\partial T'}{\partial x} + v' \left(\frac{dT_b}{dy} + \frac{\partial T'}{\partial y} \right) = \alpha \left(\frac{\partial^2 T'}{\partial x^2} + \frac{\partial^2 T'}{\partial y^2} \right) \tag{67}$$

$$\frac{\partial u'}{\partial y} - \frac{\partial v'}{\partial x} = - \frac{Kg\beta}{\nu} \frac{\partial T'}{\partial x} \tag{68}$$

In the energy equation (67) we have the option to eliminate the nonlinear terms $u'\partial T'/\partial x$ and $v'\partial T'/\partial y$ based on the assumption that in the very beginning the flow and temperature disturbances are negligibly small. Thus, in eq. (67), we retain only the first-order terms in primed (disturbance) quantities,

$$\sigma\frac{\partial T'}{\partial t} - \frac{\Delta T}{H}v' = \alpha\left(\frac{\partial^2 T'}{\partial x^2} + \frac{\partial^2 T'}{\partial y^2}\right) \tag{69}$$

A convenient way to nondimensionalize the problem is to define the new variables

$$\hat{x} = x/H, \qquad \hat{y} = y/H$$

$$\hat{u} = \frac{u'}{\alpha/H}, \qquad \hat{v} = \frac{v'}{\alpha/H} \tag{70}$$

$$\hat{T} = T'/\Delta T, \qquad \hat{t} = \alpha t/(H^2\sigma)$$

The mass, momentum, and energy equations become

$$\frac{\partial \hat{u}}{\partial \hat{x}} + \frac{\partial \hat{v}}{\partial \hat{y}} = 0 \tag{71}$$

$$\frac{\partial \hat{u}}{\partial \hat{y}} - \frac{\partial \hat{v}}{\partial \hat{x}} = -\mathrm{Ra}_H\frac{\partial \hat{T}}{\partial \hat{x}} \tag{72}$$

$$\frac{\partial \hat{T}}{\partial \hat{t}} - \hat{v} = \frac{\partial^2 \hat{T}}{\partial \hat{x}^2} + \frac{\partial^2 \hat{T}}{\partial \hat{y}^2} \tag{73}$$

The horizontal velocity $\hat{u}$ is eliminated by cross-differentiating between eqs. (71) and (72), leading to

$$\frac{\partial^2 \hat{v}}{\partial \hat{x}^2} + \frac{\partial^2 \hat{v}}{\partial \hat{y}^2} = \mathrm{Ra}_H\frac{\partial^2 \hat{T}}{\partial \hat{x}^2} \tag{74}$$

Equations (73) and (74) must be solved subject to the following isothermal impermeable wall conditions

$$\hat{v} = \hat{T} = 0 \quad \text{at } \hat{y} = 0, 1 \tag{75}$$

The initial condition to this transient problem is arbitrary; however, inspired by visual observations of Bénard cells, it makes sense to assume sinusoidal variation in $\hat{x}$ and exponential variation in $\hat{t}$,

$$\hat{T} = \theta(\hat{y})e^{p\hat{t}+ia\hat{x}}$$

$$\hat{v} = V(\hat{y})e^{p\hat{t}+ia\hat{x}} \tag{76}$$

This assumption transforms the $(\hat{T}, \hat{v})$ problem into one of determining the $\hat{y}$-profiles θ and V subject to

Momentum: $$-\alpha^2 V + V'' = -\alpha^2 \mathrm{Ra}_H \theta \tag{77}$$

Energy: $$p\theta - V = -\alpha^2 \theta + \theta'' \tag{78}$$

Boundary conditions: $\quad \theta = V = 0, \quad \text{at } \hat{y} = 0, 1 \tag{79}$

Finally, we zero in on the condition of *neutral stability* ($p = 0$, see Chapter 6) and eliminating $V(y)$ between eqs. (77) and (78), we obtain

$$\theta^{\mathrm{IV}} - 2\alpha^2 \theta'' + \alpha^4 \theta = \alpha^2 \mathrm{Ra}_H \theta \tag{80}$$

This equation admits solutions of the form

$$\theta = C \sin(n\pi\hat{y}) \tag{81}$$

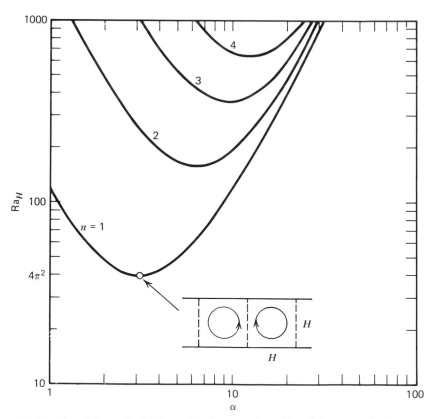

Figure 11.16 The minimum Rayleigh number for neutrally stable cellular convection in a porous layer heated from below.

where C is an arbitrary constant and n is an integer so that the boundary conditions (79) are satisfied. Combining eqs. (81) and (80) we learn that the assumed flow is neutrally stable when

$$\text{Ra}_H = \frac{(n^2\pi^2 + \alpha^2)^2}{\alpha^2} \tag{82}$$

Equation (82) is the result of the stability experiment we chose to perform, using the conduction solution (63) and (64) as a subject. It says that the assumed disturbance (n, α) is likely to exist, neither growing nor decaying, if the Rayleigh number is as high as in eq. (82). Figure 11.16 shows the dependence of Ra_H on both α and n. As Ra_H increases above zero, the first chance of convective heat transfer materializes at $n = 1$ and $\partial\text{Ra}_H/\partial\alpha = 0$, that is, when

$$\text{Ra}_H = 4\pi^2 = 39.5, \quad (n = 1, \alpha = \pi) \tag{83}$$

As is shown in Fig. 11.16, the disturbance $(n = 1, \alpha = \pi)$ represents rolls with square cross-sections, that is, rolls whose horizontal dimension is equal to the porous layer thickness H.

From a heat transfer engineering standpoint, result (83) implies that only for Rayleigh numbers less than approximately 40 is the heat transfer rate accurately predicted by the pure conduction estimate. For Rayleigh numbers much larger than 40, engineers are forced to rely on experimental and numerical measurements. Figure 11.17 shows Cheng's [1] compilation of time-averaged Nusselt number measurements reported by nine independent investigators [16–24]. Figure 11.17 is perhaps the best tool to use in engineering calculations; it shows that above $\text{Ra}_H \sim 40$ the conduction-based Nusselt number

$$\text{Nu} = \frac{q''}{k\,\Delta T/H} \tag{84}$$

is a strong function of the Rayleigh number, and that there is considerable scatter in the data.

The dependence of Nu on Ra_H (Fig. 11.17) is one of the most challenging and fascinating problems in convective heat transfer today. The large volume of activity during the past 30 years in this area has been reviewed most recently by Cheng [1]. Due to space limitations, in this section we discussed only the first step in the direction of theoretically understanding the heat transfer message of Fig. 11.17. This first step was the linearized stability experiment which, via eq. (83), anticipates correctly the sharp knee in the Nu–Ra_H curve of Fig. 11.17.

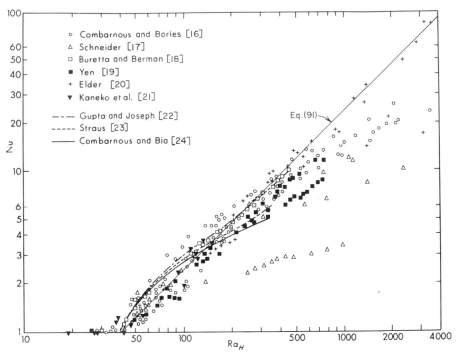

Figure 11.17   Heat transfer measurements in a porous layer heated from below [1].

The slope of the $\log \mathrm{Nu}$–$\log \mathrm{Ra}_H$ curve for $\mathrm{Ra}_H > 40$ can be predicted on the basis of a pure scaling argument. Educated by the linear stability experiment that showed the formation of a local cellular flow, we can think of a convection-dominated regime made up of rising warm plumes coexisting with descending cold plumes. Let L (unknown) be the thickness of each plume, that is, the horizontal extent of an elementary cell (Fig. 11.18). Let δ_H be the thermal boundary layer thickness across which the bottom–top ΔT takes place. (Note that in the convection-dominated regime the scale of $\partial T/\partial y$ is not $\Delta T/H$: it is $\Delta T/\delta_H$ with δ_H unknown.)

The cellular flow model contains two building blocks:

1.   *The core region*: A vertical counterflow of length H and thickness L.

2.   *The end regions*: Two identical boundary layer flows of length L and thickness δ_H.

Invoking the conservation principles in the core region, we write

$$\text{Momentum balance:} \qquad \frac{v}{L} \sim \frac{Kg\beta}{\nu} \frac{\Delta T}{L} \qquad\qquad (85)$$

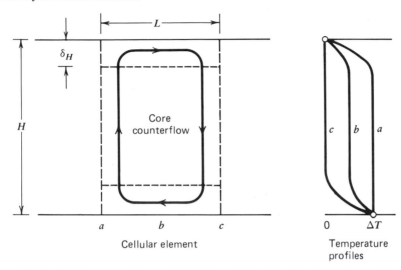

Cellular element   Temperature
                                    profiles

Figure 11.18 Cellular convection model for determining the Nusselt number scaling in the convection regime of layers heated from below.

Energy balance:
$$v\frac{\Delta T}{H} \sim \alpha\frac{\Delta T}{L^2} \tag{86}$$

Results:
$$L \sim H\,\mathrm{Ra}_H^{-1/2} \tag{87}$$

$$v \sim Kg\beta\Delta T/\nu \tag{88}$$

A relationship for δ_H is obtained by stating that the enthalpy flow vertically through the core must match the heat conducted vertically through the end region

$$\rho v L c_p\,\Delta T \sim kL\Delta T/\delta_H$$

or

$$\delta_H \sim H\,\mathrm{Ra}_H^{-1} \tag{89}$$

Therefore, in the convection-dominated regime, the Nusselt number scales as

$$\mathrm{Nu} = \frac{\text{actual heat transfer}}{\text{pure conduction}} \sim \frac{kL\Delta T/\delta_H}{kL\Delta T/H} \sim \frac{H}{\delta_H} \sim \mathrm{Ra}_H \tag{90}$$

This means that in Fig. 11.17 the slope of the $\log\mathrm{Nu}$–$\log\mathrm{Ra}_H$ curve in the convective domain must be 1 (one): this result is confirmed by the bulk of the heat transfer measurements. For heat transfer calculations then, a reasonable

Nu (Ra_H) correlation is

$$\text{Nu} = \begin{cases} 1, & \text{if } \text{Ra}_H < 40 \\ \text{Ra}_H/40, & \text{if } \text{Ra}_H > 40 \end{cases} \tag{91}$$

SYMBOLS

c_P	specific heat at constant pressure
C	constant
g	gravitational acceleration
H	vertical dimension
k	thermal conductivity of the fluid–matrix composite
K	permeability
$K_{1,2}$	core parameters, in the shallow layer solution
L	horizontal dimension
L_x	length of horizontal penetration
L_y	length of vertical penetration
Nu	Nusselt number [eq. (37)]
P	pressure
Q	heat transfer rate
q	Oseen function
q_e	value of q near the top and bottom ends of the porous layer [eq. (40)]
r_0	radius of cylindrical porous system
Ra_{H, L, r_0}	Darcy-modified Rayleigh numbers based on H, L, and r_0, respectively
t	time
t_f	time of boundary layer development [eq. (17)]
T	temperature
T_0	reference temperature
ΔT	temperature difference ($T_{\text{warm}} - T_{\text{cold}}$)
u	horizontal velocity
v	vertical velocity
V	disturbance velocity profile, dimensionless
x	horizontal coordinate
y	vertical coordinate
z	dimensionless vertical coordinate
α	thermal diffusivity $k/(\rho c_P)_f$; also, disturbance wave number [eq. (76)]
β	coefficient of thermal expansion
δ	boundary layer thickness
δ_e	the ratio (end region thickness)$/H$ [eqs. (46) and (47)]
δ_f	final boundary layer thickness [eq. (18)]

δ_H	thermal boundary layer thickness near the horizontal boundary of a porous layer heated from below [eq. (89)]
θ	disturbance temperature profile
λ	Oseen function
ν	kinematic viscosity
ρ	density
$(\rho c_P)_f$	fluid heat capacity
$(\rho c_P)_s$	solid matrix heat capacity
σ	heat capacity ratio [eq. (5)]
ϕ	porosity
ψ	streamfunction
$(\ )_b$	base solution for stability experiment
$(\ )'$	disturbance quantities
$(\hat{\ })$	dimensionless variables in the stability study of a layer heated from below
$(\ )_*$	dimensionless variables in the boundary layer analysis of the high-Ra_H regime
$(\ )_\infty$	pertaining to the core region of a porous layer in the high-Ra_H regime

REFERENCES

1. P. Cheng, Heat transfer in geothermal systems, *Adv. Heat Transfer*, Vol. 14, 1979, pp. 1–105.

2. D. Poulikakos and A. Bejan, Unsteady natural convection in a porous layer, *Phys. Fluids*, Vol. 26, 1983, pp. 1183–1191.

3. J. E. Weber, The boundary layer regime for convection in a vertical porous layer, *Int. J. Heat Mass Transfer*, Vol. 18, 1975, pp. 569–573.

4. B. K. C. Chan, C. M. Ivey and J. M. Barry, Natural convection in enclosed porous media with rectangular boundaries, *J. Heat Transfer*, Vol. 92, 1970, pp. 21–27.

5. C. G. Bankvall, Natural convection in vertical permeable space, *Wärme Stoffübertrag.*, Vol. 7, 1974, pp. 22–30.

6. K. J. Schneider, Investigation of the influence of free thermal convection on heat transfer through granular material, International Institute of Refrigeration, Proceedings, 1963, pp. 247–253.

7. S. Klarsfeld, *Revue Generale Thermique*, Vol. 108, 1970, pp. 1403–1423.

8. A. Bejan, On the boundary layer regime in a vertical enclosure filled with a porous medium, *Lett. Heat Mass Transfer*, Vol. 6, 1979, pp. 93–102.

9. P. G. Simpkins, and P. A. Blythe, Convection in a porous layer, *Int. J. Heat Mass Transfer*, Vol. 23, 1980, pp. 881–887.

10. A. Bejan, and C. L. Tien, Natural convection in a horizontal porous medium subjected to an end-to-end temperature difference, *J. Heat Transfer*, Vol. 100, 1978, pp. 191–198; also Vol. 105, 1983, pp. 683, 684.

11. K. L. Walker, and G. M. Homsy, Convection in a porous cavity, *J. Fluid Mech.*, Vol. 87, 1978, pp. 449–474.

12. C. E. Hickox, and D. K. Gartling, A numerical study of natural convection in a horizontal porous layer subjected to an end-to-end temperature difference, *J. Heat Transfer*, Vol. 103, 1981, pp. 797–802.

13. A. Bejan, Lateral intrusion of natural convection into a horizontal porous structure, *J. Heat Transfer*, Vol. 103, 1981, pp. 237–241.

14. A. Bejan, Natural convection in a vertical cylindrical well filled with porous medium, *Int. J. Heat Mass Transfer*, Vol. 23, 1980, pp. 726–729.

15. E. R. Lapwood, Convection of a fluid in a porous medium, Proc. Cambridge Philos. Soc., Vol. 44, 1948, pp. 508–521.

16. M. A. Combarnous, and S. A. Bories, Hydrothermal convection in saturated porous media, *Adv. Hydrosci.*, Vol. 10, 1975, pp. 231–307.

17. K. J. Schneider, Paper 11-4, Proc. 11th Int. Congr. Refrig., 1963.

18. R. J. Buretta and A. S. Berman, Convective heat transfer in a liquid saturated porous layer, *J. Appl. Mech.*, Vol. 43, 1976, pp. 249–253.

19. Y. C. Yen, Effects of density inversion on free convective heat transfer in porous layer heated from below, *Int. J. Heat Mass Transfer*, Vol. 17, 1974, pp. 1349–1356.

20. J. W. Elder, Steady free convection in a porous medium heated from below, *J. Fluid Mech.*, Vol. 27, 1967, pp. 29–48.

21. T. Kaneko, M. F. Mohtadi, and K. Aziz, An experimental study of natural convection in inclined porous media, *Int. J. Heat Mass Transfer*, Vol. 17, 1974, pp. 485–496.

22. V. P. Gupta and D. D. Joseph, Bounds for heat transport in a porous layer, *J. Fluid Mech.*, Vol. 57, 1973, pp. 491–514.

23. J. M. Strauss, Large amplitude convection in porous media, *J. Fluid Mech.*, Vol. 64, 1974, pp. 51–63.

24. M. A. Combarnous, and P. Bia, Combined free and forced convection in porous media, *Soc. Pet. Eng. J.*, Vol. 11, 1971, pp. 399–405.

PROBLEMS

1. Consider the heat transfer through a shallow porous layer with different end-temperatures. Show that the heat transfer rate scales as $Q \gtrsim kH\Delta T/\delta_f$ and explain the basis for the " $<$ " sign in this inequality [see eqs. (25), regime (IV)].

2. Show that in the $\mathrm{Ra}_H \to \infty$ limit, the Nusselt number of shallow layers [eq. (44)] acquires a form similar to that for the high-Ra_H regime [eq. (38)].

3. Consider the phenomenon of lateral intrusion of natural convection into an isothermal porous structure. Show that the length of lateral penetration scales as $L_x \sim H\,\mathrm{Ra}_H^{1/2}$, where H is the vertical dimension of the medium and Ra_H is the Rayleigh number based on H. Show that the heat transfer effected by the phenomenon scales as $Q \sim k\,\Delta T\,\mathrm{Ra}_H^{1/2}$: obtain this result by integrating the wall heat flux over the penetration length L_x.

4. Consider the phenomenon of vertical penetration of natural convection into an isothermal porous medium. Prove that the vertical penetration length scales as $L_y \sim L\,\mathrm{Ra}_L$, where L is the horizontal extent of the porous layer. Determine the scale of the heat flux around the region penetrated by natural convection. Integrating the heat flux over the length L_y, show that the overall heat transfer rate scales as $Q_y \sim k\,\Delta T\,\mathrm{Ra}_L$.

12

Numerical Methods
in Convection

DIMOS POULIKAKOS
Department of Mechanical Engineering
University of Illinois, Chicago

SHIGEO KIMURA
Department of Earth and Space Sciences
University of California, Los Angeles

The course offered in Chapters 1–11 stressed the availability of more than one approximate analytical method for solving a given convection problem. In addition, this course emphasized the individuality (the freedom of choice) that must be exercised by each problem-solver in choosing a particular method for a particular problem. In the present chapter, we add to this problem-solving arsenal numerical methods or computer-aided solutions to convection problems. This methodology is not only of contemporary interest—computers are revolutionizing the way we learn and practice many facets of engineering—but promises to become a much larger component of the engineering language of the future. Numerical methods deserve the student's scrutiny, because they make possible the solution to many problems that are not solvable by analytical methods. In addition, numerical methods in many cases provide *less expensive* alternatives to solving problems that might in fact be solvable based on classical analytical methods. The current status of numerical methods in heat transfer engineering and the growing interest in perfecting these methods are amply documented in the heat transfer journals (e.g., [1]).

THE FINITE-DIFFERENCE FORM OF A DIFFERENTIAL EQUATION

The first problem in the pursuit of solutions to differential equations, partial or ordinary, is placing the equations in a form that allows a numerical solution, that is, in a form that can be handled by a computer. We overcome this problem by thinking of the flow and heat transfer medium not as a continuum but as a conglomerate of discrete points in the domain of interest, points bathed by the heat and flow phenomenon of interest. These discrete points or nodes define a *grid* or a *mesh*; they represent the intersections of a *net* with which we overlay the continuum of interest. In doing so, we depart dramatically from the thinking that historically has led to analytical solutions of convection problems (see Chapters 2–5, 11, 12). Instead, we focus on a category of *experiments* based on measuring or recording information at specific points to finally piece together a picture of the entire region of interest. This is why numerical solutions are very appropriately referred to by many as *numerical experiments*.

There exists more than one method that enables us to construct a computer-compatible model of a differential equation. The objective of this section is not to list all the existing methods in a cookbook fashion, but to present two methods that, in our opinion, are effective. Both of these methods combine simplicity in formulation with numerical accuracy in results and, for these reasons, are being used routinely in all corners of the field of convection.

Truncated Taylor Series Expansions

The usual procedure for obtaining the finite-difference form of a partial differential equation with this method is to approximate all the partial derivatives in the equation by means of their Taylor series expansions, appropriately truncated depending on the desired accuracy. Obviously, the procedure for obtaining the finite-difference form of an ordinary differential equation is a simplified version of this approach.

Consider a two-dimensional region covered with a rectangular mesh as is shown in Fig. 12.1a. The mesh consists of m vertical and n horizontal lines positioned at intervals of length Δx and Δy, respectively. For simplicity, we assume that both Δx and Δy are constant. Assume that the unknown variable $\phi(x, y, \cdots)$ possesses continuous derivatives. We can approximate a first derivative of ϕ, say, $\partial \phi / \partial x$, in the following three ways:

1. Forward difference approximation. A forward Taylor expansion of ϕ in x around the point (i, j) yields (Fig. 12.1a):

$$\phi_{i+1, j} = \phi_{i, j} + \left(\frac{\partial \phi}{\partial x} \right)_{i, j} (x_{i+1, j} - x_{i, j}) + \frac{1}{2!} \left(\frac{\partial^2 \phi}{\partial x^2} \right)_{i, j} (x_{i+1, j} - x_{i, j})^2$$

$$+ \frac{1}{3!} \left(\frac{\partial^3 \phi}{\partial x^3} \right)_{i, j} (x_{i+1, j} - x_{i, j})^3 + \text{(higher order terms)} \qquad (1)$$

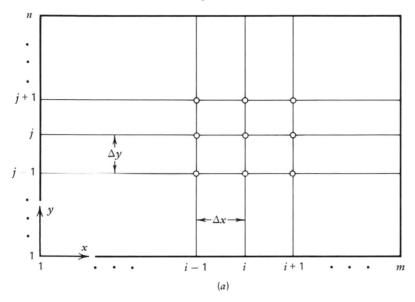

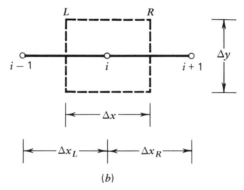

Figure 12.1   (*a*) The discretization of a two-dimensional continuous domain; (*b*) control volume.

Solving for $(\partial\phi/\partial x)_{i,j}$, we obtain

$$\left(\frac{\partial\phi}{\partial x}\right)_{i,j} = \frac{\phi_{i+1,j} - \phi_{i,j}}{\Delta x} - \frac{1}{2!}\left(\frac{\partial^2\phi}{\partial x^2}\right)_{i,j}\Delta x$$

$$- \frac{1}{3!}\left(\frac{\partial^3\phi}{\partial x^3}\right)_{i,j}(\Delta x)^2 + (\text{higher order terms}) \qquad (2)$$

where

$$\Delta x = x_{i+1,j} - x_{i,j}$$

If the desired accuracy is $O(\Delta x)$, the above equation simplifies to the *forward-difference* approximation of $\partial\phi/\partial x$:

$$\left(\frac{\partial\phi}{\partial x}\right)_{i,j} = \frac{\phi_{i+1,j} - \phi_{i,j}}{\Delta x} \tag{3}$$

Once again, it is worth keeping in mind the limitations of eq. (3) in approximating $\partial\phi/\partial x$, namely, (a) The truncation error in eq. (3) is $O(\Delta x)$, since all the terms in the Taylor series of order $(\Delta x)^2$ or greater are omitted, that is, the representation for $\partial\phi/\partial x$ is first-order accurate; (b) The truncation error decreases as Δx decreases; at the same time eq. (3) becomes a better approximation of $\partial\phi/\partial x$.

2. Backward difference approximation. A backward Taylor expansion of ϕ in x around the point (i, j) yields

$$\phi_{i-1,j} = \phi_{i,j} - \left(\frac{\partial\phi}{\partial x}\right)_{i,j}\Delta x + \frac{1}{2!}\left(\frac{\partial^2\phi}{\partial x^2}\right)_{i,j}(\Delta x)^2$$

$$- \frac{1}{3!}\left(\frac{\partial^3\phi}{\partial x^3}\right)_{i,j}(\Delta x)^3 + (\text{higher order terms}) \tag{4}$$

Following the previous procedure yields the *backward-difference* approximation of $\partial\phi/\partial x$,

$$\left(\frac{\partial\phi}{\partial x}\right)_{i,j} = \frac{\phi_{i,j} - \phi_{i-1,j}}{\Delta x} \tag{5}$$

The limitations of eq. (5) are identical to the limitations of eq. (3) and, for brevity, are not repeated here.

3. Centered difference approximation. The centered difference approximation of $\partial\phi/\partial x$ is derived by subtracting eq. (4) from eq. (1) to obtain

$$\phi_{i+1,j} - \phi_{i-1,j} = 2\left(\frac{\partial\phi}{\partial x}\right)_{i,j}\Delta x + \frac{1}{3}\left(\frac{\partial^3\phi}{\partial x^3}\right)_{i,j}(\Delta x)^3 + (\text{higher order terms}) \tag{6}$$

Next, solving for $(\partial\phi/\partial x)_{i,j}$ gives

$$\left(\frac{\partial\phi}{\partial x}\right)_{i,j} = \frac{\phi_{i+1,j} - \phi_{i-1,j}}{2\Delta x} - \frac{1}{3!}\left(\frac{\partial^3\phi}{\partial x^3}\right)_{i,j}(\Delta x)^2 + (\text{higher order terms}) \tag{7}$$

To second-order accuracy, eq. (7) can be rewritten as

$$\left(\frac{\partial\phi}{\partial x}\right)_{i,j} = \frac{\phi_{i+1,j} - \phi_{i-1,j}}{2\,\Delta x} \tag{8}$$

The truncation error in eq. (8) is $O(\Delta x)^2$, which makes expression (8) a better approximation of $\partial\phi/\partial x$ than the approximations presented in eqs. (3) and (5), in which the magnitude of the truncation error is $O(\Delta x)$.

A second-order accurate centered difference scheme for the second partial derivative of ϕ, $(\partial^2\phi/\partial x^2)$, can be easily derived by adding eqs. (1) and (4), solving for $(\partial^2\phi/\partial x^2)_{i,j}$, and omitting the terms of order $O(\Delta x)^2$ or higher. The result is:

$$\left(\frac{\partial^2\phi}{\partial x^2}\right)_{i,j} = \frac{\phi_{i+1,j} + \phi_{i-1,j} - 2\phi_{i,j}}{(\Delta x)^2} \tag{9}$$

Expressions analogous to (3), (5), and (9) can be obtained for all the partial derivatives of ϕ such as $\partial\phi/\partial y$, $\partial^2\phi/\partial y^2$, $\partial^2\phi/\partial x\,\partial y$, and so on. Thus, the centered-difference result for $\partial^2\phi/\partial x\,\partial y$ is

$$\left(\frac{\partial^2\phi}{\partial x\,\partial y}\right)_{i,j} = \frac{\phi_{i+1,j+1} - \phi_{i+1,j-1} - \phi_{i-1,j+1} + \phi_{i-1,j-1}}{4\Delta x\,\Delta y} \tag{10}$$

The derivation of eq. (10) is left as an exercise for the student. To illustrate how expressions like (3), (5), (9), and (10) can be used to "discretize" differential equations, consider the following example.

Example 1. Develop the finite-difference form of Poisson's equation.

$$\frac{\partial^2\phi}{\partial x^2} + \frac{\partial^2\phi}{\partial y^2} = q(x, y) \tag{11}$$

Solution 1. Using centered differences to approximate the partial derivatives in eq. (11) yields

$$\frac{\phi_{i+1,j} + \phi_{i-1,j} - 2\phi_{i,j}}{(\Delta x)^2} + \frac{\phi_{i,j+1} + \phi_{i,j-1} - 2\phi_{i,j}}{(\Delta y)^2} = q_{i,j} \tag{12}$$

Expression (12) can be solved for the value $\phi_{i,j}$ as a function of the value of ϕ at the four neighboring points $(i + 1, j)$, $(i - 1, j)$, $(i, j + 1)$, and $(i, j - 1)$.

$$\phi_{i,j} = \frac{1}{2}\left(\frac{(\Delta x)^2(\Delta y)^2}{(\Delta x)^2 + (\Delta y)^2}\right)\left(\frac{\phi_{i+1,j} + \phi_{i-1,j}}{(\Delta x)^2} + \frac{\phi_{i,j+1} + \phi_{i,j-1}}{(\Delta y)^2} - q_{i,j}\right) \tag{13}$$

In the special case where $\Delta x = \Delta y$ and $q = 0$, eq. (13) reduces to

$$\phi_{i,j} = \tfrac{1}{4}\left(\phi_{i+1,j} + \phi_{i-1,j} + \phi_{i,j+1} + \phi_{i,j-1}\right) \tag{14}$$

in other words, the value of ϕ at the point (i, j) is the average of the value of ϕ at the four neighboring points. Note that the case $q = 0$ corresponds to a pure conduction problem with ϕ as the temperature field (Chapter 1) or to a flow through porous medium problem with ϕ as the pressure field (Chapter 10).

We conclude our presentation of the Taylor series expansions discretization method with the following remarks. The user of this method has to bear in mind that the method assumes that the higher order derivatives of the unknown variable ϕ are not of significant magnitude. This assumption is not good when one encounters regions of steep variations of ϕ (e.g., exponential variations) and leads to sizeable errors. In addition, the Taylor series expansion method, even though relatively simple, is purely an analytical approximation and does not bring with it additional insight into the physics of the problem. Finally, the method is very accurate only in the limit $\Delta x \to 0$ and $\Delta y \to 0$. An alternative, more flexible discretization method is the *control-volume formulation* discussed next.

Control Volume Approach

Let us imagine that we divide the rectangular region of Figure 12.1a into a finite number of nonoverlapping control volumes (areas, since our region is two-dimensional). These control volumes are chosen such that each one completely engulfs only one grid point. The main idea behind this approach is to *integrate* the differential equation over each control volume using chosen (assumed) profiles for the unknown variable ϕ in order to evaluate the required integrals. Once we complete this task, we have in our hands the discretized form of the differential equation, a form that contains the values of ϕ for a group of neighboring grid points. It is worth noting that we are free to assume different profiles for ϕ to approximate different terms in the differential equation.

It is important to realize that if the differential equation expresses the conservation of some quantity like, for example, energy in the First Law of Thermodynamics for an *infinitesimal* control volume, then the discretization equation is the exact representation of the conservation of that quantity in a *finite* control volume. This is the main advantage of the control-volume formulation method: quantities such as mass, momentum, and energy are conserved exactly over any number of control volumes and, therefore, over the entire region, regardless of the coarseness of the mesh that defines the control volumes. To illustrate this method consider the following simple example.

Example 2. Derive the discretization equation of the following differential equation:

$$\frac{d^2\phi}{dx^2} = q(x) \tag{15}$$

Note that eq. (15), which describes a one-dimensional conduction heat transfer problem, is the one-dimensional simplification of eq. (11) discussed in example 1. The choice is intentional, for we plan to compare the resulting finite difference equations with those of example 1.

Solution 2.   Consider the three grid-point group shown in Fig. 12.1b. The positions of the left and right faces of the control volume are denoted by L and R, respectively. Integrating expression (15) from L to R yields

$$\left(\frac{d\phi}{dx}\right)_R - \left(\frac{d\phi}{dx}\right)_L = \int_L^R q\,dx \tag{16}$$

The time has come to select a profile for the variation of ϕ between the grid points of Fig. 12.1b. If we decide to go with a piecewise linear profile, eq. (16) can be rewritten as

$$\frac{\phi_{i+1} - \phi_i}{(\Delta x)_R} - \frac{\phi_i - \phi_{i-1}}{(\Delta x)_L} = q_i\,\Delta x \tag{17}$$

where we assumed that the value of q at point (i) prevails over the control volume surrounding it. If we further assume that points L and R are located midway between the grid-point pairs $(i-1)$, (i) and (i), $(i+1)$, respectively, $(\Delta x_L = \Delta x_R = \Delta x)$, eq. (17) takes the simple form

$$\phi_i = \tfrac{1}{2}\left[\phi_{i+1} + \phi_{i-1} - q_i(\Delta x)^2\right] \tag{18}$$

As we can easily see from eq. (13), result (18) is identical to what a centered-difference scheme would have given had it been used to discretize the differential equation (15). This finding is encouraging and adds to the credibility of both methods. However, if we had chosen a different profile for ϕ, that is, something other than a piecewise-linear shape, the two methods would have yielded different results. The student is urged to rederive result (18) by using profiles of his choice for ϕ (e.g., exponential, parabolic, etc.).

At this point we should ask ourselves: Are there any clear-cut criteria to decide when each method works best? We loosely answered this question in the opening lines of the discussion of the control volume approach, where we left the reader with the impression that the control volume method is, overall, the better method. This is our personal opinion. However, the success of any method depends greatly on the peculiarities of the problem in the hand of the numerical analyst, as well as on the analyst's talent and patience.

A generalization of the control volume method in two dimensions is presented in following sections by means of specific two-dimensional convection problems. Cases in which the Taylor-series expansion method is used to discretize a differential equation are also considered.

THE SOLUTION OF ORDINARY DIFFERENTIAL EQUATIONS

This section focuses on the solution of an ordinary differential equation of the type discussed in Chapters 2 and 4 where, with the help of a suitable *similarity variable*, we were able to transform the system of partial differential equations governing the heat and fluid flow in a boundary layer to a system of *ordinary differential equations* with the similarity variable as the only independent variable. Since the introduction of a similarity variable can, in some cases, transform the governing partial differential equations of a convection problem to ordinary differential equations, we find it necessary to provide the reader with some of the methods (tools) that are commonly used to solve an ordinary differential equation.

Depending on the boundary conditions to be satisfied by the solution, we can identify two large groups of problems in ordinary differential equations:

1. Initial value problems (IVP).

2. Two-point boundary value problems or, simply, boundary value problems (BVP).

In the first group of problems (IVP), as the name implies, all the conditions to be satisfied by the solution are specified at one point. A typical example of an IVP is the vibration of a mass–spring system (Fig. 12.2a). The motion of the point-mass m is described mathematically by the following second-order differential equation and initial conditions

$$y''(t) + \lambda^2 y(t) = 0 \qquad (19)$$

$$y(0) = 0, \qquad y'(0) = y_0 \qquad (20)$$

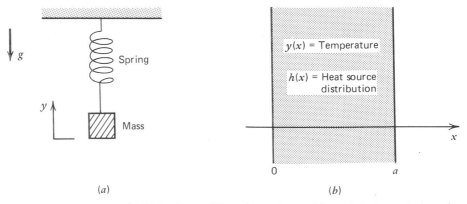

(a) $\qquad\qquad\qquad\qquad\qquad\qquad (b)$

Figure 12.2 Examples of initial value and boundary value problems: (a) the evolution of a spring-mass system; (b) the temperature distribution inside a one-dimensional conducting slab.

where t is the time and $(\ )'$ denotes differentiation with respect to time; the rest of the symbols are defined in Fig. 12.2a. The special feature of this problem (which makes it an IVP) is the fact that both conditions to be satisfied by the solution are imposed at $t = 0$, that is, at one point in time. In the example of Fig. 12.2, eqs. (20) require that at $t = 0$ the displacement is zero and that the initial velocity has a nonzero value, y_0. The solution of eq. (19) is straightforward; hence, it is omitted here for the sake of brevity.

The spring–mass system of the preceding example gives us good insight into the nature of an IVP. This type of problem is often the result of describing mathematically the evolution of a physical phenomenon in time. The phenomenon may evolve in time forever, in other words, the domain of the independent infinite, $0 \leq t < \infty$: there are no conditions to be satisfied by the solution other than the conditions at $t = 0$.

To understand the nature of the second group of problems in ordinary differential equations considered in this chapter (BVP), let us look at a specific example. Assume that we are asked to determine the steady state temperature distribution in a slab with an arbitrarily distributed heat source designated by aa function $h(x)$. The temperature on both sides of the slab is constant and equal to zero (Fig. 12.2b). Referring to the First Law of Thermodynamics (Chapter 1), the temperature in the slab is the solution of the following *linear* differential equation

$$y''(x) = h(x) \tag{21}$$

subject to

$$y(0) = y(a) = 0 \tag{22}$$

All the symbols in the above two equations are defined in Fig. 12.2b. Since conditions (22) are specified at *two* points in space ($x = 0$ and $x = a$), the problem is termed as a *two-point boundary value problem* or a *boundary value problem*. As was shown in Chapters 2, 4, 8, and 10, the similarity solution for the temperature and velocity distributions in both forced and free convection boundary layer flows results from a boundary value problem (BVP). Unfortunately, there is no available *general* numerical method for the solution of a BVP if the differential equation is *nonlinear*, which is frequently the case in convection.

A central objective of this chapter is to introduce the reader to a commonly used method for the solution of a BVP in convection, namely, the *shooting method*. Since the main idea in the shooting method is to convert a BVP to an IVP, let us first familiarize ourselves with an established algorithm for the solution of an IVP, the Runge–Kutta algorithm. In closing this section, we would like to stress that a detailed discussion of the various methods for solving initial value and boundary value problems is beyond the scope of this chapter. The interested reader will find more information on this topic in references such as [2, 3].

THE RUNGE–KUTTA METHOD

The Runge–Kutta method is one of the most commonly used algorithms for the solution of an IVP. There are several versions of the Runge–Kutta method, depending on the desired smallness of the truncation error. The fourth-order version of the Runge–Kutta method is probably the most popular. However, since the derivation of the fourth-order Runge–Kutta algorithm requires a considerable amount of algebra, we choose to derive the simpler second-order version of the method. It is worth noting that the derivation of the fourth-order algorithm is a routine generalization of the derivation of the second-order algorithm.

Consider a differential equation of the form

$$\frac{dy}{dx} = f(x, y) \tag{23}$$

In the case of an IVP, the value of $y(x)$ is assumed to be known at some starting point, that is, $y = y_i$ at $x = x_i$. Next, the value of $y(x)$ at $x = x_i + \Delta x$ is calculated by using the following recurrence formula

$$y_{i+1} = y_i + w_1 k_1 + w_2 k_2 \tag{24}$$

The *weighting functions* w_1 and w_2 in the above equation are determined based on a term-by-term matching of (24) with a second-order Taylor series expansion of y about y_i. The parameters k_1 and k_2 depend on dy/dx [eq. (23)] and are defined as

$$k_1 = f(x_i, y_i)\,\Delta x \tag{25}$$

$$k_2 = f(x_i + p\Delta x, y_i + qk_1)\,\Delta x \tag{26}$$

In the above two equations, p and q are constants to be evaluated by matching eq. (24) with the second-order Taylor series expansion

$$y_{i+1} = y_i + \frac{dy_i}{dx}\,\Delta x + \frac{1}{2!}\frac{d^2 y_i}{dx^2}(\Delta x)^2 + \cdots \tag{27}$$

It is worth stressing at this point that the main goal of the second-order Runge–Kutta method is to evaluate y_{i+1} in terms of y_i and dy_i/dx, which is known via eq. (23).

The obvious advantage of the method over a straightforward Taylor series [eq. (7)] is that it offers second-order accuracy without using $d^2 y_i/dx^2$, which is not necessarily well-behaved. Thus, we must find a way to replace the third term in eq. (27) with a combination of first-order derivatives. To do so, we first note that eq. (23) can be written as

$$\frac{d}{dx}\left(\frac{dy}{dx}\right) = \frac{\partial f}{\partial x} + \frac{\partial f}{\partial y}\frac{dy}{dx} \tag{28}$$

Combining eq. (28) with eqs. (23) and (27) yields

$$y_{i+1} = y_i + f(x_i, y_i)\,\Delta x$$

$$+ \frac{1}{2!}\left[\frac{\partial f(x_i, y_i)}{\partial x}\right.$$

$$\left. + \frac{\partial f(x_i, y_i)}{\partial y} f(x_i, y_i)\right](\Delta x)^2 \tag{29}$$

Next, using a double Taylor series expansion, we rewrite eq. (26) as

$$k_2 = f(x_i, y_i)\,\Delta x + \frac{\partial f}{\partial x} p(\Delta x)^2 + \frac{\partial f}{\partial y} qk_1\,\Delta x + \cdots \tag{30}$$

Substituting eqs. (25) and (30) into eq. (24) yields

$$y_{i+1} = y_i + w_1 f\Delta x + w_2 f\Delta x + w_2\left[\frac{\partial f}{\partial x} p(\Delta x)^2 + \frac{\partial f}{\partial y} qf(\Delta x)^2\right] \tag{31}$$

Second-order accuracy requires a term-by-term matching of eqs. (29) and (31), which yields

$$w_1 + w_2 = 1 \tag{32}$$

$$w_2 p = \tfrac{1}{2} \tag{33}$$

$$w_2 q = \tfrac{1}{2} \tag{34}$$

Since we have a system of three equations (32–34) with four unknowns (w_1, w_2, p, q), we can determine any three of the unknowns in terms of the fourth which remains to be chosen arbitrarily. The most common choice is $p = 1$ which, together with eqs. (32)–(34) yields

$$w_1 = w_2 = \tfrac{1}{2} \tag{35}$$

and

$$q = 1 \tag{36}$$

Therefore, the final recurrence formula for the second-order Runge–Kutta method is

$$y_{i+1} = y_i + \tfrac{1}{2}(k_1 + k_2) \tag{37}$$

$$k_1 = f(x_i, y_i)\,\Delta x \tag{38}$$

$$k_2 = f(x_i + \Delta x, y_i + k_1)\,\Delta x \tag{39}$$

The recurrence formula for the most commonly used fourth-order Runge–Kutta method is a result of a similar derivation that yields

$$y_{i+1} = y_i + \tfrac{1}{6}(k_1 + 2k_2 + 2k_3 + k_4) \tag{40}$$

$$k_1 = f(x_i, y_i)\,\Delta x \tag{41}$$

$$k_2 = f\left(x_i + \frac{\Delta x}{2}, y_i + \frac{k_1}{2}\right)\Delta x \tag{42}$$

$$k_3 = f\left(x_i + \frac{\Delta x}{2}, y_i + \frac{k_2}{2}\right)\Delta x \tag{43}$$

$$k_4 = f(x_i + \Delta x, y_i + k_3)\,\Delta x \tag{44}$$

The following is the sequence of steps necessary for integrating eq. (23) using the fourth-order Runge–Kutta method:

For the *given* initial value $f(x_i, y_i)$, calculate $k_1, \ldots, k_4$ via (41–44).

Substitute $k_1, \ldots, k_4$ into eq. (40) to obtain y_{i+1} which is the value of y at $x + \Delta x$.

Repeat the above process, to obtain $y_{i+2}, y_{i+3}, \ldots$, etc.

Up to this point we dealt only with first-order differential equations of the type shown in eq. (23). A reasonable question is then: Can we use the Runge–Kutta method to integrate differential equations of higher order? The answer to this question lies in the fact that a higher order differential equation can be recast into an equivalent system of first-order differential equations that must be solved simultaneously. To illustrate, consider the following example.

Example 3. Use the fourth-order Runge–Kutta algorithm to solve the equation, $d^2y/dx^2 + y = 0$, assuming that $y(0)$ and $y'(0)$ are known.

Solution 3. Define

$$Y = \frac{dy}{dx} \tag{45}$$

The original second-order differential equation can then be rewritten as

$$\frac{dY}{dy} + y = 0 \tag{46}$$

It is the system of first-order eqs. (45) and (46) that we now have to solve subject to the known initial conditions, say $y(0) = y_1$ and $Y(0) = y'(0) = Y_1$. With the help of eqs. (40–44) we can easily construct the following fourth-order Runge-Kutta algorithm in order to integrate simultaneously eqs. (45) and (46)

$$Y_{i+1} = Y_i + \tfrac{1}{6}(l_1 + 2l_2 + 2l_3 + l_4) \tag{47}$$

$$l_1 = -y_i \Delta x \tag{48}$$

$$l_2 = -\left(y_i + \frac{l_1}{2}\right)\Delta x \tag{49}$$

$$l_3 = -\left(y_i + \frac{l_2}{2}\right)\Delta x \tag{50}$$

$$l_4 = -(y_i + l_3)\Delta x \tag{51}$$

$$y_{i+1} = y_i + \tfrac{1}{6}(k_1 + 2k_2 + 2k_3 + k_4) \tag{52}$$

$$k_1 = Y_i \Delta x \tag{53}$$

$$k_2 = \left(Y_i + \frac{k_1}{2}\right)\Delta x \tag{54}$$

$$k_3 = \left(Y_i + \frac{k_2}{2}\right)\Delta x \tag{55}$$

$$k_4 = (Y_i + k_3)\Delta x \tag{56}$$

The student will find it instructive to write a small program and to obtain the numerical solution for this example, using initial conditions of his choice. The numerical solution can then be compared with the exact solution, which for our example is straightforward.

THE SHOOTING METHOD

The shooting method views the two-point BVP as an IVP. In general, the method requires the "guessing" of additional initial conditions in order to start the numerical integration. For example, if we are to solve a fourth-order ordinary differential equation, we need four initial conditions, say, $y(0)$, $y'(0)$, $y''(0)$, and $y'''(0)$ to be able to utilize available numerical algorithms such as the fourth-order Runge–Kunge method. However, some of the initial conditions, say, $y''(0)$ and $y'''(0)$ are not usually available in a BVP; instead a BVP requires the satisfaction of conditions at the other end, say, $y(a)$ and $y'(a)$. Hence, a reasonable procedure is to guess values for the unspecified initial conditions [$y''(0)$ and $y'''(0)$] and to try to adjust these values simultaneously so that the numerical solution will eventually satisfy the originally prescribed conditions at $x = a$. To illustrate the nature of the shooting method, consider the following example.

Example 4. Formulate the numerical solution for the fourth-order differential equation

$$y^{IV} + yy' = f(x) \tag{57}$$

subject to

$$y(0) = y'(0) = y(a) = y'(a) = 0 \tag{58}$$

where $f(x)$ is a known function.

Solution 4. We first note that the above is a two-point BVP. As in example 3, let us transform the fourth-order eq. (57) into an equivalent system of first-order differential equations.

$$Y_3' + yY_1 = f(x) \tag{59}$$

$$Y_2' = Y_3 \tag{60}$$

$$Y_1' = Y_2 \tag{61}$$

$$y' = Y_1 \tag{62}$$

The initial conditions for eqs. (59) and (62) are known via eq. (58). However, $Y_2(0) = y''(0)$ and $Y_3(0) = y'''(0)$ are not available and must be guessed in order for the numerical integration process to begin. The solution based on these first guesses does not necessarily satisfy the conditions at $x = a$: the degree of disagreement at $x = a$ can be expressed in a least-square sense as

$$E = y^2(a) + [y'(a)]^2 \tag{63}$$

The main goal of the shooting method is to adjust systematically the guessed values for $Y_2(0)$ and $Y_3(0)$ so that the shooting error E becomes as small as possible. The procedure followed to minimize E is shown schematically in Fig. 12.3. The point corresponding to the initial guess is marked with a cross on the $Y_2(0)-Y_3(0)$ plane, and our goal is to determine the values of $Y_2(0)$ and $Y_3(0)$ that result in $E \simeq 0$. Although conceptually simple, this procedure is difficult to follow systematically, due to the fact that the relation between E and $Y_2(0)$, $Y_3(0)$ is not known explicitly.

If the number of missing initial conditions is more than two, the task of depicting the minimization of E graphically becomes a difficult one since it

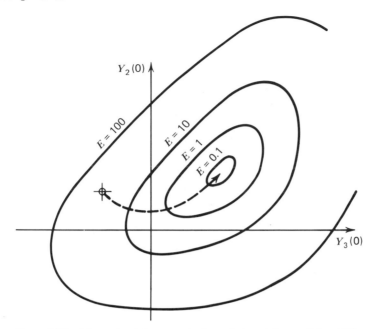

Figure 12.3 Schematic of the numerical search for minimum E [eq. (63)].

takes place in a higher order space. On the other hand, if there is only one initial condition missing and if the differential equation is linear or weakly nonlinear, we can minimize E in a relatively simple fashion by using linear interpolation. To be more specific, let us consider the heat conduction problem described by eqs. (21) and (22). In this problem, only one initial condition is missing, $y'(0)$. According to eq. (63), the E function for this case is

$$E = [y'(a)]^2 \tag{64}$$

The shooting optimization process is shown graphically in Fig. 12.4. After the first two trial integrations, the third guess for $y'(0)$ can be calculated by linearly interpolating between the first two guesses.

$$y_3'(0) = \frac{\sqrt{E_1}\, y_2'(0) + \sqrt{E_2}\, y_1'(0)}{\sqrt{E_1} + \sqrt{E_2}} \tag{65}$$

In the above expression, the subscripts denote the iteration cycle. The procedure can be repeated until E, as given by eq. (64), becomes as small as demanded by the desired accuracy of the solution.

Since the degree of nonlinearity of a problem is not always easy to determine by inspection, the linear interpolation technique is always worth trying. However, the student should bear in mind that success in solving a

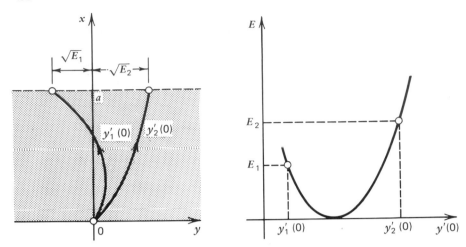

Figure 12.4 The minimization of E through linear interpolation for the heat conduction problem of Fig. 12.2b.

nonlinear problem by a *linear* interpolation technique is an exception rather than a rule [2].

We continue this section with the method first proposed by Nachtheim and Swigert [4], which is especially suitable for solving a two-point BVP with some boundary conditions specified at an initial point and others specified at limits that must be approached at large values of the independent variable. A good example of this type of problem is the equation arising in the Falkner and Skan's treatment of a laminar boundary layer flow over a flat plate (see the last section and Problem 8 of Chapter 2):

$$f''' = -ff'' + \beta(f'^2 - 1) \tag{66}$$

$$\eta = 0, \qquad f = f' = 0 \tag{67}$$

$$\eta \to \infty, \qquad f' \to 1 \tag{68}$$

The new feature in the above set of equations is the requirement to satisfy a boundary condition at infinity (eq. 68). To get around this difficulty we replace infinity with $\eta_{\max}$, where $\eta_{\max}$ is the value of the independent variable η at the edge of the boundary layer. In other words, we think that for $\eta > \eta_{\max}$ we are for all practical purposes outside the boundary layer. Since this problem is a BVP, our task is to determine the value of the missing initial condition $f''(0)$, for which the boundary condition at the edge is satisfied, that is

$$f'_{\text{edge}}[f''(0)] = 1 \tag{69}$$

where $f'_{\text{edge}} \equiv f'(\eta_{\max})$. The dependence of f'_{edge} on $f''(0)$, in this problem eq. (69), is not explicit; it is expressed via an integration of eq. (66). Hence, we must first establish an approximately valid functional form of eq. (69). We do this by first defining

$$z \equiv f''(0) \tag{70}$$

and expanding the left-hand side of eq. (69) in a Taylor series: retaining only the first two terms to preserve linearity, we obtain

$$f' + \frac{\partial f'}{\partial z} \Delta z = 1 \quad \text{at } \eta - \eta_{\max} \tag{71}$$

Solving eq. (71) for Δz yields a correction to the initial guess for z, provided that $\partial f'/\partial z$ can be evaluated at $\eta = \eta_{\max}$. To evaluate $\partial f'/\partial z$ at $\eta = \eta_{\max}$ we first differentiate eq. (66) with respect to z

$$f_z''' = -\left(f f_z'' + f'' f_z \right) + 2\beta f' f_z' \tag{72}$$

where subscript z indicates a partial derivative with respect to z. The initial conditions are

$$f_z = f_z' = 0, \qquad f_z'' = 1 \quad \text{at } \eta = 0 \tag{73}$$

Equation (72), which is sometimes called the *perturbation equation*, must be integrated at the same time with eq. (66) to yield $f_z'(\eta_{\max})$.

The magnitude of the error at $\eta = \eta_{\max}$ is estimated in the same way as in example 4 (eq. 63). In order to satisfy the asymptotic boundary condition [eq. (68)], it is desired that all the higher order derivatives of f at $\eta = \eta_{\max}$ vanish. However, if we require $f'' = 0$ at $\eta = \eta_{\max}$ the condition $f''' = 0$ at $\eta = \eta_{\max}$ is automatically satisfied by virtue of eq. (66). Therefore, it is enough to impose $f'' = 0$ at the edge of the boundary layer. A two-term Taylor expansion of this condition reads

$$f'' + f_z'' \Delta z = 0 \tag{74}$$

The error that we must minimize can then be expressed as

$$E = \left(f_z' \Delta z + f' - 1 \right)^2 + \left(f_z'' \Delta z + f'' \right)^2 \tag{75}$$

where all the dependent variables must be evaluated at $\eta = \eta_{\max}$. If E is a well-behaved function of Δz, it is likely that $\partial E/\partial(\Delta z)$ vanishes at the point of extremum. The vanishing of $\partial E/\partial(\Delta z)$ yields a correction for z for the next iteration

$$\Delta z = \frac{f_z'(1 - f') - f_z'' f''}{f_z'^2 + f_z''^2} \tag{76}$$

The derivatives in the above equation are obtained by integrating the perturbation equation (72). The procedure described above must be repeated with the corrected initial condition $z + \Delta z$ until E becomes as small as desirable.

Even though the shooting method provides satisfactory results for various types of two-point BVP's, it fails to solve problems in which the differential equations are very sensitive to the choice of the missing initial condition. In such cases, even a very accurate guess may lead to divergence in the middle of the way, that is, before the integration process reaches the other boundary. Hence, the error E (eq. 75) cannot be defined. For this type of problem, we recommend a different technique which has the advantage that it holds firmly over the entire domain. This new technique involves approximation of the differential equation at $n + 1$ discrete points, $x_0, x_1, \ldots, x_n$; each derivative is approximated by a finite-difference representation in the fashion described in detail at the beginning of this chapter. Together with the boundary conditions specified at $x = x_0$ and $x = x_n$, the finite-difference formulation leads to a system of $n - 1$ simultaneous algebraic equations, which yield the solution of the differential equation at $n - 1$ discrete points in the domain. If the original differential equation is linear, then the set of algebraic equations generated is also linear. Ferziger [7] outlines several direct methods able to handle systems of linear algebraic equations. However, if the original differential equation is nonlinear, the set of algebraic equations is also nonlinear; hence, we have to introduce some type of linearization procedure in order to solve the system iteratively. The essence of this technique resembles that of the technique used to solve partial differential equations: this topic is discussed in the next section.

THE SOLUTION OF PARTIAL DIFFERENTIAL EQUATIONS: THE VORTICITY–STREAMFUNCTION FORMULATION

The general equations governing the conservation of mass, momentum, and energy in convection were derived in Chapter 1 [eqs. (6), (20), and (39)]. A very important task is to devise numerical techniques for solving these equations subject to appropriate boundary conditions that depend on the specific engineering problems of interest. Note that the nonlinear nature of the momentum and energy equations does not permit general analytical solutions for convection problems. As stated in the introductory section of this chapter, we limit our treatment to *two-dimensional incompressible* flow situations. More general formulations of three-dimensional convection and compressible flow problems can be found in [5–7].

One of the main difficulties in determining the flow field via eq. (20) of Chapter 1 is the unknown pressure field that appears by means of pressure gradients in the momentum equation.

Since the pressure field is usually not of primary interest, it is eliminated by taking the "curl" of eq. (20). Defining the vorticity as

$$\boldsymbol{\omega} = \nabla \times \boldsymbol{v} \tag{77}$$

we obtain the vorticity transport equation

$$\rho \frac{D\omega}{Dt} = \mu \nabla^2 \omega + \nabla \times F \tag{78}$$

For example, in the (x, y), (u, v) Cartesian system (Fig. 1.2) and for a *Boussinesq-incompressible fluid* [eq. (17) of Chapter 4], the above equation reads

$$\frac{\partial \omega}{\partial t} + u \frac{\partial \omega}{\partial x} + v \frac{\partial \omega}{\partial y} = \nu \left(\frac{\partial^2}{\partial x^2} + \frac{\partial^2}{\partial y^2} \right) \omega + g\beta \frac{\partial T}{\partial x} \tag{79}$$

Using the definition of streamfunction

$$u = \frac{\partial \psi}{\partial y} \tag{80}$$

$$v = -\frac{\partial \psi}{\partial x} \tag{81}$$

we can easily rewrite eq. (77) as

$$-\nabla^2 \psi = \omega \tag{82}$$

According to the *vorticity–streamfunction* formulation we are asked to solve the system of equations (79)–(82) and the two-dimensional version of eq. (43a) of Chapter 1, subject to appropriate boundary conditions. Note that eq. (79) refers to natural convection problems where the gravitational acceleration points in the negative y direction (Chapters 4 and 5). In the case of forced convection, the last term of the right-hand side of eq. (79) vanishes.

The various methods used to discretize and solve eqs. (79)–(82) along with eq. (43a) of Chapter 1 are outlined in the next section. Figure 12.5 provides a bird's-cyc view of the successive steps that lead to the solution.

In closing this subsection, it is worth reviewing some of the limitations and disadvantages of the *vorticity–streamfunction* formulation:

1. If the pressure field is an important result, then the task of computing pressures from the vorticity field does away with the computational savings of the vorticity–streamfunction formulation.

2. The method cannot be extended to three-dimensional convection problems (the streamfunction cannot be defined in general in three-dimensional situations).

3. The value of vorticity at the boundaries is not always easy to specify.

For the above reasons, the vorticity–streamfunction formulation, even though a very effective method, should not be considered a panacea.

METHODS FOR SOLVING THE STREAMFUNCTION EQUATION

As was shown in the preceeding section, the streamfunction–vorticity formulation of convection problems requires the simultaneous treatment of three equations (streamfunction, vorticity transport, and energy) to yield the temperature and flow fields of the problem of interest. The purpose of this section is

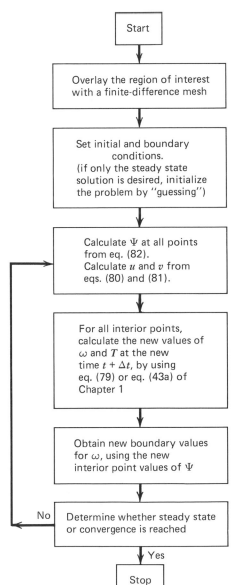

Figure 12.5 Flowchart for the numerical solution in the vorticity–streamfunction formulation. (Note: If the vorticity equation does not depend on the temperature solution, the energy equation can be solved independently after the flow field has been determined.)

to outline a few methods for solving the first of the three equations, namely, the streamfunction equation.

We begin with the observation that the streamfunction equation

$$\frac{\partial^2 \psi}{\partial x^2} + \frac{\partial^2 \psi}{\partial y^2} = q \tag{83}$$

is an elliptic (Poisson) equation: we seek to solve it in a certain domain subject to the condition that ψ is known on the boundary of that domain (this is the so-called Dirichlet condition). The finite-difference form of eq. (83) was derived in the first section of this chapter (eq. 13 with $\phi \equiv \psi$) for the rectangular region shown in Fig. 12.1a. In the special case of vanishing q and $\Delta x = \Delta y$, eq. (13) reduces to eq. (14) which states that the value of ψ at an interior point is equal to the arithmetic mean of the values of ψ at four neighboring points. Application of eq. (13) at all interior points of the domain shown in Fig. 12.1a yields a system of $(n-2) \times (m-2)$ algebraic equations with $(n-2) \times (m-2)$ unknowns; in this problem, the value of ψ on the boundary is assumed known. In principle, at least, the solution to this system of algebraic equations yields the value of ψ at all the grid points inside the domain. However, the task of solving this system is difficult (tedious and time-consuming) when the number of grid points is large, which is always the case in most convection problems where an accurate solution is required. For this reason, alternative iterative methods have been developed for solving the streamfunction equation. In what follows we discuss three of the most common iterative methods and out of these we recommend the last one.

Liebman's and Richardson's Iterative Schemes

According to Liebman's method, we initially guess the value of ψ at all interior grid points, since the value of ψ on the boundary of the domain is prescribed. These values are represented by ψ_{ij}^0, where the superscript "0" indicates the zeroth iteration and the subscripts i and j identify the point in the rectangular region of interest (Fig. 12.1a). The values of ψ for the next iteration are computed by applying eq. (13) at all interior points starting from the interior point located at the lower-left corner, namely, point (2, 2); hence,

$$\psi_{2,2}^1 = \frac{1}{2\left[1 + (\Delta x/\Delta y)^2\right]}$$

$$\times \left[\psi_{3,2}^0 + \psi_{1,2}^0 + \left(\frac{\Delta x}{\Delta y}\right)^2 \psi_{2,3}^0 + \left(\frac{\Delta x}{\Delta y}\right)^2 \psi_{2,1}^0 - (\Delta x)^2 q_{2,2}\right] \tag{84}$$

The computation at the next point [eqs. (2) and (3)], $\psi_{2,3}^1$, can be improved since $\psi_{2,2}^1$ is now available via eq. (84), etc. Following this logic we deduce a general recurrence formula for computing the value of ψ at any point (i, j) during the $(p + 1)$th iteration

$$\psi_{i,j}^{p+1} = \frac{1}{2\left[1 + \left(\dfrac{\Delta x}{\Delta y}\right)^2\right]}$$

$$\times \left[\psi_{i+1,j}^{p} + \psi_{i-1,j}^{p+1} + \left(\frac{\Delta x}{\Delta y}\right)^2 \psi_{i,j+1}^{p} + \left(\frac{\Delta x}{\Delta y}\right)^2 \psi_{i,j-1}^{p+1} - (\Delta x)^2 q_{i,j}\right]$$

$$(85)$$

The above equation is *Liebman's iterative formula*: it can be applied to all interior points. It is worth noting that in marching from left to right and from bottom to top in Fig. 12.1a, we calculate $\psi_{i-1,j}^{p+1}$ and $\psi_{i,j-1}^{p+1}$ *before* we use eq. (85) to compute $\psi_{i,j}^{p+1}$. Hence, it is only reasonable to use these "improved" values during the computation of $\psi_{i,j}^{p+1}$ [eq. (85)]. Failure to use $\psi_{i-1,j}^{p+1}$ and $\psi_{i,j-1}^{p+1}$ in calculating $\psi_{i,j}^{p+1}$ results in a slower scheme known as *Richardson's iterative formula*,

$$\psi_{i,j}^{p+1} = \frac{1}{2\left[1 + \left(\dfrac{\Delta x}{\Delta y}\right)^2\right]}$$

$$\times \left[\psi_{i+1,j}^{p} + \psi_{i-1,j}^{p} + \left(\frac{\Delta x}{\Delta y}\right)^2 \psi_{i,j+1}^{p} + \left(\frac{\Delta x}{\Delta y}\right)^2 \psi_{i,j-1}^{p} - (\Delta x)^2 q_{i,j}\right]$$

$$(86)$$

At this point we should ask ourselves: As the number of iterations increases $(p \to \infty)$ do results (85) and (86) converge to the solution of the original streamfunction equation? The answer is "yes": the proof, not included in this chapter due to space restrictions, can be found in references such as Refs. 5 and 8. In practice, since a large number of iterations is time-consuming, we accept as solution to the streamfunction equation the group of ψ values that satisfy eq. (83) within an acceptable error that depends on the desired degree of accuracy.

The Successive Overrelaxation Method (SOR)

This method was developed by Frankel [9] and Young [10] and it insures faster convergence than Liebman's method. To derive a general recurrence formula for the SOR, add and subtract $\psi_{i,j}^P$ from eq. (85). After rearranging, we obtain

$$\psi_{i,j}^{P+1} = \psi_{i,j}^P + \frac{1}{2\left[1 + (\Delta x/\Delta y)^2\right]}$$

$$\times \left\{ \psi_{i+1,j}^P + \psi_{i-1,j}^{P+1} + \left(\frac{\Delta x}{\Delta y}\right)^2 \psi_{i,j+1}^P + \left(\frac{\Delta x}{\Delta y}\right)^2 \psi_{i,j-1}^{P+1} \right.$$

$$\left. - 2\left[1 + \left(\frac{\Delta x}{\Delta y}\right)^2\right]\psi_{i,j}^P - (\Delta x)^2 q_{i,j} \right\} \tag{87}$$

According to Liebman's method, when the term in the curly brackets in the above equation becomes identically equal to zero ($\psi_{i,j}^{P+1} = \psi_{i,j}^P$), the solution to eq. (83) has been obtained. According to the SOR, the term contained by the curly brackets is multiplied by a relaxation factor $\gamma(\gamma \neq 1)$,

$$\psi_{i,j}^{P+1} = \psi_{i,j}^P + \frac{\gamma}{2\left[1 + (\Delta x/\Delta y)^2\right]}$$

$$\times \left\{ \psi_{i+1,j}^P + \psi_{i-1,j}^{P+1} + \left(\frac{\Delta x}{\Delta y}\right)^2 \psi_{i,j+1}^P + \left(\frac{\Delta x}{\Delta y}\right)^2 \psi_{i,j-1}^{P+1} \right.$$

$$\left. - 2\left[1 + \left(\frac{\Delta x}{\Delta y}\right)^2\right]\psi_{i,j}^P - (\Delta x)^2 q_{i,j} \right\} \tag{88}$$

For overrelaxed convergence, it is required that $1 < \gamma < 2$ [5]. The equation is underrelaxed if $0 < \gamma < 1$; underrelaxation slows down the convergence, however, it is useful in problems where steep gradients are present. The optimum value of parameter γ, that is, the value that yields the fastest convergence depends on the mesh size and the shape of the domain, and is usually determined by trial-and-error. For a rectangular domain of size $(m - 1) \times \Delta x$ by $(n - 1) \times \Delta y$ (Fig. 12.1a), the optimum value γ_{opt} is [5]

$$\gamma_{opt} = \frac{2\left[1 - (1 - \lambda)^{1/2}\right]}{\lambda} \tag{89}$$

where

$$\lambda = \left[\frac{\cos\left(\frac{\pi}{m - 1}\right) + \left(\frac{\Delta x}{\Delta y}\right)^2 \cos\left(\frac{\pi}{n - 1}\right)}{1 + \left(\frac{\Delta x}{\Delta y}\right)^2}\right]^2 \tag{90}$$

The optimum value of γ given by eq. (89) is based on the assumption that the problem of interest is described by eq. (83) alone. In natural convection, on the other hand, the flow and temperature fields are coupled through the temperature dependent buoyancy term in the momentum equation. In such problems the value of γ_{opt} should be determined by trial-and-error, that is, the result listed as eq. (89) should be used only as a first guess.

It can be shown [5] that the number of iterations p required for convergence is proportional to $(m - 1) \times (n - 1)$ if the SOR is used, and proportional to $[(m - 1) \times (n - 1)]^2$ if Liebman's method is used. Therefore, the SOR method is much more efficient for large meshes. More on existing methods appropriate for solving elliptic Poisson equations can be found in Refs. 5–7.

THE UPWIND FINITE-DIFFERENCE METHOD

In this section we consider the discretized form of a partial differential equation of the advection-diffusion type, with special emphasis on the vorticity transport equation (78). This equation type is very important in the field of convection, because it includes as special cases both the energy equation and the equation of conservation of species in mass transfer. Thus, the discretizing method developed for the vorticity transport equation in this section can be extended to other conservation equations found in convection: this is in fact one advantage of the vorticity–streamfunction approach to solving the flow part of a convection problem. In what follows, we first describe the numerical procedure for the time-dependent (parabolic) form of the vorticity equation, and finish with a brief discussion of the procedure for steady state (elliptic) equations of the advection–diffusion type.

Probably the most commonly used finite-difference method is *upwind-differences*. In addition to simplicity, this method possesses the highly desirable property of numerical stability. To derive the upwind differences, we will follow the control volume approach described in the introductory section of this chapter and, for simplicity, we present only the derivation for the one-dimensional vorticity equation. In dimensionless form, this equation can be expressed as

$$\frac{\partial \omega}{\partial t} = -\frac{\partial(u\omega)}{\partial x} + \frac{1}{\text{Re}} \frac{\partial^2 \omega}{\partial x^2} \tag{91}$$

where Re is a dimensionless number. Consider next a control volume around a point of position x, as shown in Fig. 12.1b. The point value of ω will refer to the average of ω over the control volume (cv). We then state that "the net accumulation of ω in the cv = the net flux of ω into cv by advection + the net flux of ω into cv by diffusion."

The net accumulation of a certain period Δt is

$$\left[\int_L^R \omega^{t+\Delta t} \, dx - \int_L^R \omega^t \, dx \right] \Delta y \tag{92}$$

According to the upwind method, the advection flux of ω into the cv depends on the flow directions at R and L. The vorticity at $(i-1)$ is convected *toward* the cv if $u_L > 0$, therefore, ω_L in the advection term takes the same value as ω_{i-1}. On the contrary, if $u_L < 0$, then the vorticity is convected out of the cv, and the value of ω_L is again set equal to the value upstream, in this case, ω_i. The velocity at L can be taken as the algebraic average of the velocities at the two nodes situated on either side of L, in other words, it is assumed that the velocity varies linearly from one node to another. A similar argument holds at the other boundary, R. In summary we have the following formula for the rate of increase of ω in the cv over a certain time period:

$$\omega_{i-1}\frac{|u_{i-1}+u_i|}{2} - \omega_i\frac{|u_{i+1}+u_i|}{2}$$

$$\text{for } u_{i-1}+u_i > 0 \quad \text{and} \quad u_{i+1}+u_i > 0$$

$$\omega_{i-1}\frac{|u_{i-1}+u_i|}{2} + \omega_{i+1}\frac{|u_{i+1}+u_i|}{2}$$

$$\text{for } u_{i-1}+u_i > 0 \quad \text{and} \quad u_{i+1}+u_i < 0$$

$$-\omega_i\frac{|u_{i-1}+u_i|}{2} - \omega_i\frac{|u_{i+1}+u_i|}{2} \tag{93}$$

$$\text{for } u_{i-1}+u_i < 0 \quad \text{and} \quad u_{i+1}+u_i > 0$$

$$\omega_{i+1}\frac{|u_{i+1}+u_i|}{2} - \omega_i\frac{|u_{i-1}+u_i|}{2}$$

$$\text{for } u_{i-1}+u_i < 0 \quad \text{and} \quad u_{i+1}+u_i < 0$$

The above four statements together with eq. (92) can be combined into a single expression, which is much better suited for computer programming.

$$-\left[\frac{1}{2}\int_t^{t+\Delta t}\{(u_R - |u_R|)\omega_{i+1} + (u_R + |u_R| - u_L + |u_L|)\omega_i\right.$$

$$\left. - (u_L + |u_L|)\omega_{i-1}\} \, dt\right]\Delta y \tag{94}$$

where $u_R = \frac{1}{2}(u_{i+1}+u_i)$ and $u_L = \frac{1}{2}(u_{i-1}+u_i)$.

Although it has some physical meaning in the context of vorticity transport, the above result is essentially based on two postulates, namely, the linear variation of velocity from node to node and the special way to choose the vorticity value at the boundary of the cv. The value of vorticity convected into or out of the cv is always evaluated at the *upstream* node; hence, the name *upwind finite-differences* for this particular discretization method.

It is also possible to approximate the advection term by centered finite-differences.

$$\frac{\partial(u\omega)}{\partial x} \cong \frac{(u_{i+1}\omega_{i+1} - u_{i-1}\omega_{i-1})}{2\Delta x} \tag{95}$$

There is nothing wrong with this alternative, besides, the centered finite difference has second-order accuracy with regard to truncation errors. However, it turns out that the use of centered finite-differences in the advection term poses severe limitations on the size of spatial mesh (i.e., on Δx and Δy) in order to achieve numerical stability. For this reason upwind finite-differences are generally preferred for the advection term.

The next step is to apply centered finite-differences to approximate the diffusion term in eq. (91). The vorticity flux due to diffusion through the cv boundaries can be approximated as

$$\frac{1}{Re}\left(\int_t^{t+\Delta t}\frac{\partial^2\omega}{\partial x^2}\,dt\right)\Delta y \cong \frac{1}{Re}\left[\int_t^{t+\Delta t}\frac{\omega_{i+1}+\omega_{i-1}-2\omega_i}{(\Delta x)^2}\,dt\right]\Delta y \tag{96}$$

Using eqs. (92)–(96), the conservation law for one-dimensional advection–diffusion becomes

$$\left[\int_L^R(\omega^{t+\Delta t}-\omega^t)\,dx\right]\Delta y =$$

$$-\left[\frac{1}{2}\int_t^{t+\Delta t}\{(u_R - |u_R|)\omega_{i+1} + (u_R + |u_R| - u_L + |u_L|)\omega_i\right.$$

$$\left. -(u_L + |u_L|)\omega_{i-1}\}\,dt\right]\Delta y$$

$$+\left[\frac{1}{Re}\int_t^{t+\Delta t}\frac{\omega_{i+1}+\omega_{i-1}-2\omega_i}{(\Delta x)^2}\,dt\right]\Delta y \tag{97}$$

For short enough intervals Δx and Δt, the integrals may be approximated as $\int_z^{z+\Delta z} f(z)\,dz \cong \bar{f}(z)\Delta z$, where $\bar{f}$ is the mean value of $f(z)$ over the Δz interval; in this way, eq. (97) reduces to

$$\frac{\omega_i^{t+\Delta t}-\omega_i^t}{\Delta t} = -\frac{1}{2\Delta x}\left[(u_R - |u_R|)\omega_{i+1}^t + (u_R + |u_R| - u_L + |u_L|)\omega_i^t\right.$$

$$\left. -(u_L + |u_L|)\omega_{i-1}^t\right] + \frac{\omega_{i+1}^t + \omega_{i-1}^t - 2\omega_i^t}{Re(\Delta x)^2} \tag{98}$$

which is the final form of finite-difference approximation for eq. (91).

The above derivation can be carried out in exactly the same manner for the two-dimensional case. Referring to the index notation shown in Fig. 12.1a, the final result for the two-dimensional case is

$$\frac{\omega_{i,j}^{t+\Delta t} - \omega_{i,j}^{t}}{\Delta t} = -\frac{1}{2\,\Delta x}$$

$$\times \left[(u_R - |u_R|)\omega_{i+1,j}^{t} + (u_R + |u_R| - u_L + |u_L|)\omega_{i,j}^{t} - (u_L + |u_L|)\omega_{i-1,j}^{t} \right]$$

$$-\frac{1}{2\,\Delta y}\left[(v_A - |v_A|)\omega_{i,j+1}^{t} + (v_A + |v_A| - v_B + |v_B|)\omega_{i,j}^{t} - (v_B + |v_B|)\omega_{i,j-1}^{t} \right]$$

$$+\frac{1}{Re}\left[\frac{\omega_{i+1,j}^{t} - 2\omega_{i,j}^{t} + \omega_{i-1,j}^{t}}{(\Delta x)^2} + \frac{\omega_{i,j+1}^{t} - 2\omega_{i,j}^{t} + \omega_{i,j-1}^{t}}{(\Delta y)^2} \right] \tag{99}$$

where $v_A = \frac{1}{2}(v_{i,j+1} + v_{i,j})$, $v_B = \frac{1}{2}(v_{i,j} + v_{i,j-1})$, and $A =$ above, $B =$ below. In the above expression, all the velocities must be evaluated at time t by differentiating the stream function: the centered finite-difference scheme is commonly used for this purpose. Since eq. (99) can be solved explicitly for $\omega_{i,j}^{t+\Delta t}$ (which is the value of vorticity at node (i, j) for a new time level $t + \Delta t$), it is convenient to recast this equation in the following form

$$\omega_{i,j}^{t+\Delta t} = a\omega_{i+1,j}^{t} + b\omega_{i-1,j}^{t} + c\omega_{i,j+1}^{t} + d\omega_{i,j-1}^{t} + e\omega_{i,j}^{t} + \Delta t\, S_{i,j}^{t} \tag{100}$$

where

$$a = \Delta t\left[\frac{1}{Re(\Delta x)^2} - \frac{1}{2\,\Delta x}(u_R - |u_R|) \right]$$

$$b = \Delta t\left[\frac{1}{Re(\Delta x)^2} + \frac{1}{2\,\Delta x}(u_L + |u_L|) \right]$$

$$c = \Delta t\left[\frac{1}{Re(\Delta y)^2} - \frac{1}{2\,\Delta y}(v_A - |v_A|) \right] \tag{101}$$

$$d = \Delta t\left[\frac{1}{Re(\Delta y)^2} + \frac{1}{2\,\Delta y}(v_B + |v_B|) \right]$$

$$e = 1 - \left[\frac{\Delta t(u_R + |u_R| - u_L + |u_L|)}{2\,\Delta x} + \frac{\Delta t(v_A + |v_A| - v_B + |v_B|)}{2\,\Delta y} \right.$$

$$\left. + \frac{2\,\Delta t}{Re(\Delta x)^2} + \frac{2\,\Delta t}{Re(\Delta y)^2} \right]$$

In eq. (100), $S_{i,j}$ stands for a possible source term, such as the buoyancy term encountered in natural convection.

Concerning the mesh size, we may ask: what is the largest possible size that preserves numerical stability and yields acceptable accuracy? Unfortunately, a generally valid answer to this question does not exist, since the answer depends not only on the individual problem but also on the individual problem solver. One way of determining and improving the accuracy of a certain solution is to first obtain a rough result with a coarse mesh and then to refine this result gradually while fixing other parameters (Δt may have to be changed for stability reasons). Plotting an overall result (such as the overall Nusselt number) versus grid size is an effective way to visualize the range of $(\Delta x, \Delta y)$ values where the grid has become fine enough, that is, the range beyond which further decreases in Δx and Δy do not yield significant changes in the overall result.

Regarding the maximum allowable time step Δt, recall that the finite-difference equations (100) can be viewed as a matrix operation whereby the matrix whose entires are $(a \cdots e)$'s is operating on ω^t in order to yield $\omega_{i,j}^{t+\Delta t}$. From this point of view it is also known that the following conditions must be satisfied by the coefficients a–e in eq. (100) in order for the numerical scheme to be stable [11].

$$a, b, c, d, e \geq 0 \tag{102}$$

Coefficients a, b, c, and d are always positive primarily due to the use of the upwind scheme, and e can be made positive if Δt is sufficiently small. This last condition determines the largest time step Δt.

$$\Delta t \leq \cfrac{1}{\cfrac{u_R + |u_R| - u_L + |u_L|}{2\Delta x} + \cfrac{v_A + |v_A| - v_B + |v_B|}{2\Delta y} + \cfrac{2}{\mathrm{Re}}\left[\cfrac{1}{(\Delta x)^2} + \cfrac{1}{(\Delta y)^2}\right]} \tag{103}$$

The student should keep in mind that condition (103) is not absolute, since the stability analysis is only applicable to linear systems, that is, to cases where coefficients a–e are constant. Therefore, in the work of solving an actual problem, it is advisable to use a time step whose magnitude is smaller than that calculated from eq. (103), say, 20 percent smaller.

Many problems in convection heat transfer are *steady state problems* that can be solved based on the same discretization of the governing equations as before, this time neglecting the time derivatives. The nonlinear system of algebraic equations obtained in this manner may be solved iteratively. The most commonly used iterative method is referred to as the *successive substitution* or *Gauss–Seidel iteration*, however, it is essentially the same as Liebman's method. The method is based on the immediate use of the new updated values wherever possible. Mathematically, this method reduces to solving the following equation for $\omega_{i,j}$:

$$\omega_{i,j}^{p+1} = A\omega_{i+1,j}^{p} + B\omega_{i-1,j}^{p+1} + C\omega_{i,j+1}^{p} + D\omega_{i,j-1}^{p+1} + ES_{i,j}^{p} \tag{104}$$

where the coefficients $A-E$ depend on the particular method used for discretization. Employing upwind differences for advection terms, Gosman et al. [12] found

$$A = \frac{\left(A_R + \frac{\Delta y}{\Delta x} \frac{1}{Re} \right)}{F}$$

$$B = \frac{\left(A_L + \frac{\Delta y}{\Delta x} \frac{1}{Re} \right)}{F}$$

$$C = \frac{\left(A_A + \frac{\Delta x}{\Delta y} \frac{1}{Re} \right)}{F}$$

$$D = \frac{\left(A_B + \frac{\Delta x}{\Delta y} \frac{1}{Re} \right)}{F}$$

$$E = -\frac{1}{F}$$

$$F = A_R + A_L + A_A + A_B + \frac{2}{Re} \left(\frac{\Delta y}{\Delta x} + \frac{\Delta x}{\Delta y} \right)$$

$$A_R = \frac{1}{8} \left\{ \left(\psi_{i+1,j-1} + \psi_{i,j-1} - \psi_{i+1,j+1} - \psi_{i,j+1} \right) \right.$$
$$\left. + |\psi_{i+1,j-1} + \psi_{i,j-1} - \psi_{i+1,j+1} - \psi_{i,j+1}| \right\}$$

$$A_L = \frac{1}{8} \left\{ \left(\psi_{i-1,j+1} + \psi_{i,j+1} - \psi_{i-1,j-1} - \psi_{i,j-1} \right) \right.$$
$$\left. + |\psi_{i-1,j+1} + \psi_{i,j+1} - \psi_{i-1,j-1} - \psi_{i,j-1}| \right\} \qquad (105)$$

$$A_A = \frac{1}{8} \left\{ \left(\psi_{i+1,j+1} + \psi_{i+1,j} - \psi_{i-1,j+1} - \psi_{i-1,j} \right) \right.$$
$$\left. + |\psi_{i+1,j+1} + \psi_{i+1,j} - \psi_{i-1,j+1} - \psi_{i-1,j}| \right\}$$

$$A_B = \frac{1}{8} \left\{ \left(\psi_{i-1,j-1} + \psi_{i-1,j} - \psi_{i+1,j-1} - \psi_{i+1,j} \right) \right.$$
$$\left. + |\psi_{i-1,j-1} + \psi_{i-1,j} - \psi_{i+1,j-1} - \psi_{i+1,j}| \right\}$$

In order for the Gauss–Seidel iteration to converge, coefficients $A-D$ must be less than or at most equal to unity [13]. This condition is satisfied by expressions (105), since $A + B + C + D = 1$. It is worth noting that the same condition is also satisfied by the coefficients in the discretized streamfunction equation (5). This convergence condition applies strictly to linear systems, that is, to cases where $A-D$ are constant during the iteration. If these coefficients vary significantly from iteration to iteration due to strong nonlinear feedback, the convergence may be impaired. One way to remedy this problem is to *underrelax* the dependent variable ω_{ij} by a certain amount in each iteration cycle. An equation similar to eq. (88) may be used for this purpose, however, this time the relaxation factor γ must be set equal to a number between 0 and 1.

OBSERVATIONS

Boundary Conditions

Determining the discretized form of the boundary conditions in a convection problem is a necessary task, which is not always straightforward. Some "easy to deal with" boundary conditions on temperature and streamfunction are:

(i)   **Specified boundary temperature.**   No problem here, we simply assign on the boundary the given value for the temperature.

(ii) **Specified boundary heat flux.**   If the boundary is perpendicular to, say, the x-axis, the boundary condition reads: $q_{\text{bdry}} = -k \times (\partial T/\partial x_{\text{bdry}})$. Therefore, the reasonable thing to do is to discretize $(\partial T/\partial x_{\text{bdry}})$ according to the first section of the chapter and solve the resulting discretization equation to obtain the boundary temperature as a function of q_{bdry} and the temperature at one or two interior grid points next to the boundary, depending on the desired accuracy.

(iii) **Impermeable boundary.**   This condition translates to $\psi = $ constant along the boundary. The most popular choice is $\psi = 0$.

Unfortunately, it is not always so simple to specify the discretized form of the boundary conditions in a convection problem. For example, how do we determine the value of vorticity on the boundaries? How do we deal with boundary conditions at infinity when we examine unbounded spaces? What is the proper choice of boundary conditions at places of *inflow* or *outflow* in the region of interest? The answers to these questions are listed in Table 12.1; however, their derivation is beyond the scope of this chapter. The interested reader will be able to find more information on this topic in the computational heat transfer and fluid mechanics literature, for example, in Ref. 5.

False Diffusion

A frequent criticism of the upwind differencing scheme for discretizing the momentum and energy equations is based on the so-called *false diffusion*.

Qualitatively speaking, it can be shown [6] that the upwind scheme becomes identical to the central differencing scheme if we reduce the diffusion coefficient (related to the viscosity or the thermal diffusivity, depending on the equation in hand) in the upwind scheme, by some amount termed as *the false diffusion*. Therefore, since the diffusion coefficient in the central differencing scheme is the real diffusion coefficient one is tempted to argue that the false diffusion misrepresents reality and constitutes an undesirable feature of the upwind differencing scheme. However, this argument is not correct, for it criticizes the upwind scheme assuming that the central differencing scheme is accurate. In reality the central differencing scheme is only accurate in weakly convective flows (e.g., low-Rayleigh numbers in natural convection). In strongly convective flows, this scheme fails and false diffusion is actually desirable because it tends to correct the wrong implications of the central differencing [6]. Other interesting features of the false diffusion effect are that it is small at weakly convective flows, for it directly depends on velocities that are also small in this case. Also, the false diffusion effect varies with the angle at which the flow crosses the grid lines, being maximum when this angle is 45° [14].

EXAMPLES OF SOLVED PROBLEMS

This last section is designed to give the reader a feel for how the numerical techniques described in this chapter work. One way to achieve this goal is to list a few solved examples and, without going into detail, share with the reader some of the experience that the solvers of these examples accumulated in the process of setting up and carrying out the numerical work. For the sake of brevity we discuss these examples in tabular form.

1. Natural convection in a triangular enclosure filled with porous medium.

Reference	D. Poulikakos and A. Bejan [15].
Problem statement	Natural convection in a triangular (A-shaped) enclosure filled with porous medium. The horizontal (bottom) wall of the cavity was hot and the sloped (top) walls cold.
Method used	The upwind differencing method was used to approximate the nonlinear convection terms in the vorticity and energy equations. The diffusion terms were discretized via central differencing. The streamfunction equation was solved with the SOR method.
Comments	The grid was uniform in x and y with $\Delta x = (\Delta y)L/H$. A grid fineness of 41×41 was enough

Table 12.1 Boundary Conditions for Streamfunction, Vorticity, and Temperature

Boundary Description	ψ	ω	T
Solid impermeable wall; specified wall temperature	$\psi_{i,J} = 0$	$\omega_{i,J} = \dfrac{-2(\psi_{i,J+1} - \psi_{i,J})}{(\Delta y)^2}$ or $\omega_{i,J} = \dfrac{7\psi_{i,J} - 8\psi_{i,J+1} + \psi_{i,J+2}}{2(\Delta y)^2}$	$T_{i,J} = T_0$ or $T_{i,J} = f(x)$
Impermeable and adiabatic wall	$\psi_{i,J} = 0$	$\omega_{i,J} = \dfrac{-2(\psi_{i,J+1} - \psi_{i,J})}{(\Delta y)^2}$ or $\omega_{i,J} = \dfrac{7\psi_{i,J} - 8\psi_{i,J+1} + \psi_{i,J+2}}{2(\Delta y)^2}$	$T_{i,J} = T_{i,J+1}$ or $T_{i,j} = \frac{1}{3}(4T_{i,J+1} - T_{i,J+2})$
Axis of symmetry	$\psi_{i,1} = 0$	$\omega_{i,1} = 0$	$T_{i,1} = T_{i,2}$
Far field conditions; the velocity U_∞, temperature T_∞, and entrainment velocity v_∞ are assumed known.	$\psi_{i,N} = \int_{y_1}^{y_N} u(y)\, dy$	$\omega_{i,N} = \dfrac{2(-\psi_{i,N-1} + \psi_{i,N} - U_\infty\,\Delta y)}{(\Delta y)^2}$ ($v_\infty = 0$ was assumed)	$T_{i,N} = T_\infty$

Inflow conditions, inflow velocity, and temperature profiles (u,T) are assumed known and v is allowed to develop

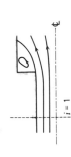

$i = 1$

$$\psi_{1,j} = \int_{y_j}^{y_N} u(y)\,dy$$

$$\omega_{1,j} = -\frac{\partial u}{\partial y}\bigg|_{1,j} + \frac{\partial v}{\partial x}\bigg|_{1,j}$$

$$= -\frac{\partial u}{\partial y}\bigg|_{1,j} - \frac{\psi_{1,j} + \psi_{3,j} - 2\psi_{2,j}}{(\Delta x)^2}$$

$$T_{1,j} = T_1$$

Outflow conditions

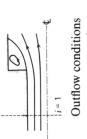

$i = M$

$$\psi_{M,j} = 2\psi_{M-1,j} - \psi_{M-2,j}$$

$$\omega_{M,j} = \omega_{M-1,j}$$

$$T_{M,j} = T_{M-1,j}$$

Concave sharp corner

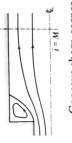

$i = 1$

$j = 1$

$$\psi_{1,1} = 0$$

$$\omega_{1,1} = 0$$

$$T_{1,1} = \tfrac{1}{2}(T_{1,2} + T_{2,1})$$

Convex sharp corner

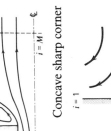

$J + 1$

$J = J$

$i = I \quad I + 1$

$\omega_{I,J}^A$

$\omega_{I,J}^B$

T_0

$q_0'' = 0$

$$\psi_{I,J} = 0$$

To calculate $\omega_{I,J+1}$,

$$\omega_{I,J}^A = \frac{-2(\psi_{I,J+1} - \psi_{I,J})}{(\Delta y)^2}$$

To calculate $\omega_{I+1,J}$,

$$\omega_{I,J}^B = \frac{-2(\psi_{I+1,J} - \psi_{I,J})}{(\Delta x)^2}$$

To calculate $T_{I,J+1}$,

$T_{I,J}^A = T_0$;

To calculate $T_{I+1,J}$,

$T_{I,J}^B = T_{I+1,J}$

449

to provide accurate solutions. The time step depended on the grid size and the Rayleigh number. Large Ra values required fine grids and small time steps. For example, the time step $\Delta t = 0.001$ was needed for Ra = 100 and a 41 × 41 grid. The results were more accurate at high values of the Rayleigh number and low values of the height/length aspect ratio, where false diffusion is not important. Representative results are shown in Fig. 12.6. Note that the Rayleigh number is based on the height of the enclosure.

2. Natural convection in a square enclosure.

Reference   S. Kimura and A. Bejan [16].
Problem statement   Natural convection in a square enclosure heated and cooled from the side: the right wall is hot, the left is cold, and the top and bottom walls are adiabatic.
Method used   The Allen–Southwell method (Problem 5) was used in the vorticity and energy equations. The streamfunction equation was solved with the SOR method.
Comments   The patterns shown in Fig. 12.7 were obtained using a uniform grid with 21 × 21 nodes. The steady state equations were solved iteratively, starting from the pure conduction solution ($\psi = 0$ everywhere, and linear temperature variation in the horizontal direction). The convergence criterion was

$$\frac{\sum_i^n \sum_j^n |\phi_{i,j}^{k+1} - \phi_{i,j}^k|}{\sum_i^n \sum_j^n |\phi_{i,j}^{k+1}|} \leq 10^{-5}$$

where ϕ stands for either T or ω, and superscript k indicates iteration order. Note that the last drawing in Fig. 12.7 shows the pattern of *heatlines* [16] discussed in the last section of Chapter 1.

3. Viscous eddy generated by a moving top wall.

Reference   O. R. Burggraf [17].
Problem statement   The top wall is moving with a specified constant speed and has a specified temperature. The other three walls are stationally and have a temperature lower than that of the top wall.

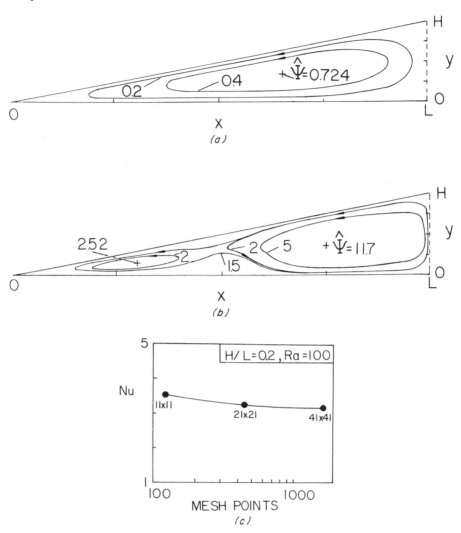

Figure 12.6 Numerical results for porous medium natural convection in a triangular enclosure with $H/L = 0.2$ [15]. (a) Steady state streamline pattern, Ra = 100; (b) Steady state streamline pattern, Ra = 1000; (c) The effect of grid fineness on the accuracy of heat transfer calculations.

Method   The vorticity and energy equations were discretized by using centered finite differences for both advection and diffusion terms; underrelaxation was used at large Reynolds numbers. The grid was uniform with 41 × 41 nodes.

Comments   The effect of buoyancy was not taken into account in this problem. Stable solutions were obtained for Reynolds numbers as high as 1000. However, as

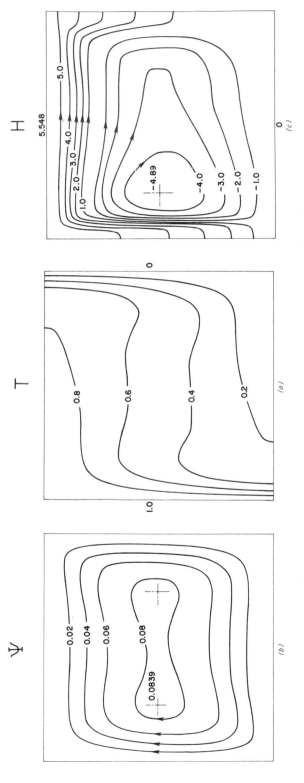

Figure 12.7 Numerical results for steady state natural convection in a square enclosure filled with Pr = 7 fluid, at Ra = 1.4×10^5 [16]: (a) temperature pattern; (b) streamline pattern; (c) heatline pattern (see Chapter 1).

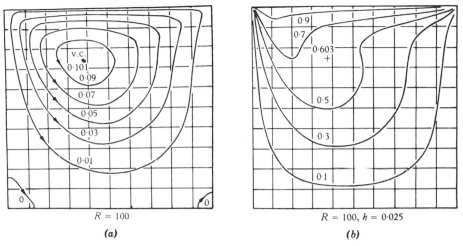

Figure 12.8 Eddy flow generated by the moving top wall of a square cavity, Re = 100 [17]; (a) streamline pattern: (b) isotherms deformed by the eddy flow.

Re increases the number of iterations increases and the grid size must be decreased simultaneously to attain acceptable accuracy. The velocity boundary conditions were treated by introducing a row of mirror-image points on the outside of the boundary, at a distance equal to the grid spacing. The streamfunction values at the boundary and at the image points were related to those at the interior points via the boundary values of velocity. The underrelaxation factor was found to be fairly insensitive to mesh size. Representative results are displayed in Fig. 12.8.

SYMBOLS

E	shooting error [eqs. (63), (64), (75)]
F	body force vector [eq. (78)]
g	gravitational acceleration
$h(x)$	heat source distribution [eq. (21)]
$k_{1,2,3,4}$	constants in the Runge-Kutta algorithm [eqs. (41)–(44)]
m	number of vertical grid lines (Fig. 12.1a)
n	number of horizontal grid lines (Fig. 12.1a)
Nu	Nusselt number
Pr	Prandtl number (ν/α)
Ra	Rayleigh number (Fig. 12)
Re	dimensionless number [eq. (91)]

t	time
$\mathbf{u}$	velocity vector
u	horizontal velocity component
v	vertical velocity component
x	horizontal Cartesian coordinate
Δx	horizontal mesh size (Fig. 12.1a)
y	vertical Cartesian coordinate
Δy	vertical mesh size (Fig. 12.1a)
β	coefficient of terminal expansion [eq. (79)] and dimensionless parameter [eq. (66)]
γ	relaxation factor [eq. (88)]
η	similarity variable [eqs. (67, 68)]
λ	auxiliary parameter [eq. (90)]
μ	viscosity
ν	kinematic viscosity ($\nu = \mu/\rho$)
ρ	density
ϕ	general function
ψ	streamfunction
ω	vorticity vector

REFERENCES

1. W. J. Minkowycz, ed., *Numerical Heat Transfer*, Hemisphere, Washington, D.C., Vol. 1 (1978)-present.

2. B. Carnahan, H. A. Luther, and J. O. Wilkes, *Applied Numerical Methods*, Wiley, New York, 1969.

3. L. Fox, *The Numerical Solution of Two-Point Boundary Problems*, Oxford University Press, Oxford, England, 1957.

4. P. R. Nachtsheim, and P. Swigert, Satisfaction of asymptotic boundary conditions in numerical solution of systems of non-linear equations of the boundary layer type, NASA TN D-3004, October 1965.

5. P. J. Roache, *Computational Fluid Dynamics*, Hermosa Publishers, Albuquerque, 1976.

6. S. V. Patankar, *Numerical Heat Transfer and Fluid Flow*, Hemisphere, Washington, D.C., 1980.

7. J. H. Ferziger, *Numerical Methods for Engineering Application*, Wiley, New York, 1981.

8. C. Y. Chow, *Computational Fluid Mechanics*, Wiley, New York, 1979.

9. S. P. Frankel, Convergence rates of iterative treatments of partial differential equations, *Math Tables and Other Aids to Computation*, Vol. 4, 1950, pp. 65–75.

10. D. Young, Iterative methods for solving partial difference equations of elliptic type, *Trans. Am. Math. Soc.*, Vol. 76, 1954, pp. 92–111.

11. P. D. Lax, and R. D. Richtmyer, Survey of the stability of linear finite difference equations, *Commun. Pure Appl. Math.*, Vol. IX, 1956, pp. 267–293.

12. A. D. Gosman, W. M. Pun, A. K. Runchal, D. B. Spalding, and M. Wolfshtein, *Heat and Mass Transfer in Recirculating Flows*, Academic Press, New York, 1969.

13. R. S. Varga, *Matrix Iterative Analysis*, Prentice-Hall International, London, 1962.

14. G. deVahl Davis, and G. D. Mallinson, False diffusion in numerical fluid mechanics, *University of New South Wales, Report 1972/FMT/1*, 1972.

15. D. Poulikakos, and A. Bejan, Numerical study of transient high Rayleigh number convection in an attic-shaped porous layer, *J. Heat Transfer*, Vol. 105, 1983, pp. 476–484.

16. S. Kimura, and A. Bejan, The "heatline" visualization of convective heat transfer, *J. Heat Transfer*, Vol. 105, 1983, pp. 916–919.

17. O. R. Burggraf, Analytical and numerical studies of the structure of steady separated flows, *J. Fluid Mech.*, Vol. 24, 1966, pp. 113–151.

18. L. C. Chow, and C. L. Tien, An examination of four differencing schemes for some elliptic-type convection equations, *Numerical Heat Transfer*, Vol. 1, 1978, pp. 87–100.

PROBLEMS

1. Consider the boundary value problem of free-convection along a vertical plate (see Chapter 4, pp. 125–130). This problem reduces to solving two differential equations

$$\frac{1}{\text{Pr}}\left(\frac{1}{2}F' - \frac{3}{4}FF''\right) = -F''' + \theta$$

$$\frac{3}{4}F\theta' = \theta''$$

Subject to the following boundary conditions

$$\eta = 0: \quad F = F' = 0, \quad \theta = 1$$

$$\eta \to \infty: \quad F' \to 0, \quad \theta \to 0$$

Develop a numerical solution following the shooting method described in the text.

2. The temperature in the boundary layer near a constant-temperature flat plate is given by the solution of the following boundary value problem (see Chapter 2, pp. 50–52).

$$\theta'' + \frac{\text{Pr}}{2}f\theta = 0$$

$$\eta = 0: \quad \theta = 0$$

$$\eta \to \infty: \quad \theta \to 1$$

Develop a finite-difference procedure for determining $\theta(\eta)$ numerically. Note that $f(\eta)$ is a function known already from the flow part of this boundary layer problem.

3. Show that the discretized version of eq. (104) for the steady state solution is identical to the finite-difference equation obtained by setting $\partial w/\partial t = 0$ in eq. (99).

4. To become acquainted with exponential finite-differences [18], refer to Fig. 12.1*a* and, for clarity, label the grid points $(i - 1, j)$, (i, j), $(i + 1, j)$, $(i, j - 1)$, $(i, j + 1)$, $(i - \frac{1}{2}, j)$, $(i + \frac{1}{2}, j)$, $(i, j - \frac{1}{2})$, and $(i, j + \frac{1}{2})$ using W, P, E, S, N, w, e, s, and n, respectively. Consider the one-dimensional vorticity equation $(\partial^2\omega/\partial x^2) = (\partial/\partial x)(u\omega)$ and the directional flux defined by $F \equiv (\partial\omega/\partial x) - u\omega$. (Note that this expression is analogous to the energy flux due to both convection and diffusion.)

(a) Assume that the vorticity flux is constant between grid points W and P and that the velocity is also constant (both the vorticity and the velocity are evaluated at w). Integrate twice to obtain

$$\omega = -\frac{F_w}{u_w} + c\exp(u_w x)$$

where c is a constant of integration.

(b) Evaluate ω at points W and P, and prove that

$$\frac{\omega - \omega_W}{\omega_P - \omega_W} = \frac{\exp[u_w(x - x_W)] - 1}{\exp[u_w(x_P - x_W)] - 1}$$

(c) Introduce two new functions

$$f_+(x) = \frac{x}{e^x - 1}, \qquad f_-(x) = \frac{-x}{e^{-x} - 1}$$

and prove that the directional flux of ω at point w is

$$F_w = \frac{f_+(u_w \Delta x)\omega_P - f_-(u_w \Delta x)\omega_W}{\Delta x}$$

where $\Delta x = x_i - x_{i-1}$

(d) Repeat the above procedure for the other boundaries and, combining the resulting expressions, show that the conservation of these fluxes in the control volume yields the exponential finite-difference equation [18]

$$F_e - F_w + F_n - F_s + S\Delta x = 0$$

where it is assumed that $\Delta x = \Delta y$ and

$$F_e = \frac{f_+(u_e \Delta x)\omega_E - f_-(u_e \Delta x)\omega_P}{\Delta x}$$

$$F_w = \frac{f_+(v_s \Delta y)\omega_P - f_-(v_s \Delta y)\omega_S}{\Delta y}$$

$$F_n = \frac{f_+(v_n \Delta y)\omega_N - f_-(v_n \Delta y)\omega_P}{\Delta y}$$

S = possible source term

5. The Allen–Southwell finite-difference formulation [18] can be illustrated using the same example as in the preceding problem. This time we assume that u is constant and equal to its value at the grid point P, and that the directional flux of vorticity $F = (\partial\omega/\partial x) - u\omega$ varies linearly in x between W and E, that is, $F = (\partial\omega/\partial x) - u\omega = Ax + B$, where A and B are constants.

(a) Integrating the directional flux show that

$$\omega = -\frac{A}{u}\left(x + \frac{1}{u}\right) - \frac{B}{u} + Ce^{ux}$$

where C is a constant of integration.

(b) Evaluate the above equation at points W, P, and E in order to determine the constants, and show that the directional flux at point w

$$F_w = \frac{\partial\omega}{\partial x} - u\omega = -A\frac{\Delta x}{2} + B$$

is equal to

$$F_w = A\Delta x\left[\frac{1}{2} - \frac{f_+(u\Delta x) - 1}{u\Delta x}\right] + \frac{f_+(u\Delta x)\omega_P - f_-(u\Delta x)\omega_W}{\Delta x}$$

where

$$A\Delta x^2 = f_+(u\Delta x)\omega_E - f_-(u\Delta x)\omega_P$$

$$-[f_+(u\Delta x)\omega_P - f_-(u\Delta x)\omega_W]$$

(c) Repeating the above analysis for points e, n, and s, show that if $\Delta x = \Delta y$ the conservation of the directional flux in the control volume requires

$$f_+(u\Delta x)\omega_E - f_-(u\Delta x)\omega_P - \left[f_+(u\Delta x)\omega_P - f_-(u\Delta x)\omega_W\right]$$

$$+ \left[f_+(v\Delta y)\omega_N - f_-(v\Delta y)\omega_P\right] - \left[f_+(v\Delta y)\omega_P - f_-(v\Delta y)\omega_S\right]$$

$$+ S(\Delta x)^2 = 0$$

This finite-difference form is the Allen–Southwell scheme [18].

6. A certain flow is bounded by a solid impermeable wall parallel to the x coordinate. Express the wall vorticity in terms of the discrete values of streamfunction, by considering the following steps.

(a) Show that the impermeable wall condition implies that the wall vorticity is $\omega_W = \partial u/\partial y|_W$, where subscript W indicates a node on the wall.

(b) Considering a Taylor series of ψ about a node on the wall, express $\partial u/\partial y|_w$ in terms of ψ and show that

$$\omega_w = -\frac{2(\psi_{W+1} - \psi_W)}{(\Delta y)^2}$$

where the subscript $W + 1$ indicates an interior node next to the wall.

7. The solid impermeable wall of the preceding problem is replaced by a porous wall along which the blowing rate (the velocity normal to the wall) is specified. Derive the boundary condition for ω suitable for the vorticity–streamfunction formulation of the numerical solution.

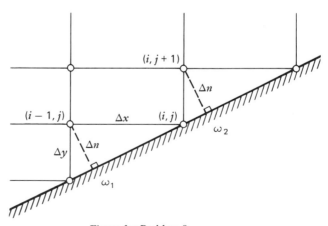

Figure for Problem 8.

8. As shown in the attached figure, a solid impermeable wall is inclined relative to the rectangular mesh superimposed on the flow. Derive an expression for the wall vorticity at point (i, j) based on the following steps.

(a) Obtain wall vorticities ω_1 and ω_2 in terms of $\psi_{i-1, j}$, $\psi_{i, j+1}$, and Δn.

(b) Interpolating along the wall, determine the wall vorticity value $\omega_{i, j}$ in terms of $\psi_{i-1, j}$, $\psi_{i, j+1}$, Δx, and Δy.

Appendix

THERMOPHYSICAL PROPERTIES

This section contains tables of thermophysical properties of those materials encountered most frequently in heat transfer applications. Space limitations, coupled with the fact that this book is not a handbook, make it impossible to exhibit the large volume of thermophysical data available for engineering calculations: for more extensive compilations, the reader is directed to the specialized handbooks (e.g., [1, 2]).

A common feature of the numerical values collected here and in the handbooks is the inexact and somewhat ephemeral character of the measured data. This characteristic must be stressed, particularly in view of the comparisons made earlier in the text between "less exact" and "more exact" analytical methods. I am grateful to J. Taborek [3] for bringing to my attention Fig. A.1, which shows how the value of one property varies with time or, better said, with the technological era in which the experimentalist lived [4]. Figure A.1 should plant a sobering thought in the minds of those who believe in the existence of "exact engineering calculations": highly successful theoretical results exist (e.g., the Nusselt numbers for fully developed laminar flow through a pipe); however, the physical quantities calculated using such results are as inexact as the thermophysical properties assumed known by the theoretician. From a teaching standpoint, this observation stresses the importance of understanding the approximate analytical methods first.

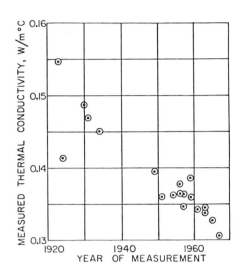

Figure A.1   The thermal conductivity of liquid toluene at 20 °C [4].

Table A.1 Thermodynamic Properties of Water at Atmospheric Pressure[a]

Temperature T [°C]	Density ρ [g/cm^3]	Specific Heat at Constant Pressure c_P [J/g K]	Specific Heat at Constant Volume c_v [J/g K]	Latent Heat of Evaporation h_{fg} [J/g]	Coefficient of Thermal Expansion β [K^{-1}]
0	0.9999	4.217	4.215	2.501×10^3	-0.6×10^{-4}
5	1.	4.202	4.202	2.489×10^3	$+0.1 \times 10^{-4}$
10	0.9997	4.192	4.187	2.477×10^3	0.9×10^{-4}
15	0.9991	4.186	4.173	2.465×10^3	1.5×10^{-4}
20	0.9982	4.182	4.158	2.454×10^3	2.1×10^{-4}
25	0.9971	4.179	4.138	2.442×10^3	2.6×10^{-4}
30	0.9957	4.178	4.118	2.430×10^3	3.0×10^{-4}
35	0.9941	4.178	4.108	2.418×10^3	3.4×10^{-4}
40	0.9923	4.178	4.088	2.406×10^3	3.8×10^{-4}
50	0.9881	4.180	4.050	2.382×10^3	4.5×10^{-4}
60	0.9832	4.184	4.004	2.357×10^3	5.1×10^{-4}
70	0.9778	4.189	3.959	2.333×10^3	5.7×10^{-4}
80	0.9718	4.196	3.906	2.308×10^3	6.2×10^{-4}
90	0.9653	4.205	3.865	2.283×10^3	6.7×10^{-4}
100	0.9584	4.216	3.816	2.257×10^3	7.1×10^{-4}

[a]Adapted from Refs. 2 and 5.

Table A.2 Transport Properties of Water at Atmospheric Pressure [a]

Temperature T [°C]	Viscosity μ [g/cm s]	Kinematic Viscosity ν [cm²/s]	Thermal Conductivity k [W/m K]	Thermal Diffusivity α [cm²/s]	Prandtl Number $Pr = \dfrac{\nu}{\alpha}$	$\dfrac{Ra_H}{H^3 \Delta T} = \dfrac{g\beta}{\alpha\nu}$ [K⁻¹ cm⁻³]
0	0.01787	0.01787	0.56	0.00133	13.44	-2.48×10^3
5	0.01514	0.01514	0.57	0.00136	11.13	$+0.47 \times 10^3$
10	0.01304	0.01304	0.58	0.00138	9.45	4.91×10^3
15	0.01137	0.01138	0.59	0.00140	8.13	9.24×10^3
20	0.01002	0.01004	0.59	0.00142	7.07	14.45×10^3
25	0.00891	0.00894	0.60	0.00144	6.21	19.81×10^3
30	0.00798	0.00802	0.61	0.00146	5.49	25.13×10^3
35	0.00720	0.00725	0.62	0.00149	4.87	30.88×10^3
40	0.00654	0.00659	0.63	0.00152	4.34	37.21×10^3
50	0.00548	0.00554	0.64	0.00155	3.57	51.41×10^3
60	0.00467	0.00475	0.65	0.00158	3.01	66.66×10^3
70	0.00405	0.00414	0.66	0.00161	2.57	83.89×10^3
80	0.00355	0.00366	0.67	0.00164	2.23	101.3×10^3
90	0.00316	0.00327	0.67	0.00165	1.98	121.8×10^3
100	0.00283	0.00295	0.68	0.00166	1.78	142.2×10^3

[a]Adapted from Refs. 2 and 5.

Table A.3 Properties of Water at the Saturation Pressure[a]

Temperature T [°C]	Density ρ [g/cm^3]	Specific Heat at Constant Pressure c_P [J/g K]	Viscosity μ [g/cm s]	Kinematic Viscosity ν [cm^2/s]	Thermal Conductivity k [W/m K]	Thermal Diffusivity α [cm^2/s]	Prandtl Number Pr
0	0.9999	4.226	0.0179	0.0179	0.56	0.0013	13.7
10	0.9997	4.195	0.0130	0.0130	0.58	0.0014	9.5
20	0.9982	4.182	0.0099	0.0101	0.60	0.0014	7
40	0.9922	4.175	0.0066	0.0066	0.63	0.0015	4.3
60	0.9832	4.181	0.0047	0.0048	0.66	0.0016	3
80	0.9718	4.194	0.0035	0.0036	0.67	0.0017	2.25
100	0.9584	4.211	0.0028	0.0029	0.68	0.0017	1.75
150	0.9169	4.270	0.00185	0.0020	0.68	0.0017	1.17
200	0.8628	4.501	0.00139	0.0016	0.66	0.0017	0.95
250	0.7992	4.857	0.00110	0.00137	0.62	0.0016	0.86
300	0.7125	5.694	0.00092	0.00128	0.56	0.0013	0.98
340	0.6094	8.160	0.00077	0.00127	0.44	0.0009	1.45
370	0.4480	11.690	0.00057	0.00127	0.29	0.00058	2.18

[a]Adapted from Ref. 2.

Table A.4 Properties of Dry Air at Atmospheric Pressure[a]

Temperature T [°C]	Density ρ [g/cm³]	Viscosity μ [g/cm s]	Kinematic Viscosity ν [cm²/s]	Thermal Conductivity k [W/cm K]	Thermal Diffusivity α [cm²/s]	Prandtl Number $Pr = \dfrac{\nu}{\alpha}$	$\dfrac{Ra_H}{H^3 \Delta T} = \dfrac{g\beta}{\alpha\nu}$ [cm⁻³ K⁻¹]
−180	3.72×10^{-3}	0.65×10^{-4}	0.0175	0.76×10^{-4}	0.019	0.92	3.2×10^{4}
−100	2.04×10^{-3}	1.16×10^{-4}	0.057	1.6×10^{-4}	0.076	0.75	1.3×10^{3}
−50	1.582×10^{-3}	1.45×10^{-4}	0.092				
0	1.293×10^{-3}	1.71×10^{-4}	0.132	2.4×10^{-4}	0.184	0.72	148.
10	1.247×10^{-3}	1.76×10^{-4}	0.141	2.5×10^{-4}	0.196	0.72	125.
20	1.205×10^{-3}	1.81×10^{-4}	0.150	2.5×10^{-4}	0.208	0.72	107.
30	1.165×10^{-3}	1.86×10^{-4}	0.160				
60	1.060×10^{-3}	2.00×10^{-4}	0.188				
100	0.946×10^{-3}	2.18×10^{-4}	0.230	3.2×10^{-4}	0.328	0.7	34.8
200	0.746×10^{-3}	2.58×10^{-4}	0.346				
300	0.616×10^{-3}	2.95×10^{-4}	0.481				
500	0.456×10^{-3}	3.58×10^{-4}	0.785				
1000	0.277×10^{-3}	4.82×10^{-4}	1.74	7.6×10^{-4}	2.71	0.64	0.163

[a]Adapted from Refs. 2 and 5.

Table A.5 The Variation of Fluid Properties with Temperature, Shown as the Ratio of the Property Evaluated at 100°C Divided by the Property Evaluated at 20°C, at Atmospheric Pressure[a]

	$\dfrac{\mu(100)}{\mu(20)}$	$\dfrac{v(100)}{v(20)}$	$\dfrac{k(100)}{k(20)}$	$\dfrac{\alpha(100)}{\alpha(20)}$	$\dfrac{c_P(100)}{c_P(20)}$	$\dfrac{Pr(100)}{Pr(20)}$
Gases						
Air (Table A.4)	1.2	1.5	1.2	1.5	1.0	1.0
Hydrogen (Pr = 0.71, at 0°C)	1.2	1.5	1.2	1.6	1.0	0.96
Helium (Pr = 0.7, at 0°C)	1.2	1.5	1.1	1.4	1.0	1.0
Water vapor (between 200°C and 100°C)	1.3	1.7	1.4	1.9	0.90	0.87
Liquids						
Water (Tables A.1–A.3)	0.36	0.30	1.1	1.2	1.0	0.25
Oil (Pr = 170, at 20°C)	0.22	0.16	0.97	0.88	1.2	0.19
Oil (Pr = 480, at 20°C)	0.098	0.10	0.95	0.84	1.2	0.13
Oil (Pr = 10^4, at 20°C)	0.021	0.025	0.94	0.85	1.2	0.026
Mercury (Pr = 0.029, at 0°C)	0.82	0.81	1.2	1.3	1.0	0.56

[a]After Ref. 6.

MATHEMATICAL FORMULAS

Error function definition and properties:

$$\text{erf}(x) = \frac{2}{\pi^{1/2}} \int_0^x e^{-m^2} \, dm$$

$$\text{erf}(-x) = -\text{erf}(x)$$

$$\text{erfc}(x) = 1 - \text{erf}(x)$$

$$\frac{d}{dx}\left[\text{erf}(x)\right]_{x=0} = \frac{2}{\pi^{1/2}} = 1.12838$$

Representative values of the error function:

x	$\text{erf}(x)$
0.	0.
0.01	0.01128
0.1	0.11246
0.2	0.2227
0.3	0.32863
0.4	0.42839
0.5	0.5205
0.6	0.60386
0.7	0.6778
0.8	0.74210
0.9	0.79691
1.	0.8427
1.2	0.91031
1.4	0.95229
1.6	0.97635
1.8	0.98909
2.	0.99532
2.5	0.99959
3.	0.99998
∞	1.

Leibnitz's integral formula:

$$\frac{d}{dx}\left[\int_{a(x)}^{b(x)} F(x, m) \, dm\right] = \int_{a(x)}^{b(x)} \frac{\partial F(x, m)}{\partial x} \, dm + F(x, b)\frac{db}{dx} - F(x, a)\frac{da}{dx}$$

CONVERSION FACTORS

Length	1 in.	= 2.54 cm
	1 ft	= 0.3048 m
	1 mile	= 1.609 km
Area	1 in.^{2}	= 6.452 cm^2
	1 ft^2	= 0.0929 m^2
	1 mile^2	= 2.59 km^2
Volume	1 in.^{3}	= 16.39 cm^3
	1 ft^3	= 0.02832 m^3
		= 28.32 liters
	1 gal. (U.S.)	= 3.785 liters
Mass	1 lbm	= 0.4536 kg
	1 oz	= 28.35 g
Force	1 lbf	= 4.448 N
		= 0.4536 kgf
	1 dyne	= 10^{-5} N
Pressure	1 psi	= 6895 N/m^2
	1 atm	= 14.69 psi
		= 1.013×10^5 N/m^2
	1 bar	= 10^5 N/m^2
	1 Torr	= 1 mm Hg
		= 133.32 N/m^2
	1 psi	= 27.68 in. H$_2$O
	1 ft H$_2$O	= 0.4335 psi
Energy	1 BTU	= 1055 J
		= 788 ft lbf
		= 3412.14 kW hr
		= 2544.5 hp hr
	1 cal	= 4.184 J
	1 erg	= 10^{-7} J
Power	1 BTU/s	= 1055 W
	1 hp	= 745.7 W
	1 ft lbf/s	= 1.3558 W

REFERENCES

1. W. Ibele, Thermophysical properties, Section 2 in *Handbook of Heat Transfer*, W. M. Rohsenow and J. P. Hartnett, eds., McGraw-Hill, New York, 1973.

2. K. Raznjevic, *Handbook of Thermodynamic Tables and Charts*, Hemisphere, Washington, D.C., 1976.

3. J. Taborek, private communication, March 1983, Washington, D.C.

4. G. M. Mallan, M. S. Michaelian, and F. J. Lockhart, Liquid thermal conductivities of organic compounds and petroleum fractions, *J. Chem. Eng. Data*, Vol. 17, October 1972, pp. 412–415.

5. G. K. Batchelor, *An Introduction to Fluid Dynamics*, Cambridge University Press, Cambridge, 1967.

6. J. Kestin and P. D. Richardson, Heat transfer across turbulent, incompressible boundary layers, *Int. J. Heat Mass Transfer*, Vol. 6, 1963, pp. 147–189.

Index